Knowledge and Networks in a Dynamic Economy

Festschrift in Honor of Åke E. Andersson

Springer

Berlin
Heidelberg
New York
Barcelona
Budapest
Hong Kong
London
Milan
Paris
Santa Clara
Singapore
Tokyo

M. J. Beckmann · B. Johansson
F. Snickars · R. Thord (Eds.)

Knowledge and Networks
in a Dynamic Economy

Festschrift in Honor of Åke E. Andersson

With 75 Figures
and 25 Tables

 Springer

Prof. Dr. Martin J. Beckmann
Brown University, Economics Department, 64 Waterman Street,
02912 Providence, RI, USA

Prof. Dr. Börje Johannsson
Jönköping International Business School, Economics Department,
P.O. Box 1026, S-55111 Jönköping, Sweden

Prof. Dr. Folke Snickars
Royal Institute of Technology, Department of Infrastructure and Planning,
S-10044 Stockholm, Sweden

Dr. Roland Thord
Thord Connector, Lützengatan 9, S-11520 Stockholm, Sweden

ISBN 3-540-64245-5 Springer-Verlag Berlin Heidelberg New York

Library of Congress Cataloging-in-Publication Data
Die Deutsche Bibliothek – CIP-Einheitsaufnahme
Knowledge and networks in a dynamic economy: Festschrift in honor of Åke E. Andersson;
with 25 tables / M.J. Beckmann ... (ed.). – Berlin; Heidelberg; New York; Barcelona; Buda-
pest; Hong Kong; London; Milan; Paris; Santa Clara; Singapore; Tokyo: Springer, 1998
  ISBN 3-540-64245-5

© Springer-Verlag Berlin · Heidelberg 1998
Printed in Germany

The use of general descriptive names, registered names, trademarks, etc. in this publication
does not imply, even in the absence of a specific statement, that such names are exempt from
the relevant protective laws and regulations and therefore free for general use.

Hardcover-Design: Erich Kirchner, Heidelberg

SPIN 10674178   42/2202-5 4 3 2 1 0 – Printed on acid-free paper

Preface

This Festschrift has been created to honor Professor *Åke E. Andersson* - a friend and deeply admired colleague of us. The book provides a limited sample of the broad ranging research of *Åke E. Andersson*. Here some of his friends and colleagues have contributed to give examples from the emerging research field "Knowledge and Networks in a Dynamic Economy". This is an area of research to which *Åke E. Andersson* has been a great inspiration to us all and in which he himself passionately has contributed as part of his prodigious output.

The volume has been sponsored by *the Stockholm County Council, the Swedish Post, the Swedish Transport and Communication Research Board* and *Telia AB*. *Temaplan AB* has delivered resources beyond the ordinary in finalizing the volume and in preparing the complete manuscript. *Marianne Bopp* and *Werner A. Müller* at *Springer-Verlag* have encouraged us in this work and also been of great support during our long endeavor in preparing this book.

The Editors would like to express their sincere thanks for making this Festschrift a reality to all the above friends and colleagues of *Åke E. Andersson* who have contributed to the volume, as well as to all supportive organizations.

Martin Beckmann
Börje Johansson
Folke Snickars
Roland Thord

Contents

Dynamics of Economical Networks

Creativity - Learning and Knowledge in Networks

The Infrastructure Arena -
Transportation and Communication Networks

Introduction

Introduction

Börje Johansson
Jönköping International Business School

Folke Snickars
Department of Infrastructure and Planning
Royal Institute of Technology, Stockholm

and
Roland Thord
Temaplan AB and Thord Connector, Stockholm

Åke E Andersson
– Biosketch and Scientific Contribution 1936 -1996

In 1996 Åke Emanuel Andersson could celebrate his golden jubilee, born 60 years earlier in the town of Sollefteå in the middle part of northern Sweden. His residence in 1996 was Uppsala, while his workplace was dispersed around the globe with a main office in Stockholm. The location dynamics of Åke started when he left Sollefteå at the age of 15 and sailed the sea with major stops in New Zealand, Australia and USA. Back in Sweden for military service and studies in Göteborg, his mobility sequence includes Vänersborg, Stockholm, Göteborg, Philadelphia, Göteborg, Laxenburg outside Vienna, Umeå and Uppsala.

From the middle of the 1950's Åke E Andersson started his college and university studies with a focus on philosophy, economic history and economics. In this period he also had time both to get married and take responsibility as chairman of the student's union. In 1962 he started a first career as teacher in economics at Göteborg's School of Business Administration. Two years later he had begun the doctoral studies at the University of Göteborg. Initially, there was no professor present at the department and the new professor Roland Artle did not arrive until 1966. Under his supervision and advice a thesis was completed 1967. The contribution was a simulation model of the economy of the Göteborg region.

Already at this time Åke E Andersson was active in decision-related regional planning. In 1966-1967 he was working as planning officer of the Älvsborg County Council, with the office in Vänersborg. In 1967 he moved to Stockholm and was employed as chief economist at Generalplaneberedningen, the city planning office of Stockholm. In 1970 he was appointed professor at the Nordic Institute for Urban and Regional Planning (Nordplan) in Stockholm. His scientific contributions from the early 1970's comprise work on housing economics and housing policy, regional development theory, and interdependencies between transportation, location and land values. His book "The Metropolitan Problem" was included as a supplement to the Long Term Perspective on the Swedish Economy in 1971. The same year he was appointed associate professor at the University of Gothenburg. During this Stockholm period he started an intense interaction with a group of young economists and mathematicians which has continued through the decades.

In 1972 Åke E Andersson returned to the Department of Economics at the University of Gothenburg, where he remained until 1978 – including a period as visiting professor at the University of Pennsylvania 1974-1975. During the first half of this period his research comprises fields such as consumption theory, income distribution issues, welfare economics, and household economics. Some of these contributions were associated with work for the Economic Commission for Europe. In regional science he worked on endogenous city formation, e.g. with Anders Karlqvist in the paper "Population and Capital in Geographical Space", and on various aspects of spatial interdependencies. The latter issue was his main focus during his visit at the University of Pennsylvania, and this was reflected in a course together with Tony Smith on interdependencies in urban models. Back in Göteborg Andersson together with Ingvar Holmberg lead an interdisciplinary research on the interaction between economic and demographic development. He also worked together with students in Göteborg on industrial dynamics.

In 1973 Åke E Andersson and Walter Isard were appointed joint editors of the North-Holland journal "Regional Science and Urban Economics". They managed to develop the journal to an influential meeting-place for new ideas in regional science and spatial economics. This work, which continued till 1980, influenced life at the Department of Economics in Göteborg – with an intense flow of visiting international scholars. This period ended 1978 when Åke E Andersson moved to the International Institute for Applied Systems Analysis (IIASA), hosted in Schloss Laxenburg outside Vienna, previously a hunting-seat of the Habsburg family.

The period at IIASA was rewarding to Åke E Andersson. It opened up new strands of research and he could establish durable interaction networks with new groups of scientists. At IIASA research perspectives included dynamical systems, development and evolutionary processes, which is reflected in his contributions from this time. His research problems from this time comprise endogenous economic growth, regional development and allocation of land to activities in a development context. Part of this work included simulation experiments with non-linear economic growth models. At this time he also became editor (together with Walter Isard) of the North-Holland book series "Studies in Regional Science and Urban Economics".

After having arrived at IIASA Åke E Andersson was appointed professor of regional economics 1979 at the University of Umeå, and he went there in 1980 to start a new chapter of his career. The time in Umeå, which lasted (with some interruptions) until 1988, was particular in several respects. Åke had already in the late 1960's been eager to implement results from research in development plans such as location decisions, design of investment profiles, and systems architecture in general. The Umeå time is especially coloured by an interaction between formulating new research ideas and outlining new policies for regional development. Umeå had until the middle of the 1960s been a very small town, and it was a great surprise when the government decided to locate the university of northern Sweden there. During his stay in Umeå Åke E Andersson elaborated on a theory of an emerging knowledge society (C-society) and the role of education and R&D in the development of such regions. In view of these ideas he organised and promoted symposia, conferences and regular chamber music festivals in the Umeå region. He was also instrumental for the design and establishment of the Centre for Regional Science Research (CERUM) at the university. His ideas of the C-society (stressing communication, cognitive capacity, creativity, and culture) are developed in the book "Creativity: The Future of Metropolis" from 1985. This contribution was later followed by a series of books on the C-society topic.

During his Umeå period Åke E Andersson returned to IIASA and stayed there between 1984 and 1985. He participated in a project on the global forest sector development and became the leader of a regional development project. Under influence of various research efforts at IIASA he brought with him back to Sweden a research programme which emphasised the role of non-linearities in creative and innovative processes, and hence in all types of development processes. When he moved to Uppsala (residence) and Stockholm (office) 1988 to become director of the Institute for Futures Studies he could further promote the analysis of phenomena comprising complex dynamics. Through the period 1988 - 1996 and onwards, the research at the institute has been based on three principles. Firstly, futures studies is not a science in itself. It has to rely on multidisciplinary collaboration between researchers, each of them with a strong foundation in a basic discipline. Secondly, it is necessary to distinguish between slow and fast processes, where phenomena with slow adjustments function like an arena or an infrastructure for changes which adjust in a fast time scale. Thirdly, fundamental interdependencies between change processes with different time scales bring about complex dynamics – which in a deep sense embody the future.

Basic values of individuals constitute a slowly changing arena for the society. This is one of the issues examined in several research activities at the institute. The economic development of functional regions as well as world regions (such as Europe, South East Asia, etc.) has been another focus. One may also recognise that in the beginning or the 1990's Åke E Andersson played a crucial role in promoting the establishment of the International Business School in Jönköping (now the workplace of one of the editors). Being a medium-sized city which lacked a university, Jönköping was an interesting test case. Can a strong injection of knowledge activities in a region bring about a renewal of the economy and the entire regional society? Is it possible to trigger the development of medium-sized C-regions? Åke E Andersson's ex ante answer was in the affirmative.

Åke E Andersson has always been intellectually on the move. He has selected his own track through the academic system. He has formed a school of thought which has brought him international recognition. He received the Japanese Honda Prize for his work on dynamic analysis in the fields of regional economics and regional planning in 1995. This prize which had been given earlier to researchers like Benoit Mandelbrot was a reward he really deserved. It was a prize he was duly proud of also for the reason that it came from another world economic region. One of his many intellectual and operational contributions is to promote research contacts between the Nordic area and Japan.

During his work-life trip from Vänersborg to Uppsala Åke E Andersson's research has been as dynamic as his space-time work trajectory. At the same time it has rested on foundations which were established early in his academic career. The three cornerstones of his scientific interests have been welfare analysis, regional economic dynamics and human capital theory. But the foundations have always been built on his thorough training in philosophy, economic history and economics from the 1960's.

Åke E Andersson is one of those persons who found a intellectual track to adhere to already early in the career. His solid readings in economics coupled with his deeply rooted interest in real-world regional problems has provided him with a flooding and lasting well of creativity. Philosophy and economic history brought him the structural perspective. His basic studies in economics made him realise the strengths and weaknesses of neoclassical economics. His early economic thinking was inspired by contacts with Norwegian research on structural economic transformation. His early interest in non-linear mathematics brought him into contact with researchers in this field at the Royal Institute of Technology, Stockholm.

He turned to regional economics partly because he saw the problems of industrial transformation in the Swedish regions, partly because he was inspired by his teacher Roland Artle who had performed novel studies in the structure of the Stockholm economy. His interest in human capital theory started with population and labour market studies. His collaboration with Walter Isard, Martin Beckmann and others contributed substantially to the development of modern regional science as a discipline which has inspired the intellectual thinking in other fields of social science, maybe more than realised until the breakthrough of regional thought in the 1990's.

Many of the ideas which have recently been substantiated by Åke E Andersson and his network of research colleagues, the members of his Nordic school of regional structural dynamics, were first developed during the early 1970's. The problems of long-term regional planning in metropolitan regions was addressed by Åke E Andersson and his colleagues in a way that foreruns the current research on creative regions based on contact-intensive activities. The current research on infrastructure and regional economic dynamics was started with research on dynamic input-output analysis for the Swedish regions. His early work on welfare theory questioned the foundations of microeconomic analysis and paved the way for later developments in stochastic choice theory.

A substantial part of this research has not been widely published in the internatio-nal literature. The contacts with the international research frontier which were developed through Åke E Andersson's stay with the Regional Science Department, University of Pennsylvania 1974-75 and his work as research leader at IIASA 1978-80 and 1984-86. Especially the IIASA stay provided a connection point between the research on regional dynamics which had been performed in the Nordic countries with Åke E Andersson as a central contributor was rewarding. The interaction with researchers as Brian Arthur, Jena-Pierre Aubin and others at IIASA produced a state of structural instability in the the the thought structure of Åke E Andersson.

The phase transition towards his most productive period both as a researcher, a teacher and a policy advisor started after this period. Thoughts which had been there already during his Stockholm and Gothenburg periods in the 1970's reemerged in the new paradigm. Åke E Andersson suddenly understood the driving forces in the global economy in a completely new light. What had been discussed and analysed with his colleagues in a series of seminars and papers during the earlier period was provided with a fundamental theoretical framework. The energetic work along this track in the end qualified Åke E Andersson for the Honda Prize. He even created the Institute for Futures Studies, of which he has been the director for a decade, mainly on this paradigm of societal development as a complex dynamic system.

The role of logistical revolutions as driving forces for the slow societal processes was a truly challenging thought. Åke E Andersson's background in economic his-tory came to help in the formulation of those thoughts. The economic theory of knowledge-creation which he developed together with Martin Beckmann, Kyoshi Kobayashi and others was another path-breaking thought structure. His experience as an public-policy advisor and his being instrumental in the creation of several research organisations in Sweden provided a body of resonance for those ideas. His contributions to infrastructure economics together with several of the authors pre-sent in the current volume has one of its foundation in the thoughts of Roland Artle of the city as an infrastructure consisting of infrastructures.

One of the most prominent traits of Åke E Andersson's scientific contributions so far in his career in the elegance and simplicity of the ideas. There are no diffuse thoughts in the papers he writes, and no void statements in his lectures and public presentations. He has a most personal charm and pedagogical skill in presenting his views to the research community and to the informed public. This has given him a unique position as a contributor to the public debate in Sweden.

A genius in the classical sense would be a person who is able to invent new artefacts or find solutions to difficult technical problems. The modern genius would be a person who is able to invent new ideas or suggest solutions to complex policy problems. In the old days the royal courts had to be convinced. The contemporary courts would maybe resemble the mass media. At the age of sixty Åke E Andersson is beginning to show the signs of such a complex human capital system. He might, indeed, be a Swedish genius.

An Outline of the Volume

This Festschrift provides a limited sample of the broad ranging research interest of *Åke E. Andersson*. Here some of his friends and colleagues has contributed to give various examples from the growing research field of "Knowledge and Networks in a Dynamic Economy". This is an area of research in which *Åke E Andersson* has been a great inspiration to us all and in which he himself passionately has contributed as part of his prodigious research output.

Regional Science Perspective for the Future

Walter Isard opens up the book by projecting what's ahead. Since among regional scientists he can claim to have more years than most of struggle to understand the dynamics of spatial development, he dares to extrapolate the past. He starts with the Big Bang and ends up in today's development of advanced networks of communication.

Synergetics and Control in Networks

Hermann Haken focuses his attention on self-organization, whereby general concepts of the interdisciplinary field of synergetics as a general theory of self-organization are invoked. Basic concepts, such as order parameters, control parameters, and the slaving principle are briefly outlined and general laws for self-organization of patterns are sketched. Then the recognition of patterns is placed in analogy with pattern formation. In the next step, decision making is put in analogy with pattern recognition, whereby remarkable insights are obtained. Finally optimization problems in settlements, such as the location of buildings to building sites are discussed.

In his contribution *John Casti* advances the thesis that to address mathematically and computationally the major problem areas of control theory, we need an extension of the classical modeling paradigm upon which control theory rests. In his paper he discusses what elements any such extension necessarily involves, and outlines one extension of this type that enables us to account for crucial life-giving activities of self-repair and replication. A number of problems in biology, economics, and engineering are discussed as potential application areas of this new paradigmatic framework.

The chapter written by *Christopher L Barrett, Roland Thord* and *Christian Reidys* considers building and using simulations of socio-technical systems. These systems have (usually, but not necessarily, human) adaptive, learning, variable, etc., components as well as important social context, and essential physical/technological aspects. Although an actual list is much longer, this kind of system is exemplified by transport systems, financial systems, communication systems and even warfighting systems.

Jean-Pierre Aubin presents an investigation on one aspect of complexity closely related to the decentralization issue, connectionist complexity. This is a mechanism to adapt to viability constraints through the emergence of links between the agents of the dynamical economy and their evolution, together with or instead of making use of decentralizing "messages". Comparisons are made with other possible dynamical decentralization mechanisms, such as price decentralization.

Dynamics of Economical Networks

In the chapter written by *Donald G Saari* he argues that the emphasis on the parts of a problem at the expense of the whole can seriously distort the resulting conclusions. To explain this theme he appeals to Arrow's well known impossibility assertion from choice theory.

Anna Nagurney and *Stavros Siokos* present a dynamic model of international financial behaviour that yields the optimal portfolio of assets and liabilities of sectors in each country along with the equilibrium prices of the financial instruments in the currencies as well as the exchange rates. The model is analyzed qualitatively and conditions are provided for both stability and asymptotical stability of the financial adjustment process. They also prove that the equilibrium solution can be formulated as the solution to a network optimization problem.

The contribution by *Robert E Kuenne* deals with the difficult task of scaling characteristics that are not at the present time directly measurable in the meaning of the theory of measurement. Examples are given by attributes which can be measured only in a nominal 0-1 manner. The procedure in the paper moves from scaling at the product or brand level to studying competition without tacit collusion to price-nonprice competition in a collusive context, and from general functional forms in closed analysis to specific forms in simulation runs.

In the chapter written by our honored colleague *Åke E Andersson*, together with *Börje Johansson*, a specific aspect of the product cycle theory has been selected by stressing the conditions that bring about location advantages for individual urban regions. Along a product cycle path the knowledge intensity is high when a product is non-standardized and the process is non-routinized. To standardise and routinise means reduced knowledge intensity. The chapter presents a class of models which may be employed to explain this regularity. For a given product and product group, the location advantage is dynamic, i.e., it is technique-specific in the sense that the characteristics of the most advantageous economic milieu are different during the various phases of a product cycle. Propositions about this process are based on the assumption that a region's supply of resources and its entire economic milieu adjust on a slow time scale. The milieu functions as an arena for decision processes which determine plant locations, technique vintages, sales patterns and labour use along the evolution of a product cycle.

In the chapter written by *Tönu Puu* the location of a firm in geographical space, given two prelocated input sources and one prelocated output market, is studied in the spirit of Alfred Weber's classical contribution. It is seen that when production technology is not one of fixed proportions, but admits substitution, then there are multiple local optima for the choice of location. Gradient process dynamics with moderate step length is seen to become chaotic and move on fractal attractors. In this presentation the process of local gradient dynamics is studied in a case of three coexisting local optima.

Creativity, Learning and Knowledge in Networks

In the paper by *Kiyoshi Kobayashi* and *Kei Fukuyama* research issues to improve human interaction potential are discussed, where specific references are made upon modeling the possibility of face-to-face communication. The increased value of time helps to explain why people need more incentives for meeting as the knowledge society becomes more extensive, covering a wider spectrum of life. This analysis explains a frequent casual observation, which otherwise appears economically puzzling; human contacts in the knowledge society are increasingly sought but decreasingly attained.

In their contribution *Björn Hårsman* and *John M. Quigley* incorporate the spatial distribution of the demand for educational qualifications and the spatial distribution of the supply of educated workers into the knowledge production framework. This empirical analysis is based upon the worktrip behaviour of commuters in metropolitan Stockholm. The analysis is undertaken using gravity models and more sophisticated models like Poisson and the negative binomial relationship of worktrip behaviour and employment potential.

In the chapter written by *Antoine Billot* and *Tony E Smith* a model of candidate selection is proposed in which there exists only limited information about the candidates' abilities. It is assumed that only the reputation earned by candidates in competition with one another is known. Competition is here formalized in terms of contests which may involve individual and/or coalitions of individuals, and in which only winners are observed.

Scientific collaboration resulting in joint publications is discussed in the contribution written by *Martin J Beckmann* and *Olle Persson*. This is done both from the temporal and the spatial side of the problem. Studies from Sweden, Taiwan and the Nordic countries presented in the paper support the notion that informal face to face communication may be an essential ingredient in research collaboration and that greater geographical distance with the additional travel cost and time involved are impediments to collaborations.

The study presented in the contribution by *Wei-Bin Zhang* examines some aspects of the thoughts of Confucius from a social and economic point of view. The focus is concentrated on issues about learning, as well as moral and social organization. Zhang emphasizes some possible implications of Confucianism for economic and political development in China.

The Infrastructure Arena - Transportation and Communication Networks

David F Batten examines some behavioural traits of today's users of logistical networks. Unexpected outcomes may arise through co-evolutionary learning by a heterogeneous population of travelers. The paper explores the dynamics of congested traffic as a nested process involving co-evolutionary behaviour at three levels in space and time: the network as a whole, the vehicular flow pattern on each link or route and the driver population's travel. The evolutionary possibilities at each of these levels are highly interdependent. For example, seemingly small changes in the behavioural ecology of drivers can have profound effects on the final state of the transportation system as a whole.

The contribution given by *Peter Nijkamp* and *Heli Koski* aims to model the adapters' behaviour in modern information and communication networks from the viewpoint of both cross-sectional and sequential interdependencies. In particular it focuses on the economic consequences of the compatibility between network technologies. The formal analysis in the paper is supported by empirical evidence from the European microcomputer market. The empirical findings indicate that the incompatibility of technological components may have substantial implications for the spread of network technologies in general.

In his chapter *Folke Snickars* analyses the interdependency among transport investment projects. The objective of the analysis is to show how the existence of negotiation interdependencies on the demand side might affect the net benefits of alternative investment projects. A further aim is to investigate how the decision-making process preceding the decision to implement the projects might proceed under different patterns.

The chapter written by *David E Boyce*, *Der-Horng Lee* and *Bruce N Janson* describes applications and extensions of a dynamic network equilibrium model to the ADVANCE network. ADVANCE is a dynamic route guidance field test designed for 300 square miles in the northwestern Chicago area. The proposed dynamic route choice model is formulated as a link-time-based variational inequality problem. Realistic traffic engineering-based link delay functions are adopted to estimate link travel times and intersection delays of various types of links and intersections. Unexpected events that cause nonrecurrent traffic congestion are analyzed with the model. Route choice behavior based on anticipatory and non-anticipatory network conditions are considered in performing the incident analysis.

The chapter written by *T.R. Lakshmanan* highlights the important and recently increasingly role of the variety of information and knowledge technologies embodied in transportation capital and infrastructure. These technologies vastly improve the speed, reliability, and safety of transportation, while sharply dropping its costs. They provide information and knowledge vital for transport operations, enhancing their responsiveness and efficiency, and enabling and indeed driving innovations in transportation. The nature of these knowledge technologies are discussed as well as the factors underlying their rapidly expanding application in the transportation system. Special attention is also given the variety of benefits of the rapid dissemination of transport IT accruing to travelers, transport enterprises and to the customers of the enterprises in the emerging Knowledge society.

Regional Science
Perspectives for the Future

Commonalities, Regional Science and the Future: A Septuagenarian Perspective

Walter Isard
Department of Economics
Cornell University

Knowing well Åke Andersson's intense concern for the future and given the very title of this book, *Knowledge and Networks in the Dynamical Economy.* I find it impossible to avoid grappling with the problem of projecting what's ahead. And since among regional scientists I can claim to have more years than most of struggle to understand the dynamics of spatial development (space-time paths), I dare to extrapolate the past. I choose to do so beginning from the first event identified by scientific man, the Big Bang, and in doing this I will draw heavily upon the analysis in my forthcoming book, *Commonalities of Art, Science and Religion: An Evolutionary Perspective* (Scheduled for July 1996 publication by Avebury Press).

The first space-time process that I find common to all the fields of study I have ventured to explore is *symmetry breaking*. Particle physicists identify the first symmetry breaking as occurring when the force of gravity split off from the single Ur-force, when the graviton (a particle yet to be experimentally verified) split off from the indistinguishable Ur-matter. Subsequently, with decrease in temperature and pressure, there were symmetry breakings when the strong and electroweak forces split apart from the residual Ur-force, when the electroweak later divided into the weak and electromagnetic forces, when concomitantly quarks, leptons and gluons formed and were able to keep their identity after collision, and later other elements such as protons, neutrons and electrons (the basic constituents of atoms). Frequently, if not characteristically, the process of hierarchic construction took place, this process being a second fundamental *commonality* – whether we observe at one extreme the set of chemical elements ordered by number of shells of electrons and electrons in each shell, or at the other extreme the nested self-gravitating systems that comprise planets, stars, star clusters, galaxies and galaxy clusters.

The process of hierarchical construction continued subsequently and is omnipresent in biological organization. To illustrate we may point to the building blocks (e.g. amino acids and nucleotides) that form macromolecules (e.g. proteins, DNA), which then form macromolecular complexes (e.g. ribosomes, membranes) and so on through bacterial cells and higher cells, tissues, organs, and organ systems.

In human society, hierarchical structure is everywhere, and ever since the Dark Ages, structures of increasing complexity have been evolving. We clearly see this trend in development in the recent rapid change in the hierarchical configuration of the world's financial community and its network of communication channels.

Associated with increasing complexity of hierarchical configurations have been increasing *specialization* and concomitantly *scale, localization, urbanization and in general spatial juxtaposition and agglomeration economies as externalities*. This has been so ever since the Big Bang (for instance, in the formation of electrons, protons, nucleons, later in the formation of nuclei, then the many types of atoms of successively increasing mass and complexity), and in the development of organisms (for example, from noncellular organic matter to cellular, then to multicellular organisms and so on).

To illustrate more specifically, the forces leading to the bondings of like chemicals, such as hydrogen bonding, parallels the force of localization economies stemming from spatial juxtaposition of like enterprises. The forces leading to multiple bondings among elements --- the formation of a double bond from a single bond, a triple bond from a double bond, etc. that lead to decreasing bond lengths and increasing bond strengths --- parallel the forces lying behind scale economies (i.e. the reduction in unit costs and the strengthening of a firm's competitive position). The forces leading to the bonding of chemicals of different kinds and in different ways to form combinations, such as the hydrocarbons, parallel the forces that lie behind society's urbanization and agglomeration economies.

Stability and instability are properties of elements and structures that are of concern when investigating economic dynamics. These properties have been present in the world ever since the Big Bang – with regard to units of primordial matter, the early and later particles of matter, cells, structures of multicellular organisms and today's complicated multilevel hierarchical structures. Those efficient structures that have survived are ones that have by and large achieved *stratified stability* (to use Bronowski's term) – stability at each level of a hierarchy as each has been attained over time, a development that counters reversibility.

Social as well as biological hierarchical structures are dependent on *information flows*, lateral as well as vertical – flows that form a network with as complicated a pattern as that of the hierarchical structure. Typically there is a top node of a structure to and from which information is transmitted. Given the limited capacity of any decision-making unit to absorb and process information, it is clear that a highly efficient structure must apply Pattee's *principle of optimum loss of detail*, a Simon's *near-decomposable frame-work* or an equivalent factor) to preclude information overload.

In studying the functioning of social hierarchical structures and associated information networks one often tends to focus on the top node and the decision making at that level – as when we study the future financial community or world government and its associated information network implied by current and future advances in technology. However, one must also consider the supporting base at the ground level and close-by levels. Here, too, as implied by chaos and associated theory, there are commonalities such as *fractal structures, self-similarity at different scales*, and *self-reproduction* whose presence reflect the "potential" for negative externalities—diseconomies (which society implicitly tends to avoid) from excessive scale of operations, localization and urbanization or put otherwise from insufficient deglomeration, spatial dispersion and diffusion of activities.

In contemplating the future, Storper and Scott (1995), echoing general hypotheses of the older central place theory, state: "virtually all economic and social processes are sustained by transactions, by which we mean the transmission and exchange of information, goods, persons and labour. . . ." (p. 506) for example, transactions between sellers and buyers in a local market, between manufacturers of a particular good and others who supply inputs of materials, components, information and services, and between managers within a firm and workers on the shop floor. These and other transactions have a wide range of attributes which "impose constraints on the geographical scale at which they may feasibly be carried out. In very general terms, the greater the substantive complexity, irregularity, uncertainty, unpredictability and uncodifiability of transactions, the greater their sensitivity to geographical distance. In all these circumstances, the cost of covering distance will rise dramatically and in some cases, no matter what the cost, the transaction becomes unfeasible at great distance . . . Thus, we find transactions ranging from the virtually costless, instantaneous, global transmission of certain kinds of information or financial capital, where distance forms virtually no barrier, to those that have great qualitative complexity and high spatially dependent costs. The latter state of affairs applies to a wide variety of cases: the group of scientists working on a new frontier technology; the manufacturers and subcontractor in an information-intensive central city garment industry complex; the customer who expects fresh bread from the bakery in the morning; the student who needs frequent contact with a particular professor.

Thus, different kinds of transactions – and the communities that congeal around them – occur at different geographical scales. These range from the highly localized (household, neighborhood), to the regional, national and international levels" (pp.506-07).

The above conclusions have been demonstrated to be valid by past work of others such as the visionary Philbrick (1957, 1973), Philbrick and Brown (1972) and later scholars who have explored the implications of the pioneering works of Christaller (1966) and Lösch (1954) on the spatial distribution of central places and clusters of different sizes. Further, changes in technology as in communications that cause great cost reductions in certain kinds of transactions lead also to new specializations which invariably require new networks of transactions some of which are highly sensitive to geographic distance as a cost factor because of substantive complexity, uncertainty and other factors. Still more, despite the communications revolution and major advances of technology anticipated in the future, inequalities in per capita incomes among population within regions and the global society are likely to persist if not increase (there is very little in the historical record to support decreasing inequality). The marginalization of many local productive and service activities and of the people engaged in them is not likely to diminish, nor the constraints imposed by poverty and near poverty on the pattern of settlement and movement of a significant fraction if not the majority of the world's population.

If one is to be realistic in researching knowledge and networks in the dynamic economy, one must take into account another commonality namely the play of chance and *noise*, and at least speculatively consider the disintegration of economic structures from presence of excessive noise as has so often characterized the fall of civilizations. (Parenthetically, there have been attempts to incorporate noise formally in dynamic models – models that have employed the Master and Fokker-Planck equations [see Isard and Liossatos (1969) and Isard and Lobo (1996)] but as yet far from satisfactory.) The play of chance and noise from the environment is pointedly demonstrated by Prigogine (1984) in his analysis of transitions from an equilibrium or existing far-from-equilibrium state to another far-from-equilibrium state. The very simple transition from molecular chaos to molecular convection that he so well presents illustrates how the exact path that a deterministic framework takes at a point of bifurcation is strictly a matter of chance.

Often noise has a major and deep-seated impact on social development which is not easily unravelled, perceived and understood. Using biological terminology, wherein a human organism is essentially a *coadapted gene complex* with culture as a bag of *coadapted meme complexes* (A meme (as defined by Dawkins, 1989, p.290) is "a unit of cultural inheritance, hypothesized as analogous to the particulate gene, and as naturally selected by virtue of its 'phenotypic' consequences on its own survival and replication in the cultural environment."), chance together with other factors functions via *mutation* of genes, memes, and their complexes to generate unique human organisms (phenotypes). As one studies the rise and fall of empires and civilization, it becomes crystal clear that in their heyday there was at the top node of the socio-political-economic hierarchy a truly unique coadapted gene complex in the form of a tremendously outstanding leader [Cyrus/Darius in the Persian (Achaemenid) empire, Augustus in the Roman empire, Pope Innocent III in the papal empire] who controlled, for their times, highly elaborate networks of communication channels. Inevitably, these truly brilliant tremendously efficient administrators were in time succeeded by less competent administrators --- administrators unable to manage the intricate and highly efficient hierarchical structures that their predecessors had carved out. This *descendance* and subsequent *decay* of leadership sparked, in cumulative *autocatalytic* and *crosscatalytic* fashion, demoralization, depersonalization, devitalization, dissension, disquietude, disbelief, despoilation of resources, and so forth leading to the dissolution and disintegration of the finely honed hierarchical systems.

In the chaotic dissembling of these structures, common social diseases were the ungodly passions of incompetent officials for power, obsession with preserving the status quo, excessive impersonality and disconcern for civil rights, rampant plotting and occurrence of assassinations, widespread transmission of misinformation, rottenness in government rule and costly bureaucracy, disastrous management of the fisc, tax evasion and corruption in the tax system, ubiquitous and paralyzing inflation, increasing impoverishment of the poor and inability to combat pestilences, intensification of urban problems, civil disobedience, riots and other disorders and a number of other degenerating elements.

What better illustration of the degradation of society and the destruction of the vast interconnection and network of communication channels established in the heyday of the Roman empire than the feudal system's isolated manorial estates of the 10th century Dark Ages?

While in this brief paper, we are not able to expound upon all the commonalities that we have explored, and others which we have not yet perceived, what insights can be gained for the decades immediately ahead? One major implication is clear today. For our times, and its technology and until recently the unheard of and unimaginable advanced network of communication channels, there exists no unique, brilliant leadership able to guide and drive development along a space-time path that could materialize in decades, let alone centuries, of prosperity and relative peace. There exists no combination of a favorable background and a unique, tremendously capable and far-sighted coadapted gene complex to create an advanced global society much concerned with the welfare of its diverse constituents. Nor can we today, in reviewing our immediate past (the 20th century and even earlier), point to long periods of world prosperity and relative peace engineered by such leadership. As a consequence, today we have not inherited a finely-honed interdependent society, brilliantly constructed, with relativel; few fissures. Hence, today we do not confront a situation with significant *potential* for major descendance and decay of leadership. We are and have been for many decades dependent upon, accustomed to and necessarily receptive of independent, suboptimal leadership of diverse calibre at the national, regional, and local level. (Such leadership, operating in relatively uncoordinated fashion, would have been eliminated in an Augustan-type realm.) This suboptimal, albeit energetic, leadership is omnipresent in our emerging multi-internetted cyberspace and it provides many points of stability for local, regional and national societies. It copes with, admittedly frequently ineffectively, fissures in the international society; and sufficiently so as to dampen (unawaringly) the crosscatalytic snowballing process of decline. Put otherwise, there does exist today, a cadre of power-seeking leaders, diligent scholars, world citizens and other motivated individuals who can be expected, many in their own independent and uncoordinated ways, to muster and organize efforts of citizens to confront excessive terrorism, civil strife, warfare and destabilizing forces – thus to preclude a modern-day replication of the Dark Ages.

Thus in my role as a regional scientist, and a well-along septuagenarian, I see all kinds of developments possible over space and time. I see continued building up of hierarchies of all sorts, whose rates of construction are unpredictable. I see mounting specialization, and increasing scale, spatial juxtaposition and agglomeration economies with respect to some of both existing and new functions – economic, political, social and cultural. At the same time, I see more measurable spatial dispersion and diffusion of other functions and population settlement. I foresee no drastic revolutionary decline or growth of society – no change any more revolutionary character than that which followed (for their times) say the railroad, automobile and airplane in transportation, and the telegraph and telephone in communications. Change will just continue via the emergence of patterns consistent with the general commonalities within the universe that have been explicit and implicit since the 10^{-12} fraction of a second after the Big Bang – but for the present in no drastic manner.

References

Bronowski, J. (1970), "New Concepts in the Evolution of Complexity: Stratified Stability and Unfounded Plans," *Zygon*, Vol. 5, No. 1, pp. 18-35.

Christaller, Walter (1966), *The Central Places of Southern Germany*, Englewood Cliffs, N.J.: Prentice-Hall.

Dawkins, Richard (1989), *The Extended Phenotype*, New York: Oxford University Press.

Isard, Walter and Panagis Liossatos (1979), *Spatial Dynamics and Optimal Space-Time Development*, Amsterdam: North Holland.

Isard, Walter (1996), *Commonalities in Art, Science and Religion: An Evolutionary Perspective*, Brookfield, Vt.: Avebury.

Isard, Walter (1996), "Perspectives of a Political Geographer on the Future," forthcoming in *Futures*.

Isard, Walter and Lobo, Jose, "Impact of Noise on Regional Development," unpublished.

Lösch, August (1954), *The Economics of Locations*, New Haven: Yale University Press.

Pagels, Heinz (1983), *The Cosmic Code*, New York: Bantam.

Pattee, H.H. "The Physical Basis and Origin of Hierarchical Control," in *Hierarchy Theory* (H.H. Pattee, ed.), New York: Braziller.

Philbrick, Allen K. (1957), "Areal Functional Organization in Regional Geography," *Papers*, Regional Science Association, Vol. 3, pp. 87-98.

Philbrick, Allen K. (1973), "Present and Future Spatial Structure of International Organization," *Papers*, Peace Science Society (International), Vol. 21, pp. 65-72.

Philbrick, Allen K. (1972), "Cosmos and International Hierarchies of Functions," *Papers*, Peace Science Society (International), Vol. 19, pp. 61-90.

Prigogine, Ilya and Isabelle Stengers (1984), *Order Out of Chaos*, London: New Science Library.

Storper, Michael and Scott, Allen J. (1995), "The Wealth of Regions: Market Forces and Policy Imperatives in Local and Global Context," *Futures*, Vol. 27, No. 5, June, pp. 505-526.

Synergetics and Control in Networks

Decision Making and Optimization in Regional Planning

Hermann Haken
Institute for Theoretical Physics and Synergetics
University of Stuttgart

1. Introduction

By means of my contribution I wish to express my high esteem of Åke E. Andersson and his fundamental contributions to regional science [1]. In my article I will make an attempt to show how my own field of research *Synergetics* [2] might be able to contribute to regional science. My starting point will be that settlements with their buidlings, such as homes, service centers, stores a.s.o. can be considered as patterns in an abstract sense [3]. At the same time, these arrangements serve specific functions so that we may also speak of *specific functional patterns*. A fundamental problem of regional planning is the question of the competition or interplay between organization (planning) versus self-organization. Synergetics may be considered as a rather general theory of self-organization. Self-organization means that a system acquires its specific structure or function without specific interference from the outside. One must bear in mind, however, that self-organization requires general conditions that are imposed on the region by the planner or by the environment. Thus self-organization is - in most cases - never just a wild growth, but a growth under certain conditions. On the other hand, the details of the evolving structures need not be fixed from the outside. It came as a big surprise to a number of scientists that self-organization may take place not only in the animate world of animals and plants, but also in the inanimate world of physics and chemistry. And this was actually the starting point of synergetics. A typical example is provided by fluid dynamics. When a fluid is heated from below, it may spontaneously form specific patterns, e.g. hexagons, in spite of the fact that the heating from below occurs quite uniformly (Fig. 1). Thus the evolving pattern shows a feature inherent in the system and not imposed on it from the outside.

Figure 1: A fluid heated uniformly from below may sponta-neously form patterns in form of hexagons. In the center of each cell the fluid rises, cools down at the surface and sinksdown at the border.

In the first part of my paper (section 2) I wish to remind the reader of basic concepts of synergetics that allow us to deal with the spontaneous formation of patterns in a great variety of systems that may range from physics over biology to regional planning. Then, in the next section 3, I want to study how humans perceive patterns, or, in other words, how they recognize specific patterns. In section 4 I will argue that making decisions in human life can be directly related to pattern recognition. We will see that a number of phenomena that we encounter in pattern recognition are repeated in decision making. Such decision making may, of course, refer to decision making with respect to specific plans of settlements. As it will turn out, this kind of decision making described here is of a rather psychological nature. Then I wish to address the same problem by making rational decisions on account of optimization, for instance how buildings must be arranged in order to increase or maximize the efficiency of a settlement. Astonishingly it will turn out that the decision making related to psychological factors and optimization are based on practically the same formalism. As a consequence it turns out that pattern formation by self-organization, pattern recognition, decision making and optimization have the same root. To ap-proach a broad audience, I will formulate these sections without mathematical formu-las. Then in section 6, I will present a brief sketch of the underlying mathematics and concepts so that the interested reader can penetrate more deeply in my approach. Section 7 then will provide the reader with an outlook.

2. Pattern Formation by Self-Organization: the Core of Synergetics

In many fields of the natural sciences, we can observe the spontaneous formation of patterns or, in other words, of structures. Prominent examples are, of course, the growth of plants and animals out of germs. Here we see a great variety of structures or forms and at the same time a variety of behavioral patterns. The spontaneous formation of patterns can also be observed in the inanimate world, such as in fluids. Some essential features of these pattern formations can be elucidated by the simple example of a fluid heated from below (Fig. 1). When the heating is only small, the fluid remains at rest. However, when the heating from below exceeds a critical magnitude, suddenly a pattern, e.g. in form of rolls, occurs (Fig. 2). With increasing heating the rolls start moving more rapidly. As the mathematical analysis reveals, close to the point where the pattern is formed, the system tests a number of different rolls or other patterns. However, when one or several patterns grow faster than any other pattern, competition sets in. This competition is won by that pattern which grows fastest. All the other patterns are suppressed. The amplitude or size of the winning pattern is called the *order parameter*. Once the order parameter is known, the whole pattern is determined. Thus instead of describing the individual motion of the individual molecules of a fluid, it is sufficient to describe the corresponding order parameter. This means that complex systems may be described by few variables, only provided they are close to instability points where a new pattern evolves. Under specific circumstances, several order parameters may have the same growth rate and thus may evolve in principle. In this case the initially strongest order parameter wins the competition and determines the eventually evolving pattern. The mathematical formalism of synergetics [2] allows us to calculate the evolving patterns provided the microscopic laws for the formation of patterns are known. As it turns out, order parameters may compete as was mentioned above, but they also may coexist, or even cooperate. An important feature of order parameters is that they change much more slowly than the individual parts of the system. This allows the individual parts of the system to follow the order parameters immediately. Such a time scale

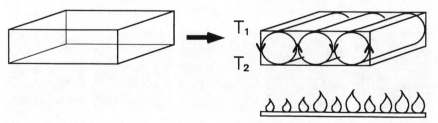

Figure 2: Spontaneous formation of rolls in a fluid heated from below if the temperature difference between the lower and the upper surface T_2, T_1 is sufficiently large.

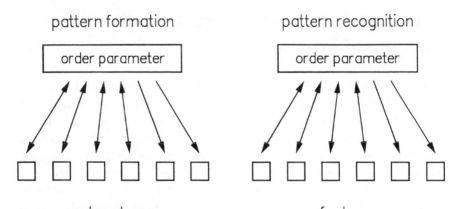

pattern formation   pattern recognition

order parameter   order parameter

subsystems   features

Figure 3: Analogy between pattern formation and pattern recognition.
L.h.s.: If part of the subsystems is in an ordered state, they call upon the order
parameter that, in turn, brings the total set of subsystems into the ordered state.
R.h.s.: If part of a pattern is given, it calls upon its order parameter that, in turn,
completes the whole pattern.

separation was recognized rather early in economy by Marshall and others and Åke Andersson has pointed at the importance of this time scale separation at many instances, where he included, for instance, also educational systems. The time scale separation is the basis of the so-called slaving principle of synergetics. It states that once the order parameters are known, it is possible to determine the states of the individual elements of a system. On the other hand, the elements generate the order parameters by means of their cooperation. This methodology has found numerous applications that we shall not dwell upon here, because it has been described elsewhere and is not quite in the center of our paper.

Figure 4: Examples of stored prototype patterns with family names encoded by letters.

Figure 5: When part of a face is used as an initial condition for the pattern reco-
gnition equations, their solution yields the complete face with the family name.

3. Pattern Recognition as Pattern Formation

We now want to show that pattern recognition by humans or advanced computers is nothing but pattern formation. First we have to explain what pattern recognition means. For instance when we see the face of a person, we want to associate with the face the name of the person or instances where we met this person, a.s.o. Thus pattern recognition is based on an associative memory. In synergetics we have shown that pattern recognition can be conceived as pattern formation [4]. To this end, look at Fig. 3. At the left-hand side the lower part shows a systems symbolically, where some of the elements are already in an ordered state, whereas the rest is not yet in that state. According to synergetics, the already ordered elements call upon their order parameter, which in turn acts on the system and forces it to adopt the fully

Figure 6a: Examples of city maps with their labels (H: Hambourg, B: Berlin, S: Stuttgart, M: Munich) are stored in the synergetic computer as prototype patterns.

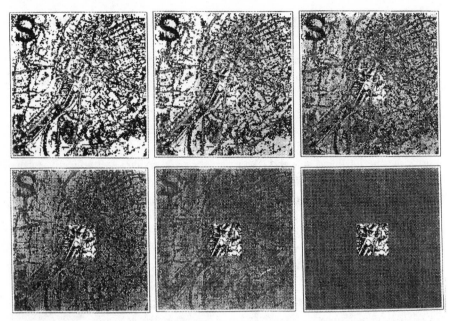

Figure 6b: Recognition, i.e. completion including label of a city map.

ordered state by means of the slaving principle. We claim that nothing else happens in pattern recognition. Part of a system, for instance part of a face, that is shown to the computer or to the observer calls upon the corresponding order parameter, which reacts on the other parts and forces the whole system into the complete pattern. Examples of this procedure are shown in Figs. 4,5. In this sense we may also state that when we have stored a number of cities or settlements in our mind, then we can complete these patterns to a full pattern even if only part of these patterns are provided (Figs. 6a,6b). There are a number of interesting instances of pattern recognition by humans. For instance when we look at the pattern of Fig. 7, we see either a vase or two faces. Interestingly enough the vase or the two faces cannot be perceived all the time, but after a while, for instance, the vase vanishes, gives place to the two faces, then the two faces vanish, a.s.o. Thus an oscillatory process sets in. This phenomenon can be explained by the assumption that attention for that specific percept, e.g. the vase, fades away and allows the brain to perceive the other interpretation, a.s.o. (Fig. 8). In a

Figure 7: Example of an ambiguous pattern: vase or two faces?

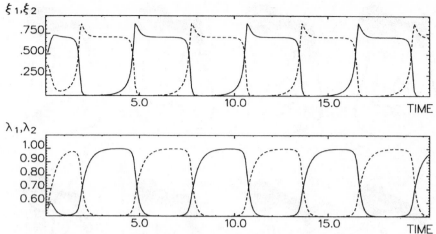

Figure 8: Oscillations of the order parameters (upper part) and of the corresponding attention parameters (lower part)

number of cases there may be a bias. For instance in Fig. 9 80 percent of male people first perceive the young woman, whereas 60 percent of females perceive the old woman first. By a detailed model we found that the period of oscillation becomes longer for the preferred interpretation as compared to the less favored interpretation. Finally we may observe the phenomenon of so-called hysteresis, which is explained by Fig. 10 and its legend. Instead of showing faces, we can, of course, show also city maps or other illustrations of settlements. But faces are perhaps still more impressive to humans.

4. Decision Making as Pattern Recognition

There is a huge amount of literature on decision making. Here we want to argue that decision making can be considered as pattern recognition (cf. also Table 1). In this psychologically based approach we first establish some similarity between a situation that is newly posed to us with a situation that has been encountered by us before. So in complete analogy to pattern recognition we first establish a similarity measure between a newly posed pattern and one stored in our mind when the pattern recognition process,

Figure 9: Ambiguous pattern: Young woman/old woman?

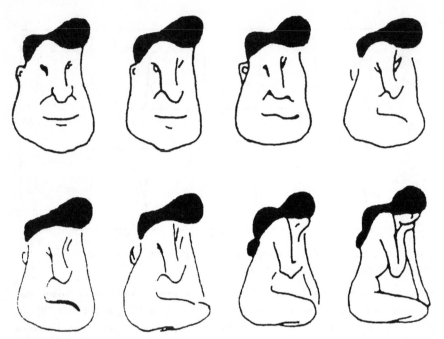

Figure 10: Example of hysteresis in pattern recognition. When the sequence of figures is looked at from the upper left to the lower right corner, the switch from a man's face to a woman occurs in the lower row, when looked at in the opposite direction, the switching occurs in the upper row.

that can be mimicked by the so-called synergetic computer, leads to a decision that is based on the largest similarity. However, other cases may occur also, namely for instance in ambiguous situations we start an oscillation between two different or even several equivalent patterns. This is quite in analogy to the oscillations between vase and faces or between young and old woman. Finally the phenomenon of hysteresis may occur in the interpretation related to decision making. We may state that we make a decision on the basis of a situation that we know in spite of the fact that the situation has changed. Quite evidently, these considerations here are of a rather psychological nature, namely we may establish a similarity between a new situation and a situation that has been known to us before. I argue that such kind of behavior occurs also in planning. The reason for this is that planning settlements or cities is an extremely involved task, which bears many possibilities. Thus one is led to resort to those solutions that have been achieved before more or less successfully. The central issue here is, of course, twofold, namely we must have a repertoire of known solutions, for instance specific patterns of settlements, and we must introduce a specific similarity measure. Thus the crucial intellectual task consists here of establishing similarities. How this can be done in quantitative terms will be outlined in chapter 6.

Table 1. Correspondence between the elements and processes of pattern recognition and those of decision making

Pattern Recognition	Decision Making
patterns	data, quantitative
pictures	qualitative, yes/no
arrangement of objects	rules, laws, regulations
visual, acoustic signals	algorithms, programs
movement patterns	flow charts, diagrams
actions	orders
(often encoded as	multi-dimensional
vectors)	in short: "data"
prototype patterns	sets of known complete "data"
learned or given	learned or given
test patterns	incomplete data
	in particular "action" lacking
similarity measure	
dynamics	
bias	$\longrightarrow$
attention, awareness	
unique identification	unique decision
or	or
oscillations between	oscillations between
two or more percepts	two or more decisions
hysteresis	do what was done last time
	even under changed circumstances
complex scenes	multiple choices
saturation of attention	failure, new attempt
	based on new decisions
	"heuristics"

5. Optimization

As it will turn out in the mathematical formulation, there are two ways to steer the competition process between different kinds of decisions or, if one wishes, plans. The first approach, i.e. the foundation of pattern recognition, is based on the assumption that all stored patterns are basically equivalent to each other, but that there is an initial bias towards a specific similarity between a new situation and a formerly known one. Under these circumstances the pattern with the strongest similarity to the newly given pattern wins the competition and the decision making process. On the other hand, we may give all the patterns that are basically possible the same opportunity, but we pay to a specific pattern a higher attention than to the other one. Then the mathematics shows that, indeed, the pattern to which the highest attention is paid wins the competition in spite of the fact that all other patterns are equivalent. This is, of course, also of psychological interest. The question arises whether we can overcome these psychologically based decisions by specific optimization procedures. Astonishingly enough, the optimization procedure that I developed is based on precisely the same processes underlying decision making. It seems that here we are unearthing relationships that have not been explored yet, but that deserve further exploration because of the challenge of combining intuition and rigorous methods in regional planning.

In order to elucidate my ideas and to be as simple and concrete as possible, I consider quite a specific problem in settlements, namely the attachment problem. Let us label sites for buildings by the index i and specific buildings, such as homes, service centers, schools, factories, a.s.o. by an index j. Then quite evidently there is a relationship between site i and building j with a specific efficiency κ_{ij}. The task is to attach the buildings to the sites in such a way that the total effficiency function $\sum \kappa_{ij}$ becomes a maximum with a specific choice of the attachments between j and i. I will cast this optimization problem into a mathematical form. As it will turn out, the solution of this attachment problem is closely related to both the pattern recognition and pattern formation problem. This may not be so surprising, because the attachment problem means that we have to find a specific pattern. What is interesting is, however, that the formalism turns out to be practically identical in both cases. Interestingly, the individual efficiency factors turn out to be just the attention parameters in pattern recognition. This allows us to reformulate the whole problem; we look for such solutions of the optimization problem so that we pay the highest attention to the highest efficiency.

This is a surprising result, because it is not trivial at all and is, on the other hand, a direct consequence of this formalism. Quite evidently, these concepts can be pushed much further, because the attachment problem can be given a much more general formulation, where also traffic connections and inter-relationships between different kinds of buildings are taken into account. While the formulation of this problem is, indeed, possible to a rather large extent, its explicit formulation would require too much space here and must be postponed to a later occasion.

6. Mathematical Formulation

According to the sections 2 – 5 I will present the mathematical approach in several steps. Basically we have to distinguish between the micro- and the macrolevel. At the microlevel we describe the state of a system by the state vector $q = (q_{1,...}, q_N)$. In physics q_j may be, for instance, the density of a fluid in a cell labelled by the index j. In settlements q_j may denote the number of buildings at a site j. When we split the index j in further indices, we may also differentiate between different kinds of buildings at the same site j. Basic to synergetics is the idea that the state vector $\mathbf{q}$ changes in the course of time because of interactions between different sites. We may also include the effect of general external conditions that are described by so-called control parameters α, which in a fluid may be the amount of heating or which in settlements may be the amount of immigration of people, a.s.o. Further we will take into account chance events by adding so-called fluctuating forces $\mathbf{F}$ that may depend on time. The general evolution equations are then given by

$$\dot{\mathbf{q}} = \mathbf{N}(\mathbf{q}, \alpha) + \mathbf{F}(t), \tag{1}$$

where $\mathbf{N}$ is a nonlinear function of the state vector $\mathbf{q}$.

The basic assumption underlying the approach of synergetics is based on the role of the control parameters. As is witnessed by many processes in a variety of systems, the behavior of a system may change dramatically if a control parameter exceeds a certain critical value α_c. If $\alpha = \alpha_0$ is below such a control parameter value, for instance the rate of immigration, we may deal with a specific solution $\mathbf{q}_0$, for instance a homogeneous distribution of buildings or of a liquid

$$\alpha_0 \ : \ \mathbf{q}_0. \tag{2}$$

But when the control parameter α exceeds a critical value, a totally new solution may set in, for instance a city may spontaneously grow and change its character dramatically. In order to take account of this effect, we make the hypothesis

$$\alpha \ : \ \mathbf{q} = \mathbf{q}_0 + \mathbf{w}(t), \tag{3}$$

where $\mathbf{w}$ is assumed to be initially small. This allows one to linearize eqs. (1). The solution of the linearization reads

$$\mathbf{w}(t) = \exp(\lambda_j t)\mathbf{v}_j, \tag{4}$$

where $\mathbf{v}_j$ is a time-independent vector that corresponds to a specific pattern. The exponent $\lambda_j \geq 0$ indicates growth and we put the index $j = u$ ("unstable") so that

$$\lambda_j \geq 0 \ : \ \lambda_u, \mathbf{v}_u. \tag{5}$$

Other configurations may decay, which is indicated by $j = s$ ("stable")

$$\lambda_j < 0 \ : \ \lambda_s, \mathbf{v}_s. \tag{6}$$

The exact solution of (1) can then be represented by the superposition

$$\mathbf{q} = \mathbf{q}_0 + \sum_u \xi_u(t)\mathbf{v}_u + \sum_s \xi_s(t)\mathbf{v}_s, \tag{7}$$

where we sum up over the unstable modes with index u and the stable modes with index s. ξ_u and ξ_s are time-dependent amplitudes. The slaving principle of synergetics – that is quite fundamental – shows that the amplitudes ξ_s can be uniquely expressed by the amplitudes ξ_u

$$\xi_s = f_s(\xi_u). \tag{8}$$

Since the number of all possible patterns $\mathbf{q}$ is much larger than the number of ξ_u, (8) implies an enormous information reduction. In a number of important cases, the behavior of the whole system is governed by one or few amplitudes ξ_u that are called the order parameters. As it turns out, the whole dynamics of the complex system, described by (1), can be reduced to the dynamics of the order parameters ξ_u that again obey nonlinear equations that can be derived explicitly in each specific case

$$\dot{\xi}_u(t) = \tilde{N}_u(\xi_u, \alpha) + F_u(t). \tag{9}$$

The abstract formalism described in eqs. (1)–(9) allows us to calculate the evolving patterns in a great variety of systems, expecially in physics, chemistry and biology, but also in sociology, etc.

We now turn to the problem of pattern recognition. In this case we choose a specific order parameter equation, namely of the form

$$\dot{\xi}_u(t) = \xi_u\left(\lambda_u - \sum_{u'} B_{u,u'}\xi_{u'}^2\right) + F_u(t). \tag{10}$$

Such an equation arises already in fluid dynamics and describes the competition between patterns. In pattern recognition, the λ_u's play the role of attention para-meters. If the λ_u's are the same for all so-called modes u, and $B_{u,u'}$ is of the form

$$B_{u,u'} = A\delta_{u,u'} + B(1-\delta_{u,u'}), B > A, \tag{11}$$

the eqs. (10) have the property that the initially strongest amplitude ξ_u wins the competition so that, eventually, $\xi_u = 1$ for the initially biggest ξ_u and $\xi_u = 0$ for all others. As a side remark we mention that the eqs. (10) can be conceived as a gradient dynamics

$$\dot{\xi}_u(t) = \frac{\partial V}{\partial \xi_u}, \tag{12}$$

where the potential V is given by

$$V = -\frac{1}{2}\sum_u \lambda_u\xi_u^2 + \frac{1}{2}\sum_{uu'} B_{u,u'}\xi_u^2\xi_{u'}^2. \tag{13}$$

Before we exploit this property, we briefly discuss what determines the biggest ξ_u in pattern recognition. To this end, we form the so-called overlap function between the stored prototype patterns $\mathbf{v}_u$ (cf. Fig. 4) and a test pattern vector $\mathbf{q}$ (cf. Fig. 5) according to the scalar product

$$\xi_u(0) = (\mathbf{q}(0)\mathbf{v}_u). \tag{14}$$

An example for stored prototype vectors $\mathbf{v}_u$ in the form of city maps and the recognition process are shown in Figs. 6. According to the dynamics described by (10), the final solutions are given by

$$\xi_u(t=\infty) = \begin{cases} 1 & u = u_c \\ 0 & u \neq u_c \end{cases},$$ (15)

where the order parameter wins with the biggest overlap (14), i.e. with the greatest similarity between the initially given vector $\mathbf{q}$ and the stored prototype patterns $\mathbf{v}_u$. In order to take into account the saturation effect of attention that occurs in ambiguous figures, we make the attention parameters time-dependent according to the differential equation

$$\dot{\lambda}_u = \lambda_{u,0} - \lambda_u - \xi_u^2.$$ (16)

In the example of two possible interpretations, the eqs. (10) jointly with (16) yield the oscillatory process as given by Fig. 8.

Finally we mention that we can include also a bias in the eqs. (10). and change them in such a way that the effect of hysteresis is taken into account (cf. Fig. 10).

So far we have given an outline of the process of pattern recognition and correspondingly of decision making. While in pattern recognition the prototype vectors $\mathbf{v}_u$ and the test pattern vectors $\mathbf{q}$ can be rather easily encoded by means of grey values of Fig. 4, decision making is a harder problem, because we have to encode abstract patterns in terms of vectors $\mathbf{v}$ and $\mathbf{q}$. For a comparison between pattern recognition and decision making consult Table 1 above.

In order to finish the sketch of the mathematical approach, let us now look at the attachment problem, where the total efficiency has to be maximized. Let us denote the different building sites of a settlement by an index i and the specific buildings by the index j. We further denote the efficiency relating building j to a building site i by κ_{ij}. Then the question is to maximize the efficiency according to

$$\sum_{ij} \kappa_{ij} \xi_{ij}^2 = \text{Max!},$$ (17)

where ξ_{ij} may take the values 1 or 0 only. Provided we have the same number of buildings as of building sites, and only one building can be built on a site, then ξ_{ij} must be a permutation matrix, for instance in the case of three buildings and three sites in the form (18). It is important to note that each row and column may contain only a single 1, and that all other matrix elements must be equal to 0.

$$\begin{pmatrix} 0 & 1 & 0 \\ 1 & 0 & 0 \\ 0 & 0 & 1 \end{pmatrix}.$$ (18)

In order to solve the optimization problem (17), we have to take into account the specific possible form of the permutation matrices which can be cast into the following constraints

$$\sum_j \xi_{ij}^2 = 1, \tag{19}$$

$$\sum_i \xi_{ij}^2 = 1, \tag{20}$$

$$\xi_{ij}^2 \xi_{kj}^2 = 0 \quad \text{for} \quad i \neq k, \tag{21}$$

$$\xi_{ij}^2 \xi_{ik}^2 = 0 \quad \text{for} \quad j \neq k. \tag{22}$$

We now take up an earlier idea [4] on how to take the constraints (19) – (22) into account. We define a total potential in the form

$$V = -\sum_{ij} \kappa_{ij} \xi_{ij}^2 + \Lambda_1 \left(\sum_j \xi_{ij}^2 - 1 \right)^2 + \Lambda_2 \left(\sum_i \xi_{ij}^2 - 1 \right)^2$$

$$+ \Lambda_3 \left(\sum_{i \neq k} \xi_{ij}^2 \xi_{kj}^2 + \sum_{j \neq k} \xi_{ij}^2 \xi_{ik}^2 \right). \tag{23}$$

As one may easily convince oneself, the form of the constraints with the multipliers Λ_j is a straight forward generalization of the constraints implicitly used in the pattern recognition process in eqs. (10) or (12, 13). Such a generalization has been formulated more recently by J. Starke [5], however, with one fundamental difference, because he does not include the term (17). He assumes that the competition between the configurations ξ_j is provided by the initial values, where the ξ_i, ξ_{ij} with the largest initial value wins the competition under the constraints (19) – (22). Here our approach is basically different, because in the present case the configurations ξ_{ij} win the competition that maximize (17) under the constraints (19) – (22). One can show that this competition process can be won only if the coefficients of the Lagrange parameters $\Lambda_1, \Lambda_2, \Lambda_3$ are big enough. The question of distinguishing between different local minima may still be open in contrast to the eqs. 12, 13, where a general theory of possible patterns and their minima of the potential function (13) exists.

7. Concluding Remarks and Outlook

In my contribution I have tried to show, how a number of concepts that are applicable to the question of planning of settlements stem from a single root. The question, of course, arises in the present concept, where self-organization comes in. In the approach outlined above, the original eqs. (1) contain, on the one hand, the internal interactions between the different components of a system and control parameters. Under the impact of control parameters describing specific sets of conditions that are based on the environment, like rivers, mountains, a.s.o. or, in our case more realistically, on specific planning actions, the system (1) acquires a certain state. It is important to note that the number of control parameters α is much smaller than the number of the components of the state vector q. In spite of this fact, because of the internal dynamics imposed on the system by the specific form of the function N, the system may acquire only a few stable states under favorable conditions imposed on the system by the control parameters. Thus only few specific patterns may evolve. The mathematical formulation allows one to determine these new evolving patterns. The evolving patterns are determined by a few order parameters ξ_u only. Then, in the next step, we have inverted the whole consideration, namely we have asked how we recognize patterns. This has led us to the concept of associative memory. The crucial role is played here by the attention parameters λ_u that may be considered as parameters controling our attention. For instance, if we set attention parameters for specific patterns equal to zero, then we are not able to perceive these patterns. The attention parameters also determine the competition between different stored patterns, when a newly pattern is given. If all patterns are given the same chance, the attention parameters decide which pattern is finally recognized.

Finally we formulated a maximization problem, which again could be cast into a form in complete analogy to the pattern recognition problem. Here the efficiencies κ_{ij} play the role of attention parameters. But because of the constraints not all configurations with the highest attention can be realized, but compromises must be found, which is expressed by a total potential function (23). Quite clearly, the mathematical approach can be extended in a variety of ways, for instance to more complicated maximization problems that take into account the interaction between different sites and also take into account still more complicated constraints than in eqs. (19) – (22).

Furthermore it may be stated that the minimization of the total potential function (23) need not be taken care of by a method of steepest descent as described by eqs. (12), but that other minimization procedures, such as genetic algorithms, a.s.o., may be employed here, too.

Finally we remark that the eqs. (10) define the algorithm of a so-called synergetic computer, which is a fully parallel computer, and represents an alternative to neural networks. The further elaboration on this approach would take, however, too much space and we refer the reader to the literature [4].

In conclusion it can be stated that precisely the same formalism allows us to treat pattern formation by self-organization, pattern recognition, decision making and optimization from a unifying point of view. I am confident that this will allow for a number of concrete applications in regional planning,which were alluded to by the example of an attachment problem.

References

[1] Among the numerous important publications of Åke E. Andersson are:
Regional Diversity and the Wealth of Nations (in Swedish), Stockholm (1984)
Creativity – Future of the Metropolis (in Swedish), Stockholm (1985)
The Future of the C-society (in Swedish), Stockholm (1988)
Creativity – Future of Japanese Metropolitan Systems (in Japanese) (1989)
Regional Development Modelling: Theory and Practice, North Holland, Amsterdam
   and New York (1982)
Advances in Spatial Theory and Dynamics, North Holland, Amsterdam and New
   York (1989)
The Cosmo-Creative Society, Springer-Verlag, Berlin, Heidelberg (1993)
Dynamical Systems, World Scientific (1993)

[2] H. Haken: Synergetics. An Introduction, 3rd. ed., Springer-Verlag, Berlin,
   Heidelberg (1983)

[3] H. Haken, J. Portugali: A Synergetic Approach to the Self-Organization of Cities
   and Settlements. Environment and Planning B: Planning and Design, vol. 22, p.
   35-46 (1995)

[4] H. Haken: Synergetic Computers and Cognition, Springer-Verlag, Berlin,
   Heidelberg (1990)

[5] J. Starke: Kombinatorische Optimierung auf der Basis gekoppelter Selektions-
   gleichungen, Ph.D. Thesis, Stuttgart (1997)

Putting Some Life Into the System

John L. Casti
Santa Fe Institute
and Institute for Econometrics, OR, and System Theory
Technical University of Vienna

1. On the "Biologization" of Control

In September 1986 a blue-ribbon panel of 52 experts met in Santa Clara, California to try to put a fence around the *terra incognita* on the map of modern control theory. For those concerned with the future direction of the field, it's instructive to reflect on the following points noted in the panel's final report [19]:

> These *large-scale systems* consist of many subsystems which have access to different information and are making their own local decisions, but must work together for the achievement of a common, system-wide goal ... each subsystem may be operating with limited knowledge of the structure of the remainder of the system, which may be reconfiguring itself dynamically due to changing requirements, changes in the environment, or failure of components.

> Some elements that appear in this research ... which have not been fully accounted for in the past are as follows: ... viewing reliability and survivability as key performance specifications, rather than issues of secondary concern.

> There is the need to include models of human decision-makers in designing distributed architectures.

> ... controllers are required to be robust with respect to modeling uncertainty and to adapt to slow changes in the system dynamics ... we lack at present a set of prescriptive methodologies that can be used to design fault-tolerant feedback control systems

> ... control scientists and engineers must face head-on the challenge of increased complexity

To control theorists, especially those concerned with applications, these observations will hardly be a source of controversy. But what is of great interest is their source.

These statements represent the collective view of a group whose composition was almost 100% engineers and applied mathematicians. Yet the remarks themselves, while describing problems arising in various applied engineering and technological areas like aircraft design, chemical process control and electronic signal processing, could have served equally well to describe the problems faced by any living thing. In short, the list of significant problem areas for 21st-century control theory outlined in [19] is nothing less than a reformulation in engineering jargon of the problems that living organisms solve routinely each and every moment of their lives.

Recognition of the nearly one-to-one match up between the problems that control engineers see as significant and the problems that living organisms discovered how to solve millions of years ago suggests the following research strategy: Let's try to reformulate the standard modeling paradigm of control theory in order to bring it into better congruence with the way living organisms function. What such an undertaking involves is extending the conventional input/output or state-variable frameworks upon which modern control theory rests to take explicitly into account processes like self-repair and replication that characterize all known living things.

Any such extension necessarily involves introducing the idea of intelligent control into the situation, since the concepts of self-repair and replication cannot possibly take place unless the system has a model of itself against which it can measure the result of carrying out these two functional activities. The balance of this article will be devoted to showing one path by which this kind of extension can be carried out in a mathematically "natural" manner. The paper concludes with a discussion of several application areas where these ideas of intelligent control might be profitably employed.

2. Cartesian vs. Intelligent Control

When he made his famous statement, "Cogito, ergo sum," René Descartes was implicitly separating the world into two parts: the self or *"sum,"* and the environment (the non-*sum*, so to speak). This Cartesian separation of the world shows up in conceptual clothing in traditional control theory, where the self is identified with the controller or decisionmaker who broadcasts controlling instructions into the "non-self," the system that is to be regulated. When translated into mathematical language, such a procedure generally leads to a differential equation of the following sort:

$$\dot{x} = f(x(t), u(t)),$$

where $x(t) \in X$ represents the state of the controlled system at time t, while $u(t) \in U$ is the control signal injected into the process by the controller. In *optimal* control theory, this signal is chosen to maximize or minimize some criterion function. For instance,

$$\max_{u} \int_0^T g(x(s)u(s))ds$$

Sometimes the controlling signal is taken to be explicitly a function of the time t, i.e., $u = u(t)$, in which case we speak of *open-loop* control. In other instances the control action is explicitly a function of the system state and only implicitly a function of t, i.e., $u = u(x(t))$, in which case we speak of *feedback* control. But in either instance, the setup represents only a slight dressing up of the Newtonian formalism passed down to control theorists via the medium of classical particle mechanics.

Whatever "intelligence" there is in the foregoing control-theoretic framework resides in the controller, not in the system being regulated; the system itself possesses no knowledge of its own behavior, and blindly blunders along taking its natural course unless acted upon by an outside observer, the controller, who can see where the system is going and take steps to modify that course if need be. If the term "intelligent control" means anything in modern parlance, it means that this picture is wrong from the very outset as a framework upon which to base a theory of "post modern" control processes.

Intelligent control systems are those in which the system and its controller are inseparably intertwined, somewhat like the two sides of a coin. One might say that an intelligent control system is one to which the reductionistic notion of the controller being separately identifiable from that which is being controlled does not apply; the controller *is* the system – and vice-versa. Of course, this is exactly the situation we find when we begin looking at biological processes. For example, the proteins in a living cell simultaneously carry out the metabolic actions of the cell as well as act to regulate those very same actions. Thus, it's impossible to point to any particular material component of the cell and say *that* "piece" is the controller and *that* one is what's being controlled. Nevertheless, the properties – metabolism, repair, replication – characteristic of living processes are precisely those we would like to associate with the sorts of systems we would deem to be "intelligent." But to create a mathematical framework that incorporates these properties forces us to extend the Cartesian/Newtonian view of the world in several rather dramatic ways, all of which ultimately rest upon the idea of injecting the concept of self-reference into the mathematical machinery. But rather than continue to speak in these vague, general terms, let me turn to a somewhat more detailed consideration of one way by which this kind of extension can be carried out.

3. Metabolism-Repair (M, R) Systems

In the papers [1–3, 18] I have built upon earlier work of Rosen [4–5] to develop such an extension, termed the "metabolism-repair" or "(M, R)"-systems. The underlying motivation for the (M, R)-systems is to consider the functional activities of a living cell – metabolism, repair, and replication – and to ask how one can extend the "metabolism only" Newtonian formalism in order to make the repair and replication aspects of the system emerge in a natural manner from the metabolic machinery alone.

If Ω is the set of cellular environments, with Γ representing the set of possible cellular metabolites, we can represent the metabolic activity of the cell by the map

$$f : \Omega \to \Gamma, \quad f \in H(\Omega, \Gamma),$$

where $H(\Omega, \Gamma)$ is the set of physically realizable cellular metabolisms. Clearly, the set $H(\Omega, \Gamma)$ is determined by various sorts of physicochemical constraints as to what kinds of cellular activities respect the usual laws of chemistry and physics. From the above, we see that there is no reason to doubt that the classical Newtonian machinery outlined above will provide a sound mathematical basis for capturing much of the cell's metabolic activity, and this has indeed turned out to be the case. But a cell does not live by metabolism alone. We need more than just a formalism for metabolism to claim that we have a good framework for modeling all of the functional activities of the cell. In particular, we need some explicit structure to account for those cellular functions that distinguish it as a living agent – the processes of repair and replication.

During the course of its metabolic activity, the cell experiences fluctuations and disturbances in both its environmental inputs and in the metabolic map f itself. The purpose of the repair operation is to stabilize the cellular activity against such perturbations. Let's imagine that the cell has a design or basal metabolism represented by the input/output behavior (ω^*, γ^*), i.e., when everything is working according to plan, the cell receives the environmental input ω^* and processes it via the metabolic map f^* to produce the cellular output γ^*. As a crude approximation to reality, the repair mechanism functions by siphoning off some of the cellular output, using this material to reconstruct a metabolic map f which is then used to process the next input. The repair map has the obvious boundary condition that if neither the design metabolic map f^* nor the design environment ω^* have been changed, then the repair map should process the output into the basal metabolism f^*. Hence, abstractly we can represent the repair operation by the map

$$P_{f^*} : \Gamma \to H(\Omega, \Gamma)$$

with the boundary condition $P_{f^*}(\gamma^*) = f^*$.

Finally, we come to the problem of replication. Just as the metabolic machinery of the cell eventually "wears out" and needs restoration by the repair component, the repair component itself runs down and needs to in some way be restored. Of course, it does no good to introduce another layer of "repairers to repairers," as the incipient infinite regress is obvious. Nature's solution is to just periodically throw the repair machinery away and replace it with a new one. This, in essence, is the process of replication and reproduction. There are several ways that one can think about formalizing this operation, the simplest being to assume the replication component of the cell functions by accepting the cellular metabolism, and then processes this metabolism into a new repair component. Abstractly, this is equivalent to postulating the existence of a map

$$\beta_{f^*} : H(\Omega,\Gamma) \to H(\Gamma,(\Omega,\Gamma)),$$

with the boundary condition $\beta_{f^*}(f^*) = P_{f^*}$. This condition emerges from the same line of reasoning used to obtain the corresponding condition for the repair map, i.e., "if it ain't broken, don't fix it."

Thus, we can summarize the structure of an abstract cell in the following diagram of sets and maps:

$$\Omega \overset{f^*}{\to} \underbrace{\Gamma \overset{P_{f^*}}{\to} H(\Omega,\Gamma)}_{} \overset{\beta_{f^*}}{\to} H(\Gamma,H(\Omega,\Gamma)),$$

$$\underbrace{\quad}_{metabolism} \underbrace{\quad}_{repair} \underbrace{\quad}_{replication}$$

with the boundary conditions $P_{f^*}(\gamma^*) = f^*,$

$$\beta_{f^*}(f^*) = P_{f^*}.$$

Remarks

(1) In the absence of the "life-giving" functions of repair and replication, corresponding to the cellular genome, the above diagram reduces to the conventional input/output diagram for a Newtonian particle system. Thus, the (M, R)-framework represents a genuine extension of the Newtonian setup suitable for accommodating the functional activities that we usually associate with the cell's genetic component.

(2) Both the repair and replication components emerge directly out of the cell's metabolic data, i.e., given the sets Ω,Γ and $H(\Omega,\Gamma)$, it's possible to directly construct the cell's genetic component, the maps P_{f^*} and β_{f^*}, as well as the set of repair maps $H(\Gamma,H(\Omega,\Gamma))$ by "natural" mathematical operations. This construction is carried out explicitly in [18] for the case of linear maps f. Thus, rather than having to make a variety of ad hoc assumptions in order to graft these essential features onto a classical Newtonian setup, the (M, R)-framework enables the cellular genotype to naturally emerge out of the metabolism alone. Since there's good reason to believe that this is exactly what may have happened on earth some four billion years or so ago, it's some measure of satisfaction to see the same steps being followed in any sort of mathematical framework purporting to represent living processes.

(3) The repair and replication components of the cell clearly represent a controller for the cell's main chemical business of metabolism. But this is a quite different type of control system than that usually seen in the engineering literature, since here there is no notion whatsoever of a controller injecting instructions into the system from the outside. Instead, the control is exerted from inside the system itself by

means of the boundary conditions on the repair and replication maps. In effect, these boundary conditions imply that the cell "knows" what it's supposed to be doing, and can utilize the repair and replication machinery to correct its own behavior if outside forces perturb it away from the design specifications. In short, this scheme gives us a means for incorporating the idea of "self-reference" into the mathematical formalism used to study living processes. The boundary conditions on P_f. and β_f. in effect serve as an internal self-model for the cell, thereby allowing us to study a variety of "anticipatory" control mechanisms absent from the standard literature. As noted in the recent monograph [12], the possibility of such self-referential and anticipatory behavior opens the door to a variety of system-theoretic considerations which are difficult, if not impossible, to deal with using the standard Newtonian-based approaches.

As just noted, the (M, R)-systems represent a new type of control process, quite different in character from those with which we are familiar from classical and modern control and system theory. In Aristotelian terms, the (M, R)-systems enable us to begin to formalize mathematically the idea of final causation, as is discussed in more detail in [1]. Since the mathematical structure of the (M, R)-systems has been developed in some detail elsewhere [18], let me devote the balance of this paper to a consideration of how these systems might be employed to represent a variety of problems arising in biology, economics and manufacturing.

4. Applications of (M, R)-Systems

The value of any theoretic framework lies in the spectrum and depth of applied questions that it encompasses. Consequently, interesting as the theoretical aspects of the (M, R)-systems are, the proof of their ultimate value resides in their ability to mathematically capture major slices of applied reality in a way that enables us to unlock some of the structure of the real world. Thus, we devote this section to providing the skeletal basis for such applications in the areas of biology, economics and manufacturing. Let's consider each of these quite different application areas in turn.

Cellular Biology

Since the (M, R)-formalism was originally inspired by the idea of an abstract cell, it seems reasonable to suppose that this area of applied biology would be a prime testing ground for checking on the utility of the entire theoretical (M, R) edifice developed above. This may indeed turn out to be the case, but at the moment there are serious obstacles to the realization of this goal.

The principle difficulty in having the (M, R)-systems make contact with cellular reality is the preoccupation of applied cellular biology with the material aspects of living cells, i.e., their physical composition and biochemistry. By now it should be patently clear that the (M, R)-framework is designed to probe the *functional* aspects of cellular reality, and says nothing about the precise physical constitution or electrochemical events taking place within the cell. Many mainline biologists would argue that if a theoretical structure doesn't address these "materialistic" considerations, then it cannot be taken seriously as a framework within which to study the workings of a cell. This argument is like saying that it's impossible to learn anything worthwhile about computing by studying the properties of algorithms, operating systems, communication networks and the like without delving into the actual physical hardware of a computing machine, such as its memory devices, switching elements and other physical components. In the computing context the absurdity of the argument is clear, and I would contend that it's equally absurd in the biological context, as well, and would hardly merit even a comment if it weren't taken so seriously by those in the biomedical profession. Nonetheless, let me examine briefly some of the steps needed to have the *functional* (M, R)-setup interface with the purely *structuralist* paradigm of experimental cellular biology.

Since the raw ingredients of the (M, R) cellular model are just the set of environmental inputs Ω, the possible outputs Γ, and the physically realizable metabolisms $H(\Omega, \Gamma)$, the basic applied problem is that of how to identify the elements of these abstract sets with their physical counterparts in a real cell. A possible way of doing this is to suppose that the components of an input sequence $\omega = \{u_1, u_2, ..., u_N\}, u_i \in R^m$, are concentrations of various chemical compounds in the cellular environment. Thus, in this situation we would have m different chemical compounds in the environment, with the components of the vector u_i representing the amount of each of these compounds that is received by the cell at time-step i. Similarly, the elements of the output sequence $\gamma = \{y_1, y_2, ...\}, y_j \in R^p$, would be the amount of various chemical metabolites produced by the cell. Finally, the elements $f \in H(\Omega, \Gamma)$ would correspond to the precise cellular chemical reactions that could transform ω into γ. Of course, the precise detailing of these quantities would require an extensive effort by not only system theorists familiar with the mathematics, but also open-minded laboratory cellular biologists. Hopefully, the arguments and constructions given above will suffice to motivate those readers with the appropriate training and interests to persevere in this effort.

Evolutionary Economics

In the short-term at least, an even better bet for fruitful application of the metabolism-repair paradigm is in the area of economic processes. Conventional uses of control and system theory for regulating economic systems suffer from the inherent defect that the decisionmaker is thought of as being someone or something *outside* the system, who broadcasts instructions into the process through some sort of "semipermeable membrane" that separates the controller from the system being controlled. *Avant garde* thinking in economic modeling questions this fiction, recognizing that for many processes there may even be no identifiable decision-maker at all, yet the process must somehow be internally regulated by some sort of "self-knowledge" of what it is supposed to be doing. This is exactly the sort of situation which the (M, R)-systems were developed to handle.

One way in which the (M, R)-formalism can be used, even in a rather classical setting, is to shed some new light on the case of input/output systems of the Leontief type. Consider the vastly over-simplified Leontief system in which the dynamics are

$$x_{t+1} = F_{x_t} + G_{u_t}, \quad x_0 = x_0,$$

where the ith component of the vector $x_t \in R^n$ represents the level of the ith product in the economic complex at time t, with the production matrix F having the form

$$F = \begin{pmatrix} 0 & 0 & \cdots & 0 & a_1 \\ a_2 & 0 & \cdots & 0 & 0 \\ 0 & a_3 & \cdots & 0 & 0 \\ \vdots & \vdots & \vdots & \vdots & \vdots \\ 0 & 0 & \cdots & a_n & 0 \end{pmatrix}, \quad a_i \geq 0.$$

The ith element of the input vector $u_t \in R^m$ represents the labor input to the ith product at time t, with the matrix G having the form

$$G = diag(g_1, g_2, \ldots, g_n), \quad g_i \geq 0.$$

Further, assume that the products $i = 1, 2, \ldots, n-1$ are intermediate products, so that the level of finished product in the economy is measured by the nth component of x_t. Thus, the economy's output at time t is

$$y_t = x_t^{(n)} = Hx_t = (0 \quad 0 \quad \cdots \quad 0 \quad 1)x_t.$$

The foregoing situation is already cast in exactly the form needed for use of the (M, R)-formalism. The "environment" ω is specified by the sequence of labor inputs $\{u_0, u_1, \ldots, u_N\}$, while the "cellular output" γ of the economy is the level of finished good $x_t^{(n)}$, i.e., $\gamma = \{x_1^{(n)}, x_2^{(n)}, \ldots\}$. As soon as we specify the basal levels of these quantities ω^* and γ^*, then we are in position to employ the entire (M, R)-framework developed above in order to construct the repair and replication components of the Leontief system. Furthermore, since the actual internal dynamics of the system are given, we can directly calculate the basal metabolism as $f^* = \{A_1^*, A_2^*, \ldots\}$, where the elements $A_i^* = H^* (F^*)^{i-1} G^*$. Once the "genetic" components of the system are in place, we can begin to ask the by now familiar questions about what kinds of changes in labor supply and/or technological coefficients can be compensated for by the system, the types of "mutations" in the technological coefficients that will be permanently imprinted upon the system, as well as begin to speak about the "growth," "decline" and "specialization" of the economy using the ideas discussed in the last section.

Much of the recent work on evolutionary economics, especially that arising as an outgrowth of the Schumpeter school [6–7], focuses upon the process of structural change in the economic process, and the role of technological innovation in serving as the engine driving the transition from one form of economic interaction to a qualitatively different type. It's of interest to examine how the notion of structural change would be characterized within the (M, R)-paradigm.

Imagine we think of the economic system as being represented by an (M, R)-system having the metabolic map $f : \Omega \to \Gamma$, together with its associated repair and replication components as developed above. Suppose the parameter vector $a \in R^k$ represents the level of technological development within the economy. Then the metabolism, as well as the induced repair and replication subsystems, are all functions of a, and we can write the metabolic map as $f_a : \Omega \to \Gamma$. Suppose now that some sort of technological innovation occurs, shifting $a \to \hat{a}$. In our earlier terminology, this is nothing more than a particular type of metabolic change $f^* \to f$, and we can employ the Metabolic Change Theorem [1, 18] to investigate those technological innovations that will be *neutralized* by the prevailing economic way of doing things. In other words, it's only those innovations that cannot be neutralized that have a chance of changing the system in a way that would correspond to our ordinary idea of a "re-structuring." But being nonrepairable is not enough to stamp an innovation as generating structural change. Somehow we need to introduce some measure of the degree to which the new economic order differs from its predecessor.

The usual way of measuring the qualitative change in the metabolism f_a as a response to changes in a is to introduce some topology on the space of maps $\{f_a\}$, saying that the change $a \to \hat{a}$ is structural if a is a point of discontinuity for the map $a \mapsto f_a$. The various notions of structural stability familiar from the literature all emerge from following this argument in different ways. The new element that the (M, R)-systems introduce into this situation is the possibility that a technological innovation $a \to \hat{a}$ may *not* result in any permanent shift of the economic order from $f_a \to f_{\hat{a}}$. As noted, such a shift will only occur if the metabolic change $f_a \to f_{\hat{a}}$ is not repairable by the system's "genetic" component. Such a possibility is literally inconceivable within the classical framework for the obvious reason that the classical mathematical framework has no provision for any sort of repair or replication components. While there is no room here to enter into details of the foregoing idea, there is good reason to think that some of the mysteries surrounding how it is that some innovations lead to major economic changes and other, seemingly equally significant innovations, are just ignored or buried by the prevailing economic order. These are matters for further investigation. Now let's turn to a related circle of questions arising in the use of (M, R)-systems to study modern manufacturing processes.

Manufacturing Systems

A much more specific area of potential application of the (M, R)-framework is for the modeling of the so-called "Factory of the Future," in which the manufacturing system is designed to be self-organizing, self-repairing, and even in some cases, self-regenerating. Let me sketch the basic ideas, referring the reader to the papers [2-3] for further details.

Every manufacturing operation can be thought of as a metabolic process in which a set of raw materials is transformed by labor, money and machines into an end product. The primary issues under discussion in manufacturing circles today involve matters of information manipulation, with the real concern being with just exactly how the processing of the raw materials into finished products should be organized to satisfy various constraints relating to flexibility, reliability, quality, adaptability, etc. of the manufacturing operation. Such considerations fall directly under the purview of the (M, R)-formalism.

Assume that Ω represents the set of sequences of available raw material inputs that can be used by the manufacturing enterprise, with Γ being the corresponding set of sequences of finished products that the factory is capable of producing. Just as in the biological case, to make contact with real manufacturing processes, we would have to find a good way of encoding these items of the real-world into the abstract elements of the sets Ω and Γ in order to invoke the (M, R)-paradigm. For example, if the factory is supposed to produce automobiles of a certain type, then this means that the real-world inputs such as the amount of steel, glass, rubber, plastic, hours of skilled labor, hours of unskilled labor, number of machine-hours for welding, drilling, etc. would all have to be encoded into the elements of the input u_t for each time t. This encoding would then specify the input sequence $\omega = \{u_0, u_1, ..., u_N\}$. A similar encoding would have to be developed for the output elements $y_1, y_2, ...,$ expressing the sequence of partial products of the overall auto manufacturing operation. This encoding would then determine the output sequence $\gamma \in \Gamma$. Then a particular type of automobile would be specified by giving a *particular* input and output sequence ω^* and γ^*. At this point, we are in a position to employ the (M, R)-setup to determine the corresponding repair and replication subsystems for this manufacturing enterprise.

Again, just as with the problem of using the abstract (M, R)-framework in cellular biology, there are a number of non-trivial questions that arise in connection with actually translating the foregoing ideas into workable schemes for real-world operations. But the main point is that the framework exists, and the tasks that need to be carried out to make contact with real-world problems are fairly clear and explicit. All that's needed is the will and time to see the exercise through to conclusion, exactly the same sort of requirements needed to use the Newtonian formalism to study the behavior of the mechanical processes of physics. The only difference is that the "Newton of Manufacturing" has not yet made his appearance, but the necessary formalism has.

With these brief indications of how one might go about using the (M, R)-formalism as a modeling framework for living processes, let me now consider a spectrum of extensions and generalizations that may be needed to effectively make use of the basic setup in a given real-world situation.

5. Networks, Nonlinearities and Evolution

Interesting as the single-cell case may be, we must consider how to "soup-up" this situation if we are to make contact with the processes of real life. Consequently, let me here outline some of the major extensions needed and give some indication of how I think they can be carried out. By and large, these extensions fall into three broad categories, each involving a weakening of one of the underlying assumptions in the models considered above.

The first condition that needs relaxing is the confinement to a single cell system. It's necessary to consider *networks* of such cells coupled together in various ways. Just as Newtonian mechanics is pretty trivial for single-particle systems, if we want to use the (M, R)-systems as a modeling paradigm for real organisms, consideration of networks is an essential step that must be taken.

The second constraint we need to address is the matter of nonlinearity. Experimental evidence is rather clear on the point that biological cells do not, in general, act in a linear manner when transforming one set of chemicals into another. Thus, it will be necessary to see how and to what degree it's possible to push through the formalism sketched above into detailed equations for various types of nonlinear metabolisms.

Finally, we have imposed no kind of optimality criterion by which the cell can decide whether or not certain types of modified metabolisms or new genetic structures are to be favored in the battle for survival. Thus if we want to use the (M, R)-structure as a framework to study evolutionary processes, we will have to superimpose some kind of selection criterion on the genetic makeups of the cells so that natural selection can take its course. It should be evident that these sorts of evolutionary considerations must be added to the general picture after dealing with the networking problem discussed a moment ago. With the above considerations in mind, let me take a longer look at each of these extensions and indicate the way in which I think they could be accomplished.

Cellular Networks

The most straightforward way in which to think about putting together a collection of individual cells into a network is to assume a situation in which at least one cell of the network receives part of its input from the environment, with at least one cell transmitting part of its output to the environment. All cells not interacting directly with the environment then receive their inputs as the outputs of other cells, and transmit their outputs to be used as the input to other cells. We assume that each cell must have at least one input and one output, and that proper metabolic functioning of the cell requires the cell to receive all of its inputs. This arrangement takes care of the metabolic activity of the network, but we must also make some sort of assumptions about the connectivity pattern involved with the individual genetic components of the network elements. The simplest is to assume that the repair component of each cell satisfies the following conditions: (i) it receives at least one environmental output as part of its input, and (ii) it must receive all of its inputs in order to function. A typical example of such a network is depicted in Figure 1, where the boxes labeled "M" represent the cell's metabolic component, while the ovals labeled "R" are the corresponding repair/replication components.

In this network, observe that if cell 1 fails, then so will cells 2–5 all of whose inputs ultimately depend upon the proper functioning of cell 1. We could call any such cell whose failure entails the collapse of the entire network, a *central component* of the network. Now note that by the hypotheses made above about the repair components, any cell whose repair component receives its own output cannot be built back into the network by the repair component. Such a cell can be termed *non-reestablishable*. Thus, cell 2 is non-reestablishable, while cell 5 is reestablishable. These elementary ideas already lead to the following important result of Rosen's [4] about (M, R)-networks.

(M, R)-Network Theorem.
Every finite (M, R)-network contains at least one non-reestablishable cell.

Corollary.
If an (M, R)-network contains exactly one non-reestablishable cell, then that cell must be a central component of the network.

The Network Theorem and its Corollary can be proved by a simple inductive argument, and show that every cellular network must contain elements that cannot be built back into the system should they fail. Further, if there are only a small number of such cells, then the failure of these cells is likely to bring down the entire network. This last is a point to ponder in connection with various types of social policies aimed at propping-up failing enterprises.

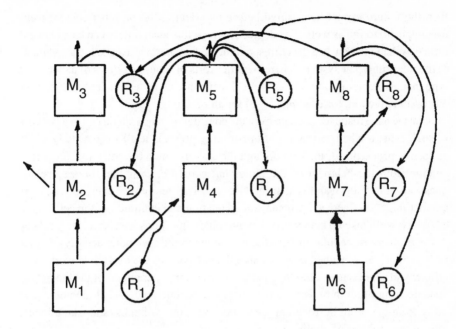

Figure 1. A Typical Cellular Network

The Network Theorem says that such propping-up cannot possibly be universally successful and, what's worse, the more successful it is, the greater is the likelihood of creating a system in which the failure of a very small number of enterprises will result in the collapse of the entire edifice.

Of course, the Network Theorem above depends on the particular assumptions made here about the need for the metabolic and repair subsystems to each have all their inputs in order to function, as well as upon the hypothesis that each repair subsystem must receive at least one environmental output as one of its inputs. In any real cellular network, these conditions may or may not hold and, if not, it will be necessary to see to what degree the kind of result obtained above can be carried over under the new conditions.

In addition to problems involving connectivity patterns, there are important time-lag effects that enter when we begin discussing networks. Besides the internal time scales associated with the metabolic, repair and replication subsystems of each individual cell, when we start connecting such cells together we quickly encounter another type of time lag, the time needed for information to pass from one cell to another. Imagine the situation in which the metabolic component of one of the cells in the network fails. If the repair system is to build this cell back into the network before the metabolic failure can be transmitted to other cells, it will be necessary for all the necessary inputs to the repair system to be available on a time scale much faster than the metabolic outputs are transmitted throughout the network. If not, then the failure cannot be repaired before its effect is felt on other cells, thereby causing them to fail as well. The degree to which this local failure can be isolated is directly related to how fast the corresponding "raw materials" can be made available to the relevant repair system. There are other sorts of time-lag effects of a metabolic type that also enter into networks, but I'll leave the interested reader to consult the discussion given in, for example, [4–5] for an account.

Once we have made the passage from the consideration of a single cell to a network of such cells, a whole new vista of opportunity presents itself for the study of the collective properties of the network and the way in which the collective properties emerge from the behavior of the individual cells. For example, there are many questions about survivability that the Network Theorem only begins to touch. In another direction, there are problems of adaptation and evolution of the cells, matters which we'll take up below. Such evolutionary questions are also closely related to the problem of cellular differentiation, in which the metabolic activity of some cells changes in a manner so as to make the behavior of the overall network more efficient, in some sense. The (M, R)-framework offers a paradigm for the theoretical consideration of all these issues, as well as much, much more. Let me now move away from questions surrounding networks and groups of cells, and take a longer look at the problem of nonlinearity.

Nonlinearities

The technical details about (M, R)-systems given in [1, 18] all rely upon the assumption that the cellular metabolism, repair and replication operations are linear functions of their inputs. From a mathematical standpoint, this is quite convenient as there is by now a very well-developed literature on linear systems which we have been able to tap into in order to provide a foundation for a technical treatment of the processes of repair and reproduction. However, there's good reason to look at this linearity hypothesis with a fishy eye, since the experimental evidence on cellular input/output behavior (what little there is of it) doesn't appear to strongly support the case for linearity. In other settings, however, as indicated in the Leontief-type economy example, the linearity hypothesis might be more easy to swallow. But the truth of the matter is that for the (M, R)-framework to provide a solid basis for modeling living processes, it's necessary to somehow extend the linear results to a wider setting.

Fortunately, the past decade or so has seen an explosion of interest in the problems of *nonlinear* system theory, and most of the machinery needed to extend the underlying vector-space setting, as well as the realization algorithms to broad classes of nonlinear systems is now in place [10-11, 16-17]. Of course, the mathematical tools needed for these extensions are quite a bit more sophisticated, in general, than the simple tools of linear algebra, although a number of important extensions to systems with some kind of linear structure (e.g., bilinear or multilinear I/O maps) can be pushed through using the linear apparatus. Since these extensions have been well documented elsewhere, here let me indicate only the basic idea in the special, but important, case of bilinear behaviors.

For linear systems, the defining relationship is that the system outputs are linear combinations of past inputs; for bilinear processes, the analogous relationship is that the outputs are linear combinations of *products of distinct* past inputs. It turns out that by a clever re-labeling of the original inputs to define a new input, it's possible to make the realization problem for bilinear behaviors become formally the same as that for linear systems, and in this way make use of the same mathematical apparatus that we have been relying upon up to now. To illustrate the basic idea, consider only the case of a scalar system ($m = p = 1$).

Let u_t be the system input at time step t, and define the new quantities

$$u_t^1 = u_t,$$

$$u_t^j = \begin{pmatrix} u_t^{j-1} \\ u_t^{j-1} u_{t+j-1} \end{pmatrix}, \quad j = 2,3,\ldots; \quad t = 0,1,2,\ldots .$$

In terms of the quantities $\{u_t^{j-1}\}$, the input/output behavior of a scalar bilinear process can be given as

$$y_t = \sum_{j=0}^{t-1} w_{t-j} u_t^j, \quad t = 1,2,\ldots .$$

Formally, this is exactly the same input/output relationship as for conventional linear processes, but with now the quantities $\{w_i\}$ and $\{u_t^j\}$ taking the place of the Markov parameters $\{A_i\}$ and the usual inputs $\{u_t\}$. By a suitable extension of the standard linear realization algorithms, it can be shown [10] that the bilinear behavior sequence $\{w_i\}$ can be realized by an internal bilinear model of the form

$$x_{t+1} = Fx_t + Gu_t + Nx_t u_t, \quad x_0 = 0,$$

$$y_t = Hx_t, \quad t = 0,1,2,\ldots .$$

For details on how to process the behavior sequence to canonically generate the elements F, G, H and N, we refer to the literature [8-9, 16-17].

What's important about the above development is that it demonstrates in explicit form that a broad class of nonlinear behaviors can be dealt with in much the same manner that we used to deal with linear processes. Thus, the lack of linearity is no real obstacle to the effective employment of our (M, R)-framework and, in fact, the same kinds of arguments that we've just sketched for bilinear processes can be carried over to much broader classes of nonlinear behaviors as detailed in [10-11].

Adaptation and Evolution

The single-cell situation treated here assumes that the cell has only a single purpose: to fulfill the prescription laid down in its design metabolism. Thus, there is no explicit criterion of performance as the entire genetic apparatus is setup solely to insure that the cell sticks to the basal metabolism, if at all possible. Such a narrow, purely survival-oriented objective is called into question when we move to the setting of cellular networks in which the cell is not just in business for itself, but also to serve the needs of the other cells. Thus, in this kind of setting it makes sense to superimpose some sort of evolutionary fitness criterion upon the basic survival condition, thereby measuring the degree to which the cell is "fit" to operate within the constraints of its "generalized" environment consisting now of not only the external physical environment, but also the other cells in the network. As soon as such a selection criterion is imposed, we then have all the ingredients necessary for an evolutionary process, since we have already mentioned ways in which the genetic subsystem can experience various sorts of "mutations."

It is by now folk wisdom in the control engineering business that choice of a performance criterion is usually the trickiest aspect of putting together the pieces of a good optimal control model for a given situation. The situation is no easier in biology, where the arguments still rage hot and heavy over competing positions on the matter of selection criteria for evolutionary processes. In fact, in biology the situation is if anything worse, since there is not even a consensus as to the level at which the selection acts, with rather vocal arguments being made for action at every level from the gene itself up to the complete organism and beyond to an entire species. Consequently, I'm afraid there is no divine mathematical insight that can be offered as to how to choose an appropriate selection criterion to use with the (M, R)-framework we've introduced here. About all that can be said is that once such a choice has been made, then a network of cells put together according to the (M, R)-prescription will provide a natural theoretical basis for the study of adaptive and evolutionary processes.

References

[1] Casti, J., "Newton, Aristotle and the Modeling of Living Systems," in *Newton to Aristotle: Toward a Theory of Models for Living Systems*, J. Casti and A. Karlqvist, eds., Birkhäuser, Boston, 1989, pp.47-89.

[2] Casti, J., "(M, R)-Systems as a Framework for Modeling Structural Change in a Global Industry," *J. Social & Biological Structures*, 12 (1989), 17-31.

[3] Casti, J., "Metaphors for Manufacturing: What Could it be Like to be a Manufacturing System?" *Tech. Forecasting & Social Change*, 29(1986), 241-270.

[4] Rosen, R., "Some Relational Cell Models: The Metabolism-Repair Systems," in *Foundations of Mathematical Biology*, R. Rosen, ed., Vol. 2, Academic Press, New York, 1972.

[5] Rosen, R., "A Relational Theory of Biological Systems," *Bull. Math. Biophysics*, 21 (1959), 109–128.

[6] *Economic Evolution and Structural Adjustment*, Batten, D., J.Casti and B. Johansson, eds., Springer, Berlin, 1987.

[7] Nelson, R. and S. Winter, *An Evolutionary Theory of Economic Change*, Harvard U. Press, Cambridge, MA, 1982.

[8] Isidori, A., "Direct Construction of Minimal Bilinear Realizations from Nonlinear Input-Output Maps," *IEEE Tran. Auto. Control*, AC-18 (1973), 626–631.

[9] Mohler, R. and W. Kolodziej, "An Overview of Bilinear System Theory and Applications," *IEEE Trans. Systems, Man & Cyber.*, SMC-10 (1980), 683–688.

[10] Isidori, A., *Nonlinear Control Systems*, 2nd Edition, Springer, Heidelberg, 1989.

[11] Casti, J., *Nonlinear System Theory*, Academic Press, New York, 1985.

[12] Rosen, R., *Anticipatory Systems*, Pergamon, Oxford, 1985.

[13] Casti, J., *Linear Dynamical Systems*, Academic Press, New York, 1987.

[14] Brockett, R., *Finite-Dimensional Linear Systems*, Wiley, New York, 1970.

[15] Fortmann, T. and C. Hitz, *An Introduction to Linear Control Systems*, Dekker, New York, 1977.

[16] Sontag, E., *Mathematical Control Theory*, Springer, New York, 1990.

[17] Nijmeijer, H. and A. van der Schaft, *Nonlinear Dynamical Control Systems*, Springer, New York, 1990.

[18] Casti, J., "The Theory of Metabolism-Repair Systems," *Applied Mathematics & Computation*, 28 (1988), 113–154.

[19] "Challenges to Control: A Collective View," IEEE Trans. Auto. Control, AC-32 (1987), 275–285.

Simulations in Decision Making for Socio-technical Systems

Christopher L. Barrett
Royal Institute of Technology, Stockholm
and Los Alamos National Laboratory

Roland Thord
Temaplan AB, Stockholm
and Thord Connector, Stockholm

and
Christian Reidys
Los Alamos National Laboratory

1. Representation of Socio-technical Systems Using Computers

This paper considers building and using simulations of socio-technical systems. These systems have (usually, but not necessarily, human) adaptive, learning, variable, etc., components as well as important social context, and essential physical/technological aspects. Although an actual list is much longer, this kind of system is exemplified by transport systems, financial systems, communication systems and even warfighting systems. We hope to help with conceptual organization of the many issues that are involved in simulation and which can complicate the use of simulations of socio-technical systems for practical purposes. Different simulation "laboratories" for common categories of real world use are suggested in the last section.

The concept of a society of individuals is largely a matter of conventions for interactions and transactions, i.e., traffic, of one sort or another. Thus vehicular-, goods-, informational-, financial-, and all other forms of traffic, the control of traffic, and the design or regulation of network resources on which traffic moves, are serious matters to society. Much public policy and industrial regulation concerns assuring and regulating networks and traffic on those networks. Traffic and the associated infrastructure are also important to scientists, engineers, and businesspersons. They may seek to understand, design or modify the actual in-place systems, develop relevant technology, or market products and services. Moreover, the larger constituency whose lives are shaped by these policies, designs, technologies, and their various consequences are obviously also interested parties to their development and character. But socio-technical systems have proven to be difficult to represent for analysis, to understand, plan, design and use for all involved and from every perspective.

Computer simulation is also really a matter of (procedural) interactions and transactions, (programming) conventions, and (data) traffic. The intuition that simulations can represent socio-technical systems and assist us to understand or design them follows largely from the observation that it seems plausible to put these computational properties into correspondence with the properties of systems outside the computer. The representation would then be a naturalistic decomposition and would be explicitly and inherently dynamical. Examination of this intuition gives useful insight regarding simulation as a means of system representation, the apparent current incomplete scientific understanding regarding the use of simulation, its enormous prospects, and perhaps its potential limitations, relative to socio-technical systems.

2. Simulation

Computer simulation is basically the art and science of using computers to calculate interactions and transactions among many separate algorithmic representations, each of which might be associated with identifiable "things" in the real world, (at least in a "world" outside the simulation program, perhaps even another simulation program.) Simulations compose these "local mappings" and thereby generate "global phenomena". One usually hopes to then be able to interpret the global phenomena as the properties of some system of real interest, which is (rather inconveniently) represented by all of the individual, interacting, parts. An immediate complication is that the composition of all the piece-parts can induce very complicated dynamics, patterns of space and time varying properties, that often defy any simple description. We can then be left with a simulation program that is (at least) nearly as complicated as the system it is intended to represent and is not particularly well understood even as a computer program. Therefore the simulation program might bring us no closer at all to a simulation model of a system, or the sense of understanding that usually implies.

Nevertheless there are compelling and hopeful possibilities which follow from noticing that actual socio-technical system dynamics are composed of local interactions from which amazing complication arises, and that this is also true of computer simulation programs. Current knowledge is mostly pre-theoretical [5, 42], but large resources are being invested on this intuition and hope [4], and larger resources invested on the basis of decisions influenced by computer simulations. This will inevitably lead to increased activity and understanding in the area of computer simulation [45]. Simulation is at a stage where nearly everything is somewhere reported to be possible. Many possible applications that now appear full of promise will surely prove empty and some of those that will eventually predominate are likely not apparent. At such times it can be useful to pause, re-examine premises, and take stock rather than to forecast.

2.1 Notions Underlying Use of Simulation for System Representation

It is often difficult to discuss simulation productively because of the idiosyncrasies of the viewpoints of the various interested audiences. It is futile to consider seriously uses of simulation when there is no agreement on what it is that might be used. Therefore in this section we leave specific reference to socio-technical systems in order to describe to some essential ideas about simulation that serve to make things somewhat more coherent. This is not a conventional presentation of concepts in simulation, but it may nevertheless provide useful perspective on simulation (but see also, e.g., [7, 17, 31, 48]).

Simulation has two basic aspects[10]. First, a simulation is a dynamics generator. A computer simulation program is a means by which to compose many distinct, data local, iterated procedures. Each procedure's dynamics is affected by the composition of procedures and, in that sense, the composition produces global (composed) system dynamics. The second aspect of simulation is mimicry of the dynamics of another system by the dynamics of the simulation program. After mimicry is established, the simulation program is often called a simulation model but this is a separate collection of issues. Attending to properties of generation first greatly helps to focus the topic of mimicry. On the other hand, it is not very helpful to consider both these issues at once and it is also less than helpful to address mimicry first.

2.1.1 Basic Generation of Dynamics in Simulation

There are three basic elements of a simulation program, focussing on the generation of dynamics: entities, entity interactions, and composition principles for combining entities in interaction (see figure 1).

Newton's kinematic laws provide an excellent illustration of these elements. If any reference is actually needed here, a classic discussion of these laws is found in [25].

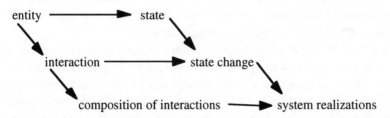

Figure 1. Elements of generation in a simulation

Entity

Newton's first law introduces a particle and a base state and is remarkable for the starkly elegant and general definition of an individual "thing" that it provides. Each individual thing is uniquely characterized by a position and inertia - a base state for simple existence. The concept of a particle is an astonishing intellectual achievement; a completely generic thing, stripped of any context or particularities. The abstraction it provides applies to classical description of everything physical. It provides a useful aesthetic standard for theorists and practitioners in simulation as regards generic agent representation.

A simulation program requires a generic entity-state concept such as a particle in physics. Usually such a thing is called an entity, object, process, agent or actor in a simulation. There is normally a collection of such things in a simulation program.

Interaction

Newton's second law introduces collision as the mechanism and momentum as the property relevant to particle interaction. It provides a beautiful concept that defines conditions for, and prescribes the form of the result of, interaction from the point of view of an individual particle in an interaction with another particle. The well-known uncertainty principle, the ill-posedness of simultaneous definition of both position and momentum of a particle, (although both are required to completely define a collision interaction), serves to remind that interaction is subtle (again, [25] has a classic exposition).

Given a collection of individual things in a simulation program, it is necessary to characterize how interactions among them affect each of them; that is, how the state of a thing changes. The "dynamics" are the properties of the variations in states of entities.

Interaction Composition Principle

Newton's third law requires that momentum be conserved during particle collisions in the system and requires every pairwise interaction to conserve momentum. This principle provides the last essential concept for the construction of physical dynamical systems from individual physical parts, a composition law.

Given a collection of individual things in interaction in a simulation program, in principle it is possible to try to update their states many different ways. So it is necessary to have a composition principle provide some kind of lawfulness over a population of interactions that are each occurring completely locally. Composition principles are manifest in conservation properties in physics and also in rationality, optimality, or equilibrium assumptions for models in economics. The applicable principles must be explicitly realized during entity state update scheduling in simulation programs [7, 10, 17, 30, 31]. The relationship between the entity representations and the update method must be, in combination, compatible with allowable compositions of entities implied by the conservation properties. Indeed, this relationship forms the representation of the conservation properties in a simulation program.

It is important to distinguish between the conceptual sufficiency of Newton's laws to characterize a system of particles and any straightforward recipe for solution of that system's detailed dynamics. For example, the so-called n-body problem is notorious, (e.g., [46]). It is instructive to recall that the extreme difficulty of analytically solving systems of even small numbers particles according to these laws was a primary motivation of numerical methods, statistical mechanics, and particle simulation in physics. In socio-technical system simulations, not only is an obvious general and completely natural composition law lacking, any presumption that its appearance would lead to straightforward representation and solution of these systems is unwarranted.

Thus, the motivation for simulation-like representations of dynamics and their difficulties is hardly new and the basic three elements of the generation method have important intellectual precedent. A diagrammatic summary of the example in this section is provided in figure 2.

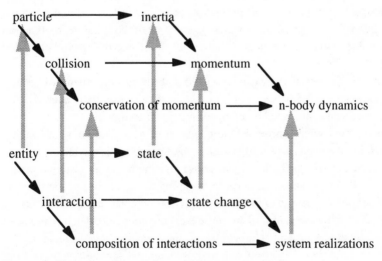

Figure 2. Elements of a simulation: kinematic example

2.1.2 Non-Cooperative Games and Player Moves, Simulations and Entity Updates

At this point it is usual to hear objections along the lines that economic/psychological/ social systems are not simple physical systems. Therefore principles which might be adequately illustrated by kinematics are manifestly uninteresting in relation to them. These other systems are, after all, comprised of much more complicated entities than Newtonian particles engaged in having collisions and conserving momentum. To begin to move away from the simplest kinematic setting and toward systems of more complex entities, we will touch on a few defining aspects of game theory. We will consider formal games because they relate to more complex agents that also can be connected conceptually to the requisites for generation of dynamics in a simulation.

The elements of a formal game are: "nature", individual players, local information, actions, strategies, individual payoffs, game outcomes, and global equilibria. Game theory is concerned with the actions of individuals that possess motives and are aware that their actions affect others. The basic jargon is introduced and idea of a game is suggested in the following description. Each entity is a player that collects information about other players and the state of "nature", which is an actor of sorts in the simulation, but has no awareness of the players. Based on this information, a player acts according to strategy which prescribes a particular sequence of actions among possible actions and sequences of actions available to that player. Based on this action and the actions of other players and nature, a given player achieves some payoff. The game is a realization of many states, action sequences, and payoffs for every player. Any projection of this manifold of values, which embodies the totality of the game, is called an outcome of the game. A combination of strategies among players that satisfies some global criterion over payoffs is called an (outcome) equilibrium.

The elements of a formal game can be related to the basic constituents of generation in a simulation, as can be seen in figure 3. Indeed, it is essentially a recipe for a simulation as can be appreciated by comparing figure 3 to figure 2.

Game theory somewhat confounds "pure" player-level generation by imbedding global criteria in the player definitions, or by making information available to each player in artificial ways, or both. Global criteria posing as (or as assumptions in) local behavioral models serve the purpose of setting the stage for the composition mathematics to work. Because of formal conveniences obtained, considerable theory is possible for cooperative games in the sense that outcomes can often be calculated directly and theorems proved. Unfortunately, these mathematical conveniences obscure entity definition. Therefore, although we claim we are moving away from particles to "agents" that are "aware" of something, interact, and can be composed into systems (games), we must acknowledge that legitimate objections to game theory render it less than fully satisfactory for our purposes. Still, something important is

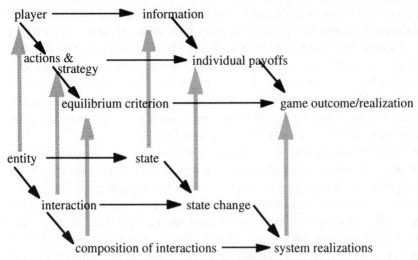

Figure 3. Elements of simulation: game theory example

learned in the considerations of game theory that can serve to help lead us away from the simplicity of simulated particles in kinetic systems toward complex individuals simulated in societies.

To address some objections to the simplest settings for formal games, one might consider noncooperative games. Noncooperative game theory is about those systems in which the order of actions taken affects the realization, i.e., outcomes, of the game and where some actions are not immediately (or ever) observed by every player (a very readable introduction is found in [41]).

Order of play matters to individual payoffs and overall game outcomes when causal dependencies exist between players. When a possible causal dependency exists in a simulation, the images of the causally dependent mappings will take different values, depending on the order in which they are updated. Conversely, causally independent mappings will not produce different values under reordering of the update. This is also true of games. The order of action in a game and the order of update in a simulation are essentially the same concept. Causal dependency between entities induces sequential order constraints in systems. This is a very general concept in relation to the appearance of sequences in nature and beyond anything to do with simulations or games. It is reflected in sequence to structure properties of biomolecules [39, 40], for example. Causal dependency in any system, physical, economic or otherwise, involves access to a resource by multiple entities or the need for one entity to access a resource that has been transformed by another entity. Dependencies inherently arrange themselves in (potentially long) transitive sequences of cause and temporal succession of outcomes (e.g., as is introduced to English speaking children as a moral issue in the nursery rhyme, "The house that Jack built.")

In a simulation program, the update mechanism, the mechanism for scheduling the order of calculating the next state for every entity in the simulation, is the center of the software framework of the program[7,29]. The point here is that the essential features of generating dynamics in kinematic systems using simulation are also pertinent to formal games composed of entities, players, with internal properties that are more complex than newtonian particles. These extended entity properties include limited information, motivations, strategies, and comprehension of other players. So perhaps game theory will benefit from simulation theory,

On the other hand, there are definitely practical reasons for students of simulation to consider the relationship of the underlying structure of computer simulation and formal games. Because different orders of actions taken in noncooperative games can cause so many different things to happen in the game outcomes, the study of games has developed tools to track the paths of actions and payoffs a game can take. An essential theoretical distinction is drawn between intentional and extensional forms of a game[41, 51]. Roughly, the intentional representation is the player-local collection of rules that characterize every player's actions, strategies, etc. The extensional form of the game is the branching graph representation of all of the moves and outcomes a game can take. Obviously these two representations are duals. A primary theoretical concern is to understand how paths in a graph of payoff outcomes are related to particular collections of player-local rules and resources.

This is by no means an easy mathematical setting. Moreover it is, essentially, the same problem that arises in a theoretical investigation of simulation[10, 37][1]. In a simulation, there is a collection of procedures acting on local information (local mappings). This collection of mappings can be updated in many orders and some of these ways produce different system outcomes. The various outcome paths can be represented by graphs and a major issue is the relationship of the collection of mappings, the order of the composition of the mappings, and the state dynamics generated. We will come back to this in section 2.1.6.

An intentional system representation and a corresponding extensional representation have a common underlying "model normal form". A non-game example of a model normal form relating the intentional and extensional forms of a system representation might serve to make the point. Consider a lattice gas representation (intentional form) of a fluid and all the state trajectories it could generate (extensional form) during a particular simulated time period with given initial conditions. Neither the local rules at the lattice vertex sites or the histories of state transitions at a site themselves necessarily look much like a fluid. However in suitable circumstances, it is possible to show that the form of the Navier-Stokes equations is a common model form for both intentional and extensional forms of the simulation system's representation. Thus, what we are calling the model normal form in this case would be equations that characterize the phenomenology of fluids and which also implies both the intentional and extensional form of a simulation program. This is very close to being able to say what it means that the simulation program in question is a simulation model of a fluid. Thus, in addition to referring to a set of entities that are more complicated than particles, formal game-derived concepts lead naturally to the important ideas of intentional, extensional, and model-normal forms which have obvious relevance to simulation. Further elaboration of entity properties and capabilities is still necessary because the constructions of game theory are probably not flexible enough for representation of many important properties of individual humans in socio-technical systems (e.g., learning actors[27], boundedly rational actors [1,2,3], etc.). To include these things in a simulation and to be able to believe in or even understand what happens when they are somehow included involves going beyond usual physical and microeconomic representation methods. However, most of the problems in such elaboration involve issues of model adequacy (e.g., see[29]), which is, as we have said, beyond the scope of the generation of dynamics to which we have limited our concerns thus far. So, for example, before learning or adaptation can be introduced to discussion of socio-technical simulation, mimicry must be better related to generation of dynamics.

2.1.3 Solvers and Generators, Mimicry and Models

To begin to be able to say when one generator is the same as another in any sense is the first step to being able to seriously discuss mimicry and thereby pass from simulation programs to simulation models. Concerns about the adequacy of simple entity models to represent individual human behaviors and the effect they may or may not have on a socio-technical system simulations requires sophisticated consideration of local and composed modeling issues.

The use of a computer for almost any system representation is occasionally referred to as a simulation. Frequently the term is used even when the program referred to is a numerical calculation of a given phenomenological model; that is, where an analytical model is given and a numerical solution method such as finite differences, finite element, etc., has been set up to calculate its solutions. These methods might better be called solvers (the use of this term is due to de Rhonde [42]) than simulations.

A derived solver is verified when it can be shown to converge to it's underlying analytical model of a phenomena. The predictive validity of the calculation is a matter of the fit of the logical entailments of the verified solver of the analytical model to measurement. Obviously there can be many verifiable solvers for an analytical model of a phenomenon. Observations and calculated results must fulfill the usual commutative diagram for models (see figure 4 for a simplified version), such as can be found in[18]. Simulation programs, on the other hand, are a naturalistic representation of the constituents of a system rather than a phenomenological representation. Simulation models are simulation programs with a requirement to be valid in the sense of figure 4. That is, the logical entailments of the calculation of the interactions must be behaviors that result in a particular solution of overall phenomenology that satisfies figure 4.

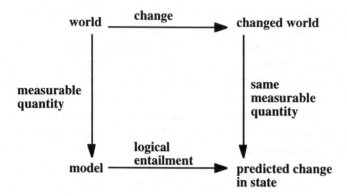

Figure 4. Commutativity diagram for predictive model validity

But we cǎn be more careful and at the same time say more than that. If different generators- from an intentional point of view- produce the same extensional form, they can be said to simulate the same system -- or each other, the question isn't really any different. The point is that we can equate one generator to another and never explicitly know the common system model. Moreover, if different generators have the same model normal form relating their respective intentional and extensional representations, they must simulate the same system. This is a primary concern for theory in simulation and will come up in section 2.1.6.

Predictive validity of a model is, of course, not a substitute for an explanation of the modeled system. Usually a simulation is considered something of an experimental environment for the controlled manipulation of the pieces of a system. In other words, a simulation allows a scientific explanatory analysis of a system that may not yet have an analytical model. This amounts to sorting out what depends on what- what the causes are. In fact it is sometimes possible to get a pretty good sense of understanding of a system from the point of view of causation and never achieve particularly good predictions, and therefore never achieve very good predictive validity, with a simulation. This is a seemingly bottomless source of confusion and discord at the moment[19] but the search for explanation of these complicated systems is also a motivation for the use of simulation and a very good reason to keep solvers and simulations conceptually distinguished.

In contrast, the causal analysis occurs in the development of an analytical model. There the sense of understanding is given by the derivation of the model, by the formal entailments of the various mathematical representations that yield the analytical form itself. The use of computers for calculation of solutions to an analytical model offers a valuable, but very different conceptual contribution, than simulation of a system that may be understood in terms of that analytical model.

In the end though, in every simulation, something is modeled, and entity resolution stops at that level. No causal inferences can be made at a level of refinement lower than the entity models. So if we, say, need to know how trader strategies evolve, we need a different simulation than if we want to know how a market evolves for a collection of trading strategies. And if we say that we don't believe the evolution of the composed market simulation unless our traders themselves evolve, we are changing the level of modeled entities which will change the causal structure dealt with the simulation program. But ultimately, the entities in a financial market simulation need not have their neurons, cell metabolism, molecular processes, or subatomic activities all represented in order to represent, say, traders. We aggregate the model of the local phenomenology of a single trader and that is enough. There are other aggregations than those we have discussed here, e.g., population ensemble properties. We will not go into ensemble averaging and ergodicity, etc., but should at least mention that we must deal with a range of within population variations such as the fact that every trader is actually a little different and that there is a very particular collection of them today, etc. These properties are also not usually going to be precisely represented in the collection of entities in the simulation. There are many ways that modeling occurs in simulations, but it is focussed on the individual pieces rather than the composed phenomenon. As a result, in the previous outline of a formal game cum simulation program, we saw simulation-like structures, but we took issue with the models of the individuals, the importance of the things that got left out, but mostly with the way that global models and information corrupt the "locality" of the entities, etc., because that affects the level of causal analysis possible using the simulation.

Therefore, in the next section we will consider what more complex local entity models are like.

2.1.4 Consequence-driven Systems

Consequence-driven systems (this term is due to Bozinovski [16]) are the class of complex entities that adapt their behaviors and also might alter their structure based on experience and local goals (figure 5). Forms of adaptation can occur in the dynamics of structurally unchanging fixed parameter systems[12], in system parameters, and in the "genotype" of structurally changing systems[28]. Study of hierarchical evolutive systems[23], anticipatory systems [43, 44], simulation in intelligent control and decision support[9], and many other areas all encounter aspects of the broad range of issues implied by consequence-driven systems. This is also a seemingly inexhaustible source of confusion and discord. Simulations of systems comprised of such entities are evidently possible since they are being produced, although it is very difficult to understand at the moment what does and does not need to actually be in the simulation program in order to represent either the individual entities or the resulting system dynamics that they generate[20].

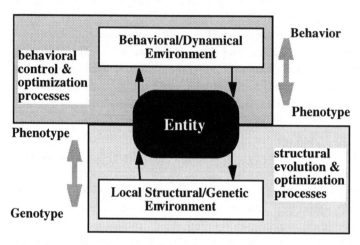

Figure 5a. General consequence-driven agent

However, all we need to mention here is that as long as local rules can be defined for entities and inter-entity causality can be reflected by state update scheduling mechanisms, then simulation techniques can in principle be applied to consequence driven systems. Indeed, the only representational medium that shows any promise at all with systems composed of consequence-driven systems is computer simulation. It is true that explanation and understanding has generally been more readily achieved than detailed predictive validity [19], probably because the state space of these simulations is very big and the dynamics are "complex"[2, 36], outcomes are nonunique relative to entity properties, and the entity "black boxes" are nonunique relative to their behaviors. A look at figure 5.b. and e.g., [9] will illustrate why this last nonuniqueness can be a problem to a modeler. Basically, state interpretive reasoning and reasoning for control actions involve induction and prediction and can compensate for one another in intelligent control setups without perfect world- or self- knowledge. It must be acknowledged that this makes the system identification problem for modeling at levels of internal processes of intelligent entities rather nasty and can open the door to unjustified overinterpretation and unscientific metaphor passing as empirical fact and mathematical analogy.

As we have said, in formal games learning and adaptation of the individual seems poorly represented and a truly "bottom up" character seems compromised. They perhaps prescribe too much, their pervasive payoff equilibria assumptions in the end

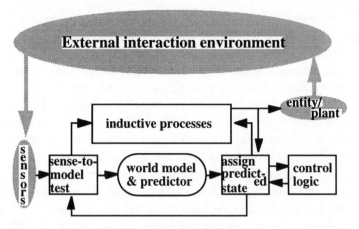

Figure 5b. General intelligently controlled agent

rendering them more like global models with fitted pieces than true dynamics generating simulations of systems of individuals. But if we push the entities a little farther and give them various adaptive, etc., capacities might simulation somehow cease to make sense? That is, are the formal conveniences of game theory necessary if the entity properties get more complicated than kinetic particles? It happens that, although we may leave the domain of formal games by incorporating complex entities, we do not exceed the scope of simulation nor the requisites for generation of dynamics. Although, we shouldn't be surprised if a model of a simulation might occasionally turn out to be interestingly expressed in a way that would permit the construction of an equivalent formal game, the complexity of behavior in large interacting systems will certainly require more than game theory, or vague appeals to evolutionary mechanisms in adaptive entities for that matter, to comprehend.

2.1.5 "Complex Dynamics" and Emergence

The "science of the complex" has been widely reported over the last decade or so and described from its relevance in social science many times, e.g., [13, 22]. Simulation has been a central feature [20] in this arguably new way of doing science. It isn't necessary to review this extensive literature here.

There are just three points that we would like to raise in regard to simulation. First, that in order to generate the fine structure of systems that exhibit complex dynamics, computational speed in the form of algorithmic efficiency becomes a functional requirement. No matter how fast computers may get, fast running compact representations are going to run faster than less elegant ones. That may be a truism, but it is also true that unnecessary elaboration of entity representations in simulation makes scientific interpretation correspondingly more difficult. Where possible the explicit representation should be absolutely spare. This aesthetic is found in computational statistical mechanics, but one must be very cautious in applying that armamentarium to simulation of socio-technical systems. There has been some interesting application of these kinds of principles to the computation of traffic in transport systems in the TRANSIMS project at Los Alamos National Laboratory [33, 34, 36, 50] and there is every reason to believe that the ideas that allow very simple generator algorithms to generate complicated traffic properties will also allow efficient representation in many socio-technical system settings. But how can that square with the above discussions of adequate representation of consequence-driven entities?

Second, a topic that has occupied a lot of attention in complex systems is the idea of emergent dynamics, or simply, emergence, e.g., [6, 36, 48]. This is normally viewed from a the point of view of the macroscopic dynamics, though perhaps on a rather fine scale, of the composed system. The general observation is that remarkably simple generator algorithms give rise to interesting global properties that are clearly more than the "sum of the parts" of the interacting simple computational devices. Again, the literature is vast and popular so we won't discuss it here. However, in [12] emergence of elaborate entity-level structures as well as macroscopic system dynamics is reported. The setting again is in road traffic in TRANSIMS. In an integer cellular automaton (CA) traffic micro simulation that exhibits phase transitions and complex phenomenology such as $1/f$ noise in the computed traffic flows[36], it is shown that it is possible to view the dynamics of individual vehicles from the point of view of their being operated by a derivative feedback adaptively compensated controller. In other words, the simple CA generator rules are producing complicated emergent local as well as global dynamical structures. Moreover this local dynamical structure- in essence an emergent driver entity- exhibits a kind of adaptation in the control of its behavior. Although the "emergent driver" does not learn from experience or mutate its genotype, it does drive differently in different traffic conditions. This means that a classical "nature-nuture" relationship is being generated in the simulation: the drivers make the traffic that forms their environment and they adapt their behaviors to a dynamcial environment they themselves create. It should be reemphasized that none of this is explicitly represented. If it were, the software for the driver model would be very slow running and, more importantly, very difficult if not impossible to calibrate. Viewed as an emergent phenomenon, the driver controller representation can be used as a model normal form, altered to yield a particular driver property such as closer following behavior, and then translated back to an explicit computational form of the generator function that will induce the new emergent controller. This "up and down emergence" may offer the possibility to produce interpretable simulations that compose the dynamics of truly complex entities and nevertheless are computationally tractable. However, this exercise requires all of the concepts in generation, interaction, composition, intentional, extensional and model normal forms we have outlined here as well as substantial knowledge of principles of statistical physics, theoretical computer science, control theory, cognitive science, and what we will introduce in the next section, theoretical principles of simulation. Of course, the practitioner will also need to be an expert in the particular applied area pertinent to the socio-technical system in question, for example, transportation systems.

As far as incorporating experiential effects on the "genotypic" structures of entities goes, there are also reasons [10, 35, 39] to believe that the sequence to structure emphasis we have outlined in the previous sections will be directly related to our ability to extend both theory and application to include them. Of course the celebrated genetic algorithm construct [28] has been applied for such purposes[3], but it is arguable that the simulation theory needs to underlie that as well, that these are "simply" more examples of simulation in action.

Third and last, complex systems have the inconvenient property that they can sometimes generate elaboration upon elaboration in patterns in state space, literally forever. This makes measurement a problem. For example taking measurements in a traffic system poses the problem of how to use them if you believe the traffic flow dynamics are chaotic. A lot of effort has gone into summarizing these properties with scaling laws and other methods e.g., see[6, 36]. In some circumstances and for some purposes these ideas have become rather well studied and are commonly applied in many disciplines. However, if we consider scaling laws for example, they won't help us much if the problem is to verify that we have an accurate (complex) simulation of a (complex) real world socio-technical system. Nor will they help us too much if we want to know if one simulation is computing the same system as another approach to representation taken by another simulation. The elaboration of state and phase space of complex systems poses a serious challenge to a basic theory of simulation: to determine congruence of simulations outside of state/phase space.

2.1.6 Elementary Theory of Simulation.

At present a highly developed theory of simulation does not exist. What theory does exist primarily relates to topics roughly at the level of figure 1. But a truly elementary theory must underlie those concepts and motivate them naturally; that is, the concepts in figure 1 should appear as propositions, lemmas, and theorems. The idea of the relationship of a basic theory of simulation to essential constructs of simulation can be illustrated by comparing figure 1 to figure 6. There the theory lies under the constructs and is not about those constructs directly. Initial work in simulation theory of this kind or its mathematical elements is found in [10, 37].

Earlier we discussed the relationship between causal dependency and sequence. Causes, whatever their genesis in an actual system might be, appear as dependencies among mappings in a simulation. These dependencies cause the image of a mapping to change as the order of the execution of the mappings changes. Thus, causal dependency in a target system is reflected in compatible orderings of rule execution in the simulation. It is therefore important to realize that watching the trace of the system generation in the allowable (causally compatible) update orders yields important information about the causal structure of the simulation. Indeed all of the causal information about the target system resident in the simulation itself is encoded in the order of update.

In [10] computer simulations are viewed abstractly as compositions of local mappings that produce global system dynamics. The theory can be developed without reference to the particulars and details of the mappings in question other than the fact that they can depend on each other. The analysis focuses on the basic dependency structure which is represented as a graph whose vertices are being updated by local maps, where locality is defined by adjacency in the graph. A sequential update graph is then developed that preserves the information about update orderings that will yield the same global dynamics. Several results are proven in this setup that pertain to the question of when it can be concluded that two simulations are the computing the same system by looking only at these structures and not making any measurements of system state at all. The last result in the paper establishes the conditions that must exist to for one simulation to encode the same causal dependencies among its mappings as another simulation. This result can be interpreted as saying that it specifies when two simulation are dynamically congruent- represent the same model normal form- if roughly the following conjecture holds: If two simulations generate the same causal dependencies and no matter what other differences exist between the representations, then they can represent the same dynamical system.

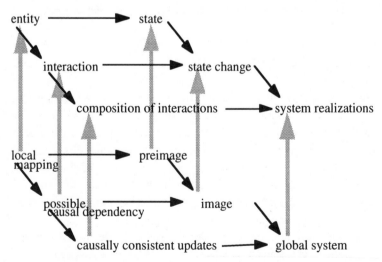

Figure 6. Elements of Simulation: Underlying theoretical properties

There are also the very real problems of assigning mappings and names to entities in the real world, in essence providing an appropriate semantics to the simulation program. But some aspects of appearances of similarity of any simulation program to another system lie outside the simulation program and this theory. To some degree, this aspect of simulation is brought structurally into the program by programmer/ modeler who is, in turn, guided by the use to which the system will be put. In a sense this is not a lot different than the difference between the information content of a message and the semantic content of the same message. In that sense there is little new in what we say here, there is a fairly universally accepted distinction between a symbol and its referent, syntax and semantics. But in simulations we are comparing not only the denotations and explicitly encoded logical relations, but the implicit emergent dynamical relations that appear in both the "real" system and a program-med generator we call a simulation program. The dynamics that appear in each of these systems are real in some sense. Anyway, we must include the use environment and the modeler for a full account of simulation and it isn't so clear how to get that in the mathematics.

Thus the rest, the extent to which the modeler and observers of the simulation draw the connection to another system and are informed by that is the entire topic of the next section. The interpretation that occurs outside the simulation program can be thought of as its use environment as well as its defining environment. In any case, for simulations to be used for decision making in socio-technical systems and to be believed, they must embody the character of their intended use.

3. The Impact of Purpose of Use on Simulation

Professor Andersson recently asked, "If simulation is really such a big deal, then why isn't Bill Gates making billions of dollars on it?" Rather than engaging his typical provocation, we simply agree it is clear that the use of simulations will direct the development of simulation science and technology. Conversely, if there are no particularly compelling uses, the science and technology will not develop. If for no other reason than that, we will outline four classes of use of simulations in regard to socio-technical systems. But in addition, because of the issues in the discussions above, it is fair to ask how we will ever sort out the conceptual and theoretical problems and actually use simulation effectively over the range of interests in these systems.

Although formally speaking it is convenient to prescribe that generation precede modeling issues, in practice the desire to model a system almost always precedes the making of a simulation. In fact, a phased "prescription" for simulations is proposed in [42]: modeling phase, simulation construction phase, computation phase. The motivation for the model - the use to which it will be applied - crucially affects the form and character of the model the simulation will be expected to embody. What we choose to model captures the sense of what is important in the simulation we construct.

Structural aspects of use context are programmed and impose themselves on the dynamics in such a way that their semantic aspects, if you will, the reason the simulation structurally represents a system in one way or another that no program could "know", are automatically reflected in the behavior of the program. This can be used. In most socio-technical systems, at the lowest level in the system things get rather "kinetic" and network constraints are powerful "dynamical filters". From vehicles on roads to bytes flying between the New York and Tokyo trading houses, the physical movement of traffic dominates at the lowest levels of resolution of dynamics in the systems. At that level there are almost always heavy constraints on what can happen because of network properties and entity physics. This is not to say that the dynamics is simple down there, it is replete with chaotic regimes and phase transitions and so on and on, in general. However, buy orders arrive in packet streams over wires and through (de-) encryption algorithms after time delays and in some order, cars cannot pass through each other or instantaneously reverse direction or disappear, and so on. What this allows is realistic computation of system outcomes that are possibly predicated on internal and personal representations a good deal less well understood, constrained, or consistent, e.g., personal preferences and opinions. In the end after all the best guesses, rough modeling, and calculation of that essential class of human intervention in a socio-technical system we can calculate consequences at successively lower levels, until we hit the traffic. We can take the behavioral consequences of the simulation in contact with constraints and feed it back into the local models of the more psycho-social-economic variety, alter them individually in some way, run the thing back down to traffic, and iterate this process over and over. So we can get the computed generation of low level dynamics to help guide local modeling without so much reliance on data collection or equilibria notions being imposed locally. Thus we are saying that use context in its structural form can (a.) be seen by users to bring the semantics of the modeled system to the simulation program and (b.) by programmers to refine the local generator algorithms in a semantically consistent way.

In TRANSIMS this kind of iteration has been done recently with regional vehicular traffic microsimulations[8,11,14, 35], and two comments about it pertain here. First, the route choice decisions of individuals get modified such that entire populations' traffic starts looking pretty realistic. This means that the iteration of the traffic micro simulation and the individual-by-individual route selection simulation has indeed allowed something to be learned by its entities, which appear as individuals in both simulations. That offers the possibility for even some of the evolutionary aspects of complex entities (figure 5.a). to appear in simulated individual travellers, in addition to the previously mentioned adaptation in driver control functions, although the implementation retains the very minimal but fast running local generator algorithms. Second, an entirely new aspect of complex individual definition arises- internal inconsistency. Internal inconsistency is a grave problem for computer programs of course. But in the TRANSIMS study, the simulated travellers were partitioned into subpopulations for an equity analysis of hypothetical infrastructure changes and associated funding decisions. Different subpopulations can perceive the impact of the changes in entirely opposite ways. There is nothing new there, it is the essence of the celebrated Arrow Paradox [32] and also it basically is why equity analysis is done. But it is also obviously true that in possible partitions of a population that a single simulated individual will reside in more than one subpopulation, and that some of these subpopulations could be in conflict from the point of view of the impact of a change to, for example, transport infrastructure. Thus our simulated individuals can exhibit internal contradiction, at least in some sense, but remain locally computationally consistent. But using feedback from the network dynamics until the traffic is "realistic" and the associated interpretation of the refinement of routing choice models as well as the partition of subpopulations and any associated ideas of incorporating contradiction in human behavioral modeling is something that clearly comes from outside the simulation program in addition to the reason for structural definition of the program. The example makes the point that it may be possible to represent a surprising range of complex entity properties with elegant and fast running simulation programs by appropriate exploitation of the inherent properties of the use environment for a simulation program.

It seems reasonable to try to propose a general, use oriented taxonomy of simulation-based environments for the study of socio-technical systems. The characteristics of the particular simulation technologies and methodologies for somebody to "make billions" with should be made clearer by including the use and users in the basic characterization of the simulations they actually need.

3.1 Strategy and Policy Laboratories

The issues here are largely goal- and priority setting, usually over long periods of time. The use of simulation is to provide an experimental environment, a laboratory, in which to (a.) characterize the effects and consequences of particular actions and policy as measured against a goal and (b.) characterize the functional properties of subsystems, that if obtained, will result in the desired consequences and in consonance with policy. The details of the means to achieve the functions, for example the technical or economic means with which to realize a functional property is not the primary function of Strategy Laboratory and therefore the simulations will likely aggregate above the details of those means. The users of this environment will be largely nontechnical management or policy makers and elected officials. Therefore the character of the environment will be heavily graphics intensive and will either be "turn key" software or staffed facilities, or both, depending on the intensity and complexity of the computing problems in the laboratory. Training environments for decision makers at this level, "war room" functions, and so on are necessary requirements for the Strategy Laboratory.

3.2 Design and Implementation Laboratories

Given that a strategy has been defined in the sense of functions and policies and goals, the use of simulations in the Design Laboratory is to guide development of subsystems that exhibit the functional requirements. This may involve generating and checking large scale system interactions, but much of this work is also done at the item-subsystem level and the simulation environments may be rather diverse. A major problem is that assuring consistency and common methods must be continually managed, but that can also be detrimental if enforced too rigidly. A major role of this laboratory is feedback to the Strategy and Policy Laboratories when functional requirements of components subsystems are not possible, etc. This allows refined strategy and policy and improves the forecasting process for strategic planning. Thus another issue is compatibility of results in the Design Laboratory with both Strategy Laboratory software on the one hand and with manufacturing methods on the other. The user population will largely be technical people although distinctions in Design- and Implementation Laboratories may be necessary to accommodate the diverse requirements of communicating with both Strategy Laboratory and manufacturing industry. Training environments for each of the subsystem simulation laboratories Design Laboratory is of course a necessary support function and must be provided, but also training environments for the iterations among Laboratories will offer unique benefits.

3.3 Control Laboratories

When subsystems are made and in place, they must be made to work. Simulation-based Control Laboratories are already part of everyday life in design and operation of control systems. Moreover, such things as flight simulators are examples of training applications in the Control Laboratory setting. The innovation that will most distinguish this technology over the next few years is incorporation of the means to implement sophisticated distributed control and self organization principles in practice. Moreover, the connection of Control Laboratory to Design Laboratory has interesting implications. The user environment is technical, operations oriented individuals.

3.4 Consensus Laboratories

Given strategy, policy, and goal options as well as options for implemented means to achieve them, consensus decision making is an essential aspect of socio-technical systems. The simpler uses of simulations to convey options to a voting populace, support studies for interest advocates, and so on, are already a reality of public life in socio-technical system funding and deployment as is exemplified by the fact that NASA regularly uses simulations of space system concepts in public relations. Consensus Laboratories offer a remarkable opportunity to convey not only the possible solutions to the range of interested parties, but also change the character of the decision process itself. A remarkable glimpse of this kind technology is a currently available personal computer game[49] from Sweden which is a low fidelity interactive simulation of the effects of investment decisions on public transportation. "SL, The Game" costs approximately $US 8.00 at SL (Storstockholms Lokaltrafik AB) ticket offices. It offers an entertaining way of helping those interested in the public investment in transport to get a feeling for how access, availability, user fees, taxes, environmental damage, convenience, and time of travel are affected by choice of building location, facility requirements, mode mix, etc. It is a remarkable program and it is essentially an example of a Consensus Laboratory concept. There are many other Consensus Laboratory settings imaginable. They are characterized by the ability to represent in an understandable way a range of options and the functional properties that the facility and technology builders can provide as well as interactions among them and consequences of decisions made. Training environments are an interesting issue from many perspectives.

Consensus Laboratory is arguably one of the most important large scale public uses of information technology in the next century, transcending both democratic ideals and the concept of news and freedom of information. It potentially provides access to something a good deal richer than simple information about decisions being made. Intriguing possibilities of education and training also arise.

And, Åke, it has entertainment possibilities.

Note

1. There is a similar use of intentional/extensional forms of representation in formal languages. The generative grammar, lexicon, etc., comprise the intentional form of a language [21, 24, 26]. The collection of well formed formulae, i.e., sentences is the extensional form of the language. In the closely related mathematical theory of computation, an equivalent formulation relates an algorithmic form of a program to the extensional set of states that it can produce. Also, games and game trees appear in some of the earliest work in Artificial Intelligence, e.g., see [47], and graphs of this kind are basic representation in Dynamic Programming [15].

References

[1] Arthur, W. B. Inductive reasoning and bounded rationality, American Economics Association Papers and Proceedings, 84 (1994)

[2] Arthur, W. B. Increasing returns and path dependencies in the economy: Ann Arbor, University of Michigan Press, 1994

[3] Arthur, W. B., Holland, J., Le Barron, B., Palmer R., and Taylor, P., Asset pricing under inductive reasoning in an artificial stock market, Santa Fe Institute working paper, 1996

[4] ASCII: Accelerated Strategic Computing Initiative, http://www.llnl.gov/asci-alliances/ , 1997

[5] Axelrod, Robert, Advancing the art of simulation in the social sciences, Santa Fe Institute Report 97-05-048, 1997

[6] Bak, P., Tang C., and Wiesenfeld, Self organized criticality, Physical Review A, 38(1), pp 364-374, 1988

[7] Barghodia A., Chandy, K., and Liao, W., A unifying framework for distributed simulation, ACM transactions on modeling and simulation, 1(4), pp 348-385, 1991

[8] Barrett, C., Berkbigler, K., Smith L., Loose, V., Beckman, R., Davis, J., Roberts, D., Williams, M., An Operational Description of TRANSIMS, Los Alamos Unclassified Report, LA-UR:95-2393, 1995

[9] Barrett, C. and Donnell, M., Real time expert advisory systems: Considerations and imperatives, North-Holland, Information and Decision Technology 16(1990).

[10] Barrett, C. & Reidys, C., Elements of a Theory of Computer Simulation I: Sequential CA over Random Graphs, Journal of Applied Mathematics and Computation, North Holland, in press.

[11] Barrett, C., Eubank, S., Nagel, K., Rasmussen, S., Riordan, J., and Wolinsky, M., Issues in the Representation of Traffic with Multi-Resolution Cellular Automata, Los Alamos Unclassified Report, LA-UR:95-2658

[12] Barrett, C., Wolinsky, M., and Olesen, M., Emergent Local Control Properties in Particle Hopping Traffic Microsimulations, in Proceedings of Traffic and Granular Flow, Julich, Germany, 1995

[13] Batten, F. Co-Evolutionary Learning on Networks. Paper prepared for a Festschrift in Honour of Åke E. Andersson. Knowledge and Networks in a Dynamical Economy, 1996.

[14] Beckman R., et. al., TRANSIMS Dallas/ Ft. Worth case study report, Los Alamos unclassified technical report available at http://www-transims.tsasa.lanl.gov, 1997

[15] Bellman, R., and Dreifus, S., Applied dynamic programming, Princeton University Press, 1962

[16] Bozinovski, S., Consequence Driven Systems, Debarska Macedonia, GOCMAR publishers, 1995.

[17] Bratley, P., Fox, B., and Schrage L., A guide to simulation, 2nd ed., New York, Springer-Verlag, 1987

[18] Casti, J., Alternative Realities, Wiley, 1990.

[19] Casti, J., Can you trust it?, Complexity, v2 (5), 1997.

[20] Casti, J., Virtual Worlds, Wiley, 1996

[21] Chomsky, N., Rules and Representations, Behavioral and Brain Sciences, v3, pp1-15, 1980.

[22] Devaney, R., An Introduction to Chaotic Dynamical Systems, 2nd ed., Addison-Wesley, 1989

[23] Ehresmann, A., VanBremeersch, Hierarchical Evolutive Systems: A Mathematical Model for Complex Systems, Bull. Math. Biol., v49(1)

[24] Fodor J.,Representations, Cambridge, Mass, MIT Press, 1981

[25] Feynmann, R., Leighton, R.B., and Sands, M., The Feynmann Lecture Series on Physics, v1-3, Addison Welsey, 1966.

[26] Ginsburg, S., The Mathematical Theory of Context-free Languages, McGraw-Hill:New York, 1966.

[27] Holland, J., Adaptation in natural and artificial systems, Ann Arbor, University of Michigan Press, 1975

[28] Holland, J., Genetic Algorithms, Scientific American, July 1992, pp 66-72

[29] Hurwicz, L., What is the Coase Theorem? Japan and the World Economy vol 7 pp 49-74, Elsevier, 1995.

[30] Jefferson, D., Virtual Time, ACM proceedings on programming languages and systems, 7(3), pp 404-425, 1985

[31] Kumar, R., and Garg, V., Modeling and control of logical discrete event systems, Kluwer Academic Publishers, 1995

[32] MacKay, A., Arrow's Theorem: The paradox of social choice, New Haven, Yale University Press, 1980

[33] Nagel, K., High-Speed Microsimulations of Traffic Flow, Ph.D. Thesis, University of Cologne, Germany, 1994

[34] Nagel, K., Particle Hopping Models and Traffic Flow Theory, Physical Review, E, American Physical Society, 1995

[35] Nagel, K., and Barrett, C., Using microsimulation feedback for trip adaptation for realistic traffic in Dallas, International Journal of Modern Physics, C., 8(3), pp 505-526, 1997

[36] Paczuski, M. and Nagel, K., Self Organized Criticality and 1/f Noise in Traffic, in Proceedings of Traffic and Grannular Flow, Julich, Germany, 1995.

[37] Reidys, C., Random induced subgraphs of generalized n-cubes, Advances in Applied Mathematics, 19, no.AM970553, pp. 360-377, 1997

[38] Reidys, C., and Barrett, C., Mutation and molecules: toward mutation-based computation, in Proceedings of the International Optimization and Simulation Conference, Singapore, Sept 1-4, 1997

[39] Reidys, C. and Fraser, S., Evolution on random structures, in Proceedings of the 1st World Congress on Systems Simulation, Singapore, 1997

[40]Reidys,C.M., Stadler, P.F., and Schuster, P.K., Generic properties of combinatory maps and neutral networks of (RNA) secondary structures, Bulletin of Mathematical Biology, 59(2), pp. 339-397, 1997

[41] Rasmussen, Eric, An Introduction to Game Theory, Blackwell, Cambridge, Mass, 1991.

[42] de Ronde, J., Mapping in High Performance Computing, Ph.D. Dissertation, Department of Computer Science, University of Amsterdam, The Netherlands, 1998.

[43] Rosen, R., Feedforwards and global system failure: a general mechanism for senescence, J. Theoretical Biology, 74, pp 579-590, 1978

[44] Rosen, R., Hierarchical Organization in Automata Theoretic Models of the Central Nervous System, In: Information Processing in the Nervous System, pp 21 -35, New York: Springer, 1969.

[45] Rubbia, C., Report of the high performance computing and networking advisory committee. Technical report, Commission of the European Communities, 1992.

[46] Saari, D. and Xia, Z., Off to infinity in finite time: Notices of the American Mathematical Society, pp 538-546, 1995

[47] Samuel, A., Some studies in machine learning using the game of checkers, IBM Journal of Research and Development 3(3), pp 210-229, 1959

[48] Sloot, P.M.A., Shoeneveld, A., de Rhonde, J.F, and Kaandorp, J.A., Large Scale Simulations of Complex Systems, Part I: Conceptual Framework, Santa Fe Institute Report 97-07-070, 1997.

[49] SL, The Game, computer game distributed by Storstockholms Lokaltrafik AB, produced by GazolineGraphics, 1997

[50] TRANSIMS, Los Alamos National Laboratory, Los Alamos NM. see http://www-transims.tsasa.lanl.gov

[51] von Neumann, J., and Morgenstern, O., The Theory of Games and Economic Behavior, New York: Wiley, 1944.

Minimal Complexity and Maximal Decentralization

Jean-Pierre Aubin
Doctoral School for Mathematical Decision Making (CEREMADE)
Université de Paris-Dauphine

1 Introduction

We shall investigate one aspect of "complexity" closely related to the decentralization issue, *connectionist complexity*, as one mechanism to adapt to viability constraints through the emergence of links between the agents[1] of the dynamical economy and their evolution, together with – or instead of – using decentralizing "messages". We shall compare it with other dynamical decentralization mechanisms, such as price decentralization.

1.1 Connectionist Mechanisms

We shall illustrate these ideas in the framework of the simple model of dynamical resource allocation. We denote by $Y := \mathbf{R}^l$ the commodity space. Assume for instance that, if there were no scarcity constraints to comply with, the dynamical behavior of each consumer $i = 1, \dots, n$ would be fully decentralized, and the evolution of its consumption $x_i(t) \in Y$ at time $t \geq 0$ would be governed by a differential equation of the form

$$\forall \ i = 1, \dots, n, \ x_i'(t) = g_i(x_i(t))$$

There is no reason why scarcity constraints

$$\forall \ t \geq 0, \ \sum_{i=1}^{n} x_i(t) \in M$$

of a collective nature would be satisfied by such independent behaviors.

We set $X := Y^n$. One first solution which comes up to the mind is to "connect" these dynamics through a connection matrix $W := (w_i^j) \in \mathcal{L}(X,X)$ and to govern the evolution of consumptions according to an evolution law of the form

$$\forall \ i = 1,\ldots,n, \ x_i'(t) = \sum_{j=1}^n w_i^j(t) g_j(x_j(t))$$

where $w_i^j \in \mathcal{L}(Y,Y)$. Naturally, we obtain a fully decentralized economy when the connection matrix is the identity matrix.

We propose mathematical metaphors designing evolution laws governing the evolution of connection matrices linking the autonomous dynamics of each agent in order to sustain the viability constraints prescribed by the environment of the isolated system.

Our point is the following: Starting with a disconnected system – for which the nondiagonal elements of the connection matrix are all equal to zero – the viability can be maintained whenever each agent of the system is linked with other through an evolving connectionist matrix, dictating cooperation among the agents. The larger the number of independent constraints, the more numerous the links between any agent with the others. In the mathematical version of the metaphor, a class of evolution laws governing the evolution of these connection matrices can be designed from the disconnected dynamical system and the set of constraints.

We shall single out for instance the problem of maximal decentralization which requires to find connecting matrices W (t) as close as possible to the identity matrix 1 – slow evolution.

This attempt to sustain the viability of the system by connecting the dynamics of its agents may be a general feature of "complex systems"[2]. Is not complexity meaning in the day-to-day language the labyrinth of connections between the components of a living organism or organization or system? Is not the purpose of complexity to sustain the constraints set by the environment and its growth parallel to the increase of the web of constraints?

Economic history has shown an everlasting trend toward a highly connected network of labyrinthine – connectionist – complexity. In this sense, complexity arose with the apparition of life and seems necessary to the pursuit of its evolution at all levels of organization.

We regard here connectionism – a less normative and more neutral term than cooperation whenever the system arises in economics or biology – as an answer to adapt to more and more viability constraints, which implies the emergence of links between the components of a system and their evolution. A system is disconnected (or autonomous, free, autarchic, decentralized, etc.) if the connection matrix is the identity (or unit) matrix. In some loose words, the distance[3] between the connection matrix – which is the matrix linking each component of the system to the others – and the identity matrix should capture and measure the concept of an index of (connectionist) complexity. The larger such a connectionist complexity index, the more complex – or labyrinthine or intricate – the connectionist feature of the system.

1.2 Decentralization Mechanisms

Compounded with this issue of connectionist complexity – that is naturally involved in neural networks – is the question of whether these links relating one agent to another can be subsumed by regulation parameters – which we call in short regulees – and whether and when they provide the same evolutions of the state of the system. Such regulees are messages sent to the consumers, and the simplest ones are the prices.

In a decentralized mechanism – actually, we should say in an "a-centralized mechanism", since we do not assume the existence of a "center" – the information on the alloaction problem is split and mediated by, say, a "message" which summarizes part of the information. In our case, we use the "price" p as a main example of message (actually, the message with the smallest dimension). Knowing the price p, consumers are supposed to know how to choose their consumption bundle, without

- knowing the behavior of their fellow consumers
- knowing the set of scarce resources

There are many other decentralized models, such as "rationing" mechanisms which involve shortages (and lines, queues, unemployment), or "frustration" of consumers, or "monetary" mechanisms, or others.

Naturally, there is no "pure" decentralization, since the choice of the decentralization message is in some sense centralized. The prices help consumers to make their choice in a decentralized way, but the difficulty is postponed to explain the evolution of prices.

In order to comply to viability constraints, we investigate successively

1. the regulation by subtracting prices to the original dynamical behavior

$$x'(t) \;=\; g(x(t)) - p(t)$$

2. the regulation by connecting the agents of the original dynamics through connecting matrices

$$x'(t) \;=\; W(t)g(x(t))$$

3. a combination of both regulation procedures

$$x'(t) \;=\; W(t)g(x(t)) - p(t)$$

The striking result we obtain in this framework is that these modes of regulation are equivalent by providing the same evolutions of commodities. In particular, this is the case of evolutions which minimize at each instant "complexity indices" of the form

$$\forall\; \alpha \;\in\; [0,1], \; \alpha \big\|1 - W(t)\big\|^2 + (1 - \alpha)\big\|p(t)\big\|^2$$

which govern evolutions of commodities independent of the choice of the weight a.

1.3 Differentiating Constraints

We shall need to "differentiate" viability constraints, and for that purpose, to implement the concept of tangency to any subset. The adequate choice to obtain viability theorems is to use the concept of contingent cone[4] introduced for the first time by Georges Bouligand in the thirties (and which happens to be the cornerstone to set-valued analysis).

Definition 1.1.
When K is a subset of X and x belongs to K, the contingent cone $T_K(x)$ to K at x is the closed cone of elements v satisfying

$$\begin{cases} v \in T_K(x) \text{ if and only if } \exists\, h_n \to 0+ \text{ and } \exists\, v_n \to v \\ \text{such that } \forall\, n,\ x + h_n v_n \in K \end{cases}$$

It is very convenient to use the following characterization of this contingent cone in terms of distances: the contingent cone $T_K(x)$ to K at x is the closed cone of elements v such that

$$\lim_{h \to 0+} \inf \frac{d(x) + hv, K}{h} = 0$$

We also observe that
if $x \in \text{Int}(K)$, then $T_K(x) = X$
and that

$$\text{if } K := \{\bar{x}\},\ T_{\{\bar{x}\}}(\bar{x}) = \{0\}$$

We recall that a function $x(.) : I \to X$ is said to be viable in K if and only if

$$\forall\, t \geq 0,\ x(t) \in K$$

Then a differentiable viable function in K satisfies

$$\forall\, t \geq 0,\ x'(t) \in T_K(x(t))$$

Let us mention that the contingent cone coincides with the tangent space $T_K(x)$ of differential geometry when K is a "smooth manifold". Also, when K is convex, one can prove that the contingent cone coincides with the tangent cone $T_K(x)$ to K at $x \in K$ of convex analysis, which is the closed cone spanned by $K - x$:

$$T_K(x) = \overline{\bigcup_{h>0} \frac{K - x}{h}}$$

In this case, the tangent cone is convex.

One can prove that the contingent cone $T_K(x)$ is convex whenever K "is sleek at x", which means that the set-valued map $T_K(.)$ is lower semicontinuous at x. Convex subsets are sleek at every elements.

There is a calculus of contingent cones which allows us to "compute them", and thus, to "apply" the viability theorems[5]. We recall in particular that if $K := h^{-1}(M)$ where $h : X \to Y$ is a continuously differentiable map such that $h'(x)$ is surjective[6] and M is closed and convex (or, more generally, sleek), then

$T_K(x) = h'(x)^{-1}T_M(h(x))$

We shall also need the following lower semicontinuity criterion (Proposition 1.5.1 of Set-Valued Analysis, [6]):

Proposition 1.2.
Consider two set-valued maps T and P from X to Y and Z respectively and a (single-valued) map c from $X \yen Z$ to Y satisfying the following assumptions:

$$\begin{cases} i) & T \text{ and } P \text{ are lower semicontinuous with convex values} \\ ii) & c \text{ is continuous} \\ iii) & \forall\, x, \; p \mapsto c(x,p) \text{ is affine} \end{cases}$$

We posit the following condition:

$$\forall\ x \in X,\ \exists \gamma > 0,\ \delta > 0,\ c > 0,\ r > 0\ \text{ such that }\ \forall\ x' \in B(x,\delta)\ \text{we}$$

$$\text{have}\quad \gamma B_Y \subset c(x', P(x') \cap r B_Z) - T(x') \qquad (1)$$

Then the set-valued map $\Pi : X \rightsquigarrow Z$ defined by

$$\Pi(x) := \left\{ p \in P(x) \mid c(x,p) \in T(x) \right\} \qquad (2)$$

is lower semicontinuous with nonempty convex values.

2   Slow Regulation

2.1   Regulation by Slow Prices

As the simplest example, we look for ways of coordinating the decisions of the consumers by subtracting prices to their dynamical priceless behavior

$$x'(t) = g(x(t)) - p(t)$$

in order that viability constraints of the form

$$\forall t \geq 0,\ h(x(t)) \in M$$

are always satisfied.

We define the **regulation map** Π_M by

$$\Pi_M(x) := \left\{ p \mid h'(x)(g(x) - p) \in T_M(h(x)) \right\}$$

The viability theorem implies that the prices $p(t)$ regulating viable solutions are given by the **regulation law**

$$p(t) \in \Pi_M(x(t))$$

In the case of the problem of allocations of scarce resources

$$\forall \ t \geq 0, \ \sum_{i=1}^{n} x_i(t) \in M$$

for instance, we see that

$$\Pi_M(x) := T_M\left(\sum_{i=1}^{n} x_i\right) - \frac{1}{n}\sum_{i=1}^{n} g_i(x_i)$$

For building feedback prices, we can for instance think of explicitly selecting some prices of the regulation map, for instance, the price $\varpi^\circ(x) \in \Pi_M(x)$ with minimal norm. Viable solutions obtained with this feedback price are called **slow viable solutions**.

We recall in the appendix that when $B \in \mathcal{L}(X,Y)$ is surjective, its orthogonal right inverse is equal to

$$B^+ = B^*(BB^*)^{-1}$$

that we can supply Y with the final norm μ^B defined by $\mu^B(z) := \|B^+ z\|$ and that we denote by π_K^B the projector of best approximation onto the closed subset K for this final norm. In this case, Proposition 3.3 below implies that the unique solution $\bar{x}$ to the minimization problem

$$\inf_{Bx \in K+v} \|x - u\|$$

is equal to

$$\bar{x} = u - B^+(1 - \pi_M^B)(Bu - v)$$

We finally recall that when M is sleek (and in particular, convex), then its tangent cones $T_M(y)$ are convex (see Theorem 4.1.8 of SET-VALUED ANALYSIS, [6, Aubin & Frankowska]). In this case, since the polar cone to the contingent cone $T_M(y)$ is the normal cone $N_M(y)$, we also know that we can write the solution to the minimization problem

$$\inf_{Bx \in T_M(y)+v} \|x - u\|$$

is equal to

$$\bar{x} = u - B^+(1 - \pi_{T_M(y)}^B)(Bu - v) = u - B^*\pi_{N_M(y)}^{B^*}(BB^*)^{-1}(Bu - v)$$

where $\pi_{N_M(y)}^{B^*}$ denotes the projector onto the normal cone $N_M(y)$ when the dual Y^* is supplied with the "dual final norm" $\|B^*q\|$.

Slow solutions to this dynamical economy to this problem are given by

Proposition 2.1.
*Let us assume that the map g is continuous and bounded and that the resource map
h is continuously differentiable and satisfies the uniform surjectivity condition:*

$$\forall \ x \ \in \ K, \ h'(x) \ \text{is surjective} \ \& \ \sup_{x \in K} \left\| h'(x)^+ \right\| \ < \ +\infty \qquad (3)$$

*and that M is closed convex (or more generally, sleek). Then the slow solution of the
dynamical economy*

$$x'(t) = g(x(t)) - p(t) \qquad (4)$$

subjected to the viability constraints

$$\forall \ t \ \geq \ 0, \ h(x(t)) \ \in \ M$$

is the solution to the differential equation

$$x'(t) = g(x(t)) - \varpi°(x(t))$$

where

$$
\begin{cases}
\varpi°(x) \ = \ h'(x)^+ \left(1 - \ \pi^{h'(x)}_{T_M(h(x))} \right) h'(x)g(x) \\
\qquad = \ h'(x)^* \pi^{h'(x)^*}_{N_M(h(x))} (h'(x)h'(x)^*)^{-1} h'(x)g(x)
\end{cases} \qquad (5)
$$

Proof:
Proposition 1.2 implies that the regulation map P_M is lower semicontinuous because
the set-valued map $T_M(.)$ is lower semicontinuous and that the surjectivity assumption
(3) implies assumption (1). Slow viable solutions do exist and are obtained by the
price $\varpi°(x)$ of minimal norm among the prices satisfying

$h'(x)p \in T_M(h(x)) - h'(x)g(x)$. Proposition 3.3 implies formula (5).

2.2 Regulation by Connection Matrices

Instead of decentralizing the dynamical economy by prices according to the dynamical
economy

$$x'(t) = g(x(t)) - p(t)$$

in order to comply to viability constraints of the form

$$\forall t \geq 0, \ h(x(t)) \in M$$

we look for the problem of sustaining viability of the evolution by connecting each

consumer with the other ones through a connection matrix $W := \left(w_i^j \right) \in L(X, X)$:

$$x'(t) = W(t)g(x(t)) \qquad (6)$$

Naturally, we obtain a fully decentralized economy when the connection matrix is
the identity matrix. But this is not always the case, and consumers may take into
account the behavior of the other consumers in their dynamics. Then the new

regulation parameter is no longer a price $p \in X^*$, but a **connection matrix** $W \in \mathcal{L}(X, X)$.

We introduce the new regulation map

$$R_M(x) := \left\{ W \in \mathcal{L}(X, X) \mid h'(x)Wg(x) \in T_M(h(x)) \right\}$$

The viability theorem implies that the connection matrices regulating viable solutions obey the regulation law $W(x) \in R_M(x(t))$. Recall that whenever $p \in X^*$ and $y \in Y$, we denote by $p \otimes y \in \mathcal{L}(X, Y)$ the rank one linear operator defined by $x \mapsto (p \otimes y)(x) := \langle p, x \rangle y$, the matrix of which is

$$\left(p^i y_j \right)_{\substack{i=1,\dots,m \\ j=1,\dots,n}}$$

Here, both finite dimensional vector spaces X and Y are identified with their duals. We then observe the following

Proposition 2.2

Let $\Gamma : X \mapsto X$ be any continuous map such that $\inf_{x \in K} \langle \Gamma(x), g(x) \rangle > 0$, connection matrix of the form

$$W(x) := \mathbf{1} - \frac{\Gamma(x)}{\langle \Gamma(x), g(x) \rangle} \otimes \varpi(x)$$

the entries of which are equal to

$$w_{ij}(x) = \delta_{i,j} - \frac{\Gamma(x)_i \varpi(x)_j}{\langle \Gamma(x), g(x) \rangle}$$

belongs to $R_M(x)$ if and only if $\varpi(x)$ belongs to $\Pi_M(x)$ and the viable solutions to the differential equations $x' = W(x)g(x)$ and $x' = g(x) - \varpi(x)$ do coincide.

Proof:

Indeed, we observe that

$$W(x)g(x) = g(x) - \frac{\langle \Gamma(x), g(x) \rangle}{\langle \Gamma(x), g(x) \rangle} \varpi(x) = g(x) - \varpi(x)$$

so that the two differential equations $x' = W(x)g(x)$ and $x' = g(x) - \varpi(x)$ are the same. Furthermore, to say that $W(x) \in R_M(x)$ amounts to saying that

$$h'(x)W(x)g(x) = h'(x)g(x) - h'(x)\varpi(x) \in T_M(h(x))$$

so that the viability conditions are the same.

In particular, the regulation by connection matrices of the form

$$W(x) := \mathbf{1} - \frac{g(x)}{\|g(x)\|^2} \otimes \varpi(x)$$

is equivalent to a price decentralization mechanism whenever $g(x) \neq 0$ for every $x \in K$.

The problem of maximal decentralization is then to find connection matrices as close as possible to the identity matrix $\mathbf{1}$. It happens that such connection matrices maximizing decentralization are of the above form.

Proposition 2.3
Let us assume that the map g is continuous and bounded and that the resource map h is continuously differentiable and satisfies the uniform surjectivity condition:

$$\forall \, x \, \in \, K, \, h'(x) \text{ is surjective } \& \, \sup_{x \in K} \frac{\left\| h'(x)^+ \right\|}{\left\| g(x) \right\|} < +\infty \quad (7)$$

and that M is closed convex (or more generally, sleek). Then the scarcity set $K := h^{-1}(M)$ is viable under the connection economy (6) if and only if for every $\forall x \in K$, the image $R_M(x)$ of the regulation map is not empty.

The solution with maximal decentralization subjected to the viability constraints

$$\forall t \geq 0, \, h(x(t)) \, \in \, M$$

is governed by the differential equation

$$x'(t) = W°(x(t))g(x(t))$$

where

$$\left\{ \begin{array}{l} W°(x) \; = \; 1 - \dfrac{g(x)}{\left\| g(x) \right\|^2} \otimes h'(x)^+ \left(1 - \Pi^{h'(x)}_{T_M\,(h(x))} \right) h'(x)g(x) \\[4mm] \qquad = \; 1 - \dfrac{g(x)}{\left\| g(x) \right\|^2} \otimes h'(x)^* \pi^{h'(x)^*}_{N_M\,(h(x))} (h'(x)h'(x)^*)^{-1} h'(x)g(x) \end{array} \right.$$

$$(8)$$

Furthermore, the slow viable solutions to dynamical economy (4) regulated by prices and the viable solutions to the connection economy (6) under maximal decentralization coincide.

Proof:
We can apply the same proof as the one of Proposition 2.1 where the connection economy is a dynamical economy for the dynamical system $c(x,W) := Wg(x)$ and where the role of the price p is played by the connection matrix W. Then, using the definition 3.5 of tensor products of linear operators of the appendise, the regulation map can be written in the form:

$$R_M(x) := \left\{ W \in L(X,X) \, \big| \, (g(x) \otimes h'(x))W \in T_M(h(x)) \right\}$$

so that viable solutions are regulated by the regulation law

$$(g(x(t)) \otimes h'(x(t))W(t) \in T_M(h(x))$$

Since the map $g(x) \otimes h'(Wx)$ is surjective from the space $L(X, X)$ of connection matrices to the resource space Y, we observe that

$$\|(g(x) \otimes h'(x))^+\| = \frac{\|h'(x)^+\|}{\|g(x)\|}$$

so that the assumptions of Proposition 1.2 are satisfied thanks to (7). Hence the regulation map R_M is lower semicontinuous.

Therefore, the viable connection matrices closest to the identity matrix 1 are the solutions W° minimizing the distance $\|W - 1\|$ among the connection matrices satisfying the above constraints.

We infer from Theorem 3.8 that since the map $g(x) \otimes h'(Wx)$ is surjective from the space $L(X, X)$ of connection matrices to the resource space Y, the solution is given by formula (8).

Remark: Mixed Regulation by Prices and Connection Matrices
Let us consider the mixed regulation mode using both connection matrices W and prices p:

$$x'(t) = Wg(x(t)) - p(t)$$

subjected to the same viability constraints

$$\forall t \geq 0, \quad h(x(t)) \in M$$

We still supply X with the Euclidian norm, identify X with its dual. Then the regulation map takes the form

$$R_M(x) := \{(W, p) \in L(X, X) \times X \mid h'(x)Wg(x) - h'(x)p \in T_M(x)\}$$

We supply the space $L(X, X) \times X$ with the norm $V_{\alpha, \beta}$ defined by

$$V_{\alpha, \beta}(W, p)^2 := \alpha\|W\|^2 + \beta\|p\|^2$$

Then the slow solutions are regulated by the pair $(W_{\alpha, \beta}(x), \varpi_{\alpha, \beta}(x))$ in the regulation map which minimizes the distance $V_{\alpha, \beta}$ $(W - 1, p)$ in $R_M(x)$. By Theorem 3.9 below, the solution is given by

$$\begin{vmatrix} W_{\alpha, \beta}(x) = 1 - \dfrac{\beta g(x)}{\alpha + \beta\|g(x)\|^2} \otimes h'(x)^+\left(1 - \pi_{T_M(h(x))}^{h'(x)}\right)h'(x)g(x) \\[4mm] \varpi_{\alpha, \beta}(x) = \dfrac{\alpha}{\alpha + \beta\|g(x)\|^2} h'(x)^+\left(1 - \pi_{T_M(h(x))}^{h'(x)}\right) h'(x)g(x) \end{vmatrix}$$

We then observe that, setting $\varpi^\circ(x) := h'(x)^+\left(1 - \Pi_{T_M(h(x))}^{h'(x)}\right) h'(x)g(x)$,

$$\begin{cases} W_{\alpha,\beta}(x)g(x) - \varpi_{\alpha,\beta}(x) \\ = g(x) - \dfrac{\beta\|g(x)\|^2}{\alpha + \beta\|g(x)\|^2}\varpi^\circ(x) - \dfrac{\alpha}{\alpha + \beta\|g(x)\|^2}\varpi^\circ(x) \\ = g(x) - \varpi^\circ(x) \end{cases}$$

Therefore, for every pair (α, β), the slow viable solutions, which are solutions to the differential equations

$$x'(t) = W_{\alpha,\beta}(x(t))g(x(t)) - \varpi_{\alpha,\beta}(x(t))$$

are the slow solutions to

$$x'(t) = g(x(t)) - \varpi^\circ(x(t))$$

3 Appendix: Right-Inverses and Tensor Products

3.1 Orthogonal Right-Inverses and Tensor Products

We denote by $(e^j)_{j=1,\dots,n}$ a basis of X and by $(f^i)_{i=1,\dots,m}$ a basis of Y.

Let X^* and Y^* denote the dual spaces of X and Y respectively. The bilinear form $\langle p, x\rangle := p(x)$ on $X^* \times X$ is called the duality product. We associate with the basis $(e^j)_j$ its dual basis $(e_j^*)_{j=1,\dots,n}$ of X^* defined by $\langle e_j^*, e^l\rangle = 0$ when $j \neq l$ and $\langle e_j^*, e^j\rangle = 1$. Hence the jth component x_j of x in the basis $(e^j)_j$ is given by $x_j = \langle e_j^*, x\rangle$, the jth component p^j of $p \in X^*$ in the dual basis $(e_j^*)_j$ is given by $p^j = \langle p, e^j\rangle$ and $\langle p, x\rangle := \sum_{j=1}^n p^j x_j$ and $\langle q, y\rangle := \sum_{i=1}^m q^j y_i$.

We recall that the transpose $W^* \in L(Y^*, X^*)$ is defined by

$$\forall q \in Y^*, \ \forall x \in X, \ \langle W^*q, x\rangle = \langle q, Wx\rangle$$

Let p belong to X^*. We shall identify the linear form $p \in L(X, \mathbf{R}) : x \in X \mapsto \langle p, x\rangle$ with its transpose $p \in L(\mathbf{R}, X^*) : \lambda \mapsto \lambda p \in X^*$. In the same way, we shall identify $x \in X$ with both the maps $p \in X^* \mapsto \langle p, x\rangle$ and $\lambda \in \mathbf{R} \mapsto \lambda x \in X$.

We shall begin by defining and characterize the orthogonal right-inverse of a surjective linear operator. This concept depends upon the scalar product i defined on the finite dimensional vector-spaces X, which is then given once and for all. We shall denote by $L \in L(X, X^*)$ its **duality map** defined by

$$\forall x_1, \ x_2 \in X, \ \langle Lx_1, x_2\rangle := l(x_1, x_2)$$

(The matrix $(l(e^i, e^j))_{i,j=1,\ldots,n}$) of the bilinear form l coincides with the matrix of the linear operator L).

We shall denote by

$$\lambda(x) := \sqrt{l(x,x)}$$

the norm associated with this scalar product.

The bilinear form $l_*(p_1, p_2) := \langle p_1, L^{-1} p_2 \rangle = l(L^{-1} p_1, L^{-1} p_2)$ is then a scalar product on the dual of X^*, called the **dual scalar product**.

Let us consider a surjective linear operator $A \in L(X, Y)$. Then, for any $y \in Y$, the problem $Ax = y$ has at least a solution. We may select the solution $\bar{x}$ with minimal norm $\lambda(x)$, i.e., a solution to the minimization problem with linear equality constraints

$$A\bar{x} = y \ \& \ \lambda(\bar{x}) = \min_{Ax=y} \lambda(x)$$

The solution to this problem is given by the formula

$$\bar{x} = L^{-1} A^*(AL^{-1} A^*)^{-1} y$$

Definition 3.1
*If $A \in L(X, Y)$ is surjective, we say that the linear operator $A^+ := L^{-1} A^*(AL^{-1} A^*)^{-1} \in L(Y, X)$ is the **orthogonal right-inverse** of A (associated with the scalar product l on X).*

Indeed, A^+ is obviously an right-inverse of A because $AA^+ y = y$ for any $y \in Y$. We observe that $1 - A^+ A$ **is the orthogonal projector onto the kernel of** A.

3.2 Projections onto Inverse Images of Convex Sets

Hence, we can write explicitly the solution $\bar{x}$ to the quadratic optimization problem with linear equality constraints

$$A\bar{x} = y \ \& \ \frac{1}{2}\lambda(\bar{x} - u)^2 = \min_{Ax=y} \frac{1}{2}\lambda(x - u)^2$$

Proposition 3.2
Let us assume that $A \in L(X, Y)$ is surjective. Then the unique solution $\bar{x}$ to the above quadratic minimization problem is given by

$$\begin{cases} \bar{x} = u - L^{-1} A^* \bar{q} = u - A^+(Au - y) \\ \text{where } \bar{q} := (AL^{-1} A^*)^{-1}(Au - y) \end{cases}$$

*is the **Lagrange multiplier** i.e., a solution to the dual minimization problem*

$$\min_{q \in Y^*} \left(\frac{1}{2} \lambda_* (A^* q - Lu)^2 + \langle q, y \rangle \right)$$

We observe easily that $\xi(q) := u - L^{-1} A^*(q)$ minimizes over X the function $x \mapsto \frac{1}{2} \lambda (x - u)^2 + \langle q, Ax \rangle$ and that the Lagrange multiplier minimizes $q \mapsto \frac{1}{2} \lambda_* (\xi(q))^2 + \langle q, y \rangle$. Consider now a closed convex subset $M \subset Y$ and the minimization problem and elements $u \in X$ and $v \in X$:

$$\inf_{Ax \in M + v} \lambda(x - u)$$

When $A \in L(X, Y)$ is surjective, we can supply the space Y with the **final scalar product**

$$m^A(y_1, y_2) := l(A^+ y_1, A^+ y_2)$$

its associated **final norm**

$$\mu^A(y) := \lambda(A^+ y) = \inf_{Ax = y} \lambda(x)$$

Its duality mapping is equal to $(AL^{-1}A^*)^{-1} \in L(Y, Y^*)$ and its dual norm is equal

to $$\mu_*^A(p) = \lambda_*(A^* p)$$

In this case, we denote by π_M^A the projector of best approximation onto the closed convex subset M when Y is supplied with the norm μ^A and by $\pi_{M^-}^{A^*}$ the projector of best approximation onto the polar cone

$M^- := \left\{ q \in Y^* \mid \forall y \in M, \langle q, y \rangle \le 0 \right\}$ when Y^* is supplied with the norm μ_*^A.

Proposition 3.3

Assume that $A \in L(X, Y)$ is surjective and that $M \subset Y$ is closed and convex. Let $u \in X$ and $v \in X$ be given. Then the unique solution $\bar{x}$ to the minimization problem

$$\inf_{Ax \in M + v} \lambda(x - u)$$

is equal to

$$\bar{x} = u - A^+ (1 - \pi_M^A)(Au - v)$$

When $M \subset Y$ is a closed convex cone, the solution can also be written in the form

$$\bar{x} = u - L^{-1} A^* \pi_{M^-}^{A^*} (AL^{-1}A^*)^{-1} (Au - v)$$

Proof

Indeed, we can write

$$
\begin{cases}
\lambda(\bar{x} - u) = \inf_{Ax \in M + v} \lambda(x - u) \\
= \inf_{y \in M + v} \inf_{Ax = y} \lambda(x - u) \\
= \inf_{y \in M + v} \lambda(A^+(y - Au)) \\
\text{(thanks to Proposition 3.2)} \\
= \inf_{z \in M} \lambda(A^+(z - (Au < -v))) \\
= \lambda(A^+(\pi_M^A(Au - v)) - (Au - v))
\end{cases}
$$

Hence $\bar{x} = u - A^+(1 - \pi_M^A)(Au - v)$.

When M is a closed convex cone, the second statement follows from the following

Lemma 3.4

Let $M \subset Y$ be a closed convex cone, $w \in Y$. If $A \in L(X,Y)$ is surjective and if Y is supplied with the final norm m^A, then

$$
(1 - \pi_M^A) \, w = (AL^{-1}A^*\pi_{M^-}^{A^*}((AL^{-1}A^*)^{-1}w)
$$

and thus

$$
A^+(w - \pi_M^A w) = L^{-1}A^*\pi_{M^-}^{A^*}(AL^{-1}A^*)^{-1}w
$$

Proof

Let us denote by $K := (AL^{-1}A^*)^{-1}$ the duality map associated with the final scalar product m^A. The projection $\pi_M^A w$ is characterized by

$$
\forall y \in M, \ m^A(w - \pi_M^A w, y - \pi_M^A w) = m_*^A(K(w - \pi_M^A w), Ky - K\pi_M^A w) \le 0
$$

When M is a closed convex cone, we can take successively $y = 0$ and $y = 2\pi_M^A w$ for deducing that this characterization is equivalent to

$$
\begin{cases}
m^A(w - \pi_M^A w, \pi_M^A w) = m_*^A(K(w - \pi_M^A w), K\pi_M^A w) = 0 \\
\forall \ y \ \in \ M, \ m^A(w - \pi_M^A w, y) = \langle K(w - \pi_M^A w), y \rangle \ \le \ 0
\end{cases}
$$

The second statement means that

$$
K(w - \pi_M^A w) \in M^-
$$

so that the first statement means that $K(w - \pi_M^A w)$ is the projection of Kw onto M^- when the dual Y^* is supplied with the dual final scalar product m_*^A since the negative polar cone M^- is a closed convex cone. Hence

$$
K(w - \pi_M^A w) = \pi_{M^-}^{A^*} Kw
$$

Consequently,

$$
A^+(w - \pi_M^A w) = L^{-1}A^*(AL^{-1}A^*)^{-1}(w - \pi_{M^-}^A w) = L^{-1}A^*\pi_{M^-}^{A^*}(AL^{-1}A^*)^{-1}w
$$

3.3   Tensor Products of Linear Operators

Let us introduce the following notations: We associate with any $p \in X^*$ and $y \in Y$ the linear operator $p \otimes y \in L(X,Y)$ defined by

$$p \otimes y \; : \; x \mapsto (p \otimes y)(x) \; := \; \langle p, x \rangle y$$

the matrix of which is

$$\left(p^i y_j \right)_{\substack{i=1,\ldots,m \\ j=1,\ldots,n}}$$

Its transpose $(p \otimes y)^* \in L(Y^*, X^*)$ maps $q \in Y^{**}$ to

$$(p \otimes y)^*(q) = \langle q, y \rangle p$$

because $\langle q, \langle p, x \rangle y \rangle \; = \; \langle \langle q, y \rangle p, x \rangle$. Hence, **we shall identify from now on the transpose** $(p \otimes y)^*$ **with** $y \otimes p$.

In the same way, if $q \in Y^*$ and $x \in X$ are given, $q \otimes x$ denotes the map $y \in Y \mapsto \langle q, y \rangle x \in X$ belonging to $L(Y,X)$ and its transpose $p \in X^* \mapsto \langle p, x \rangle q \in Y^*$ belonging to $L(X^*, Y^*)$ is identified with $x \otimes q$.

We observe that the map $(p, y) \in X^* \times Y \mapsto p \otimes y \in L(X,Y)$ is bilinear.

The operators $\left(e_j^* \otimes f^i \right)_{\substack{i=1,\ldots,m \\ j=1,\ldots,n}}$ **do form a basis of** $L(X,Y)$ **because we can**

write $x \; = \; \sum_{j=1}^{n} \langle e_j^*, x \rangle e^j$ and

$$Wx \; = \; \sum_{i=1}^{m} \langle f_i^*, Wx \rangle f^i = \sum_{i=1}^{m} \sum_{j=1}^{n} \langle f_i^*, W(e^j) \rangle \langle e_j^*, x \rangle f^i$$

$$= \; \left(\sum_{i=1}^{m} \sum_{j=1}^{n} \langle f_i^*, W(e^j) \rangle e_j^* \otimes f^i \right)(x)$$

where the entries

$$w_i^j \; := \; \langle f_i^*, W(e^j) \rangle$$

of the matrix W are the components of W in this basis.

It is useful to set

$$X^* \otimes Y \; := \; L(X,Y)$$

to recall this structure.

Definition 3.5

Let us consider two pairs (X, X_1) and (Y, Y_1) of finite dimensional vector-spaces. Let $A \in L(X_1, X)$ and $B \in L(Y, Y_1)$ be given. We denote by $A^* \otimes B$ the linear operator from $L(X,Y)$ to $L(X_1, Y_1)$ defined by

$$\forall W \in L(X,Y), \ (A^*\otimes B)(W) \ := \ BW \ A$$

We observe that when $W = p \otimes y$, we have

$$(A^*\otimes B)(p \otimes y) = A^* \ p \otimes By$$

In particular, $x \otimes B \in L(L(X,Y),Z)$ maps W to BWx.

We also observe that

$$(A^*\otimes B)^* \ = \ A \otimes B^* \tag{9}$$

For example, the transpose of $x \otimes B$ is equal to $(x \otimes B)^* \ = x \otimes B^*$. It maps any $q \in Z^*$ to the linear operator $x \otimes B^* \ q \in L(X^*,Y^*)$.

Let $A_1 \in L(X_2,X_1)$ and $B_1 \in L(Y_1,Y_2)$. Then, it is easy to check that

$$(A_1^* \otimes B_1)(A^* \otimes B) \ = \ (A_1^* A^*)\otimes(B_1 B)$$

Then the following formulas are straightforward:

$$\begin{cases} i) & \text{If } A \text{ and } B \text{ are invertible, then} \\ & \cdot \ (A^* \otimes B)^{-1} = (A^{-1})^* \otimes B^{-1} \\ ii) & \text{If } A \text{ is left – invertible and } B \text{ is right – invertible,} \\ & \text{then } A^* \otimes B \text{ is right – invertible} \\ iii) & \text{If } A \text{ is right – invertible and } B \text{ is left – invertible,} \\ & \text{then } A^* \otimes B \text{ is left – invertible} \\ iv) & \text{If } A \text{ and } B \text{ are projectors, so is } A^* \otimes B \end{cases}$$

Example: Gradients of Functionals on Linear Operators

Let us consider three spaces X, Y and Z, an element $x \in X$, a differential map $g \ : \ Y \mapsto Z$.

Proposition 3.6

The derivative of the map $H \ : \ W \mapsto H(W) \ := g(Wx)$ at $W \in L(X,Y)$ is defined by

$$H'(W) \ = \ x \otimes g'(Wx)$$

let E be a differentiable functional from Z to $\mathbf{R}$. We set $\Psi(W) \ := \ E(g(Wx))$. The gradient of ψ at $W \in L(X,Y)$ is defined by

$$\Psi'(W) \ = \ x \otimes g'(Wx)^* E'(g(Wx)) \in L(X^*,Y^*)$$

3.4 Best Approximation of Matrices

3.4.1 Scalar Products on $\mathcal{L}(X,Y)$

If we supply the spaces X and Y with scalar products $l(.,.)$ and $m(.,.)$, we shall denote by $L \in \mathcal{L}(X,X^*)$ and $M \in \mathcal{L}(Y,Y^*)$ their **duality mappings** and by

$$\lambda(x) := \sqrt{l(x,x)} \quad and \quad \mu(y) := \sqrt{m(y,y)}$$

the norms associated with these scalar products.

It is clear that if $(e^j)_{j=1,\ldots,n}$ is a l-orthonormal basis of X (i.e., satisfying $l(e^j,e^l) = 0$ when $j \neq l$ and $l(e^j,e^j) = 1$, then the basis $(e_j^* := Le^j)_j$ is the dual basis of $(e^j)_j$ because $\langle e_j^*, e^l \rangle = l(e^j, e^l)$. The matrix of the duality mapping L in these basis is the identity matrix.

Let us consider now an i-orthonormal basis (e^j) of X and a m-orthonormal basis (f^i) of Y. Let U and V be two operators of $\mathcal{L}(X,Y)$. We observe that

$$\sum_{j=1}^n m(Ue^j, Ve^j) = \sum_{i=1}^m l_*(U^* f_i^*, V^* f_i^*) = \sum_{i,j} \langle f_i^*, Ue^j \rangle \langle f_i^*, Ve^j \rangle$$

so that this expression is independent of the choice of the l- and m-orthonormal basis of X and Y.

We thus associate with the two scalar products l and m *the scalar product* $l_* \otimes m$ on $\mathcal{L}(X,Y)$ defined by

$$(l_* \otimes m)(U,V) := \Sigma_{i,j} \langle f_i^*, Ue^j \rangle \langle f_i^*, Ve^j \rangle$$

$$= \Sigma_{j=1}^n m(Ue^j, Ve^j) = \Sigma_{i=1}^m l_*(U^* f_i^*, V^* f_i^*)$$

For matrices $p_i \otimes y_i$, this formula becomes

$$(l_* \otimes m)(p_1 \otimes y_1, p_2 \otimes y_2) := l_*(p_1, p_2)m(y_1, y_2)$$

We observe that **the duality mapping from $\mathcal{L}(X,Y)$ to $\mathcal{L}(X^*,Y^*)$ of the scalar product** $l_* \otimes m$ is $L^{-1} \otimes M$ because

$$(l_* \otimes m)(U,V) = \Sigma_{i,j} \langle f_i^*, Ue^j \rangle \langle f_i^*, Ve^j \rangle$$

$$= \Sigma_{i,j} \langle f^i, MUL^{-1} e_j^* \rangle \langle f_i^*, Ve^j \rangle$$

$$= \langle MUL^{-1}, V \rangle = \langle (L^{-1} \otimes M)U, V \rangle$$

If we consider the matrices (u_i^j) and (v_i^j) of the operators U and V for the orthonormal basis $(e^j)_j$ and $(f^i)_i$, then

$$(l_* \otimes m)(U,V) = \sum_{i=1}^m \sum_{j=1}^n u_i^j v_i^j$$

We shall denote by

$$(\lambda_* \otimes \mu)(U) := \sqrt{(l_* \otimes m)(U,U)}$$

the norm associated with this scalar product.

3.4.2 Orthogonal Right-Inverse of $A \otimes B$

Proposition 3.7
Let us assume that $A \in \mathcal{L}(X_1, X)$ is injective and that $B \in \mathcal{L}(Y, Y_1)$ is surjective. Then $A * \otimes B$ is surjective and its orthogonal right-inverse is given by the formula:

$$(A* \otimes B)^+ = (A*)^+ \otimes B^+$$

Then $\overline{W} := B^+ W_1 A^-$ is the solution to the operator equation $BWA = W_1$ with minimal norm (for the norm $\lambda_* \otimes \mu$ on $\mathcal{L}(X,Y)$).

Proof
In order to prove that $(A * \otimes B)^+ = (A*)^+ \otimes B^+$, we apply the formula for the orthogonal right-inverse: Let us set $\mathcal{A} := A* \otimes B$ and recall that the duality map $J \in \mathcal{L}(\mathcal{L}(X,Y), \mathcal{L}(X*,Y*))$ is equal to $L^{-1} \otimes M$, so that its inverse is equal to $J^{-1} = L \otimes M^{-1}$.
   Hence

$$\mathcal{A}^+ = J^{-1} \mathcal{A} * (\mathcal{A} J^{-1} \mathcal{A}*)^{-1}$$

But we see that

$$\mathcal{A} J^{-1} \mathcal{A}* = A * LA \otimes BM^{-1}B*$$

so that

$$\mathcal{A}^+ = LA(A* LA)^{-1} \otimes M^{-1}B(BM^{-1}B*)^{-1}$$

It is enough to observe that $M^{-1}B(BM^{-1}B*)^{-1} = B^+$ and that

$$LA(A* LA)^{-1} = (A*)^+ = \left((A* LA)^{-1} A * L\right)^* = (A^-)**$$

so that the formula holds true.

3.4.3 Orthogonal Projection on Subsets of Matrices

Proposition 3.3 imply the following formula for the optimal solution to the quadratic minimization problem under constraints

$$(\lambda_* \otimes \mu)(\overline{W} - U) = \inf_{BWx \in K+v} (\lambda_* \otimes \mu)(W - U)$$

where $U \in \mathcal{L}(X,Y)$ is given and where $K \subset Z$ is a closed convex subset. Recall that when $B \in \mathcal{L}(Y,Z)$ is surjective, we can supply Z with the final norm μ^B defined by $\mu^B(z) := \mu(B^+z)$ and that we denote by π_K^B the projector of best approximation onto the closed subset K for this final norm.

Theorem 3.8

Let $B \in L(Y,Z)$ be a surjective operator, $K \subset Z$ be a closed convex subset and $x \in X$ be different from 0. Then the matrix

$$\overline{W} \ := \ U - \frac{Lx}{\lambda(x)^2} \otimes B^+(1 - \pi_K^B)\,(BUx - v)$$

is the best approximation of U in the subset of linear operators W satisfying the constraints

$$BWx \ \in \ K + v$$

If $K \subset Z$ is a closed convex cone, then the optimal matrix $\overline{W}$ can also be written in the form

$$\overline{W} \ := \ U - \frac{Lx}{\lambda(x)^2} \otimes M^{-1} B^* \pi_{K^-}^{B^*}\,(BM^{-1}B^*)^{-1}(BUx - v)$$

Proof

We know that the right inverse of the map $W \mapsto BWx$ is equal to the map $z \mapsto \frac{Lx}{\lambda(x)^2} \big| \otimes B^+ z$. Therefore, the final norm on the vector space Z associated with this operator is equal to $\mu(B^+ z)\lambda(x)$. Consequently, the orthogonal projector onto K associated with this norm is equal to the projector π_K^B associated with the final norm $\mu(B^+ z)$.

Consequently, by Proposition 3.3, we infer that the solution to

$$(\lambda_* \otimes \mu)(\overline{W} - U) \ = \ \inf_{BWx \in K + w} (\lambda_* \otimes \mu)(W - U)$$

is equal to $U - (x \otimes B)^+(1 - \pi_K^B)(BUx - v)$.

Theorem 3.9

Let $B \in L(Y,Z)$ be a surjective operator, $K \subset Z$ be a closed convex subset and $x \in X$ be different from 0. Let us supply the product space $L(Y,Z) \times Y$ with the norm $v_{\alpha,\beta}$ defined by

$$v_{\alpha,\beta}(W,v)^2 \ := \ \alpha(\lambda_* \otimes \mu)(W)^2 + \beta\mu(w)^2$$

Then the pair $(\overline{W}, \overline{w})$ defined by

$$\begin{cases} \overline{W} \ = \ U - \dfrac{\beta Lx}{\alpha + \beta\lambda(x)^2} \otimes B^+((1 - \pi_K^B)(BUx - v)) \\[2ex] \overline{w} \ = \ \dfrac{\alpha}{\alpha + \beta\lambda(x)^2} B^+((1 - \pi_K^B)(BUx - v)) \end{cases}$$

is the best approximation of (U,u) in the subset of pairs (W,w) satisfying the constraints

$$BWx - Bw \in K + v$$

Proof

We have to compute the orthogonal right inverse of the linear operator $A \in L(L(X,Y) \times Y, Z)$ defined by

$$A(W,w) := BWx - Bw$$

the transpose of which is defined by

$$A * q = (x \otimes B * q, -B * q)$$

Since the duality map is equal to $J_{\alpha,\beta} = \alpha(L^{-1} \otimes M) \times \beta M$, we infer that

$$AJ_{\alpha,\beta}^{-1} A^* q = \left(\frac{\langle Lx, x \rangle}{\alpha} + \frac{1}{\beta} \right) BM^{-1}B^* q$$

and that

$$(AJ_{\alpha,\beta}^{-1} A^*)^{-1} z = \frac{\alpha\beta}{\alpha + \beta\lambda(x)^2} (BM^{-1}B^*)^{-1} z$$

Therefore, the orthogonal right-inverse is equal to

$$A^+ z = J_{\alpha,\beta}^{-1} A^* (AJ_{\alpha,\beta}^{-1}A^*)^{-1} z = \left(\frac{\beta Lx}{\alpha + \beta\lambda(x)^2} \otimes B^+ z, -\frac{\alpha}{\alpha + \beta\lambda(x)^2} B^+ z \right)$$

Now, we supply Z with the final norm $v_{\alpha,\beta}^A$ equal to

$$v_{\alpha,\beta}^A(z)^2 = v_{\alpha,\beta}(A^+ z) = \left(\alpha(\lambda_* \otimes \mu)^2 + \beta\mu^2 \right) (A^+ z)$$

In other words, the final norm

$$v_{\alpha,\beta}^A(z) = \sqrt{\frac{\alpha\beta}{\alpha + \beta\lambda(x)^2}} \mu(B^+ z)$$

is proportional to the final norm $\mu^B(z) := \mu(B^+ z)$ associated with the surjective operator $B \in L(Y,Z)$.

Consequently, by Proposition 3.3, we infer that the solution to

$$\alpha(\lambda_* \otimes \mu)(\overline{W} - U)^2 + \beta\mu(\overline{w} - u)^2) = \inf_{BWxBw \in K+v} (\alpha(\lambda_* \otimes \mu)(W - U)^2 + \beta\mu(w - u)^2)$$

is equal to

$$(\overline{W}_{\alpha,\beta} \overline{w}_{\alpha,\beta}) = (U,u) - A^+ (1 - \pi_K^B)(BUx - v)$$

This yields that

$$\begin{cases} \overline{W}_{\alpha,\beta} = U - \dfrac{\beta Lx}{\alpha + \beta\lambda(x)^2} \otimes B^+ (1 - \pi_K^B)(BUx - v) \\[3mm] \overline{w}_{\alpha,\beta} = u + \dfrac{\alpha}{\alpha + \beta\lambda(x)^2} B^+ (1 - \pi_K^B)(BUx - v) \end{cases}$$

We also observe that

$$\overline{W}_{\alpha,\beta}x - \overline{w}_{\alpha,\beta} = Ux - u - B^+((1-\pi_K^B)(BUx - Bu))$$

does not depend of α and β.

Notes

1. The emergence of the cyberspace may give a new life to connectionist mechanisms, since any two agents would soon be able to communicate at each instant without the mediation of a price or other messages providing the necessary information.

2. Physicists have attempted to measure "complexity" in various ways, through the concept of Clausius's entropy, Shannon's information, the degree of regularity instead of randomness, "hierarchical complexity" in the display of level of interactions, "grammatical complexity" measuring the language to describe it, temporal or spatial computational, measuring the computer time or the amount of computer memory needed to describe a system, etc.

The problem is that living systems being open, there is no way to describe them entirely and consequently, to demonstrate the truth of any proposition. We are left with metaphors, and among them, mathematical ones, to validate a consensual definition of one meaning –actually, one submeaning, so to speak – of complexity.

Therefore, scientific activity begins by dividing a system into two classes, the system under study and its environment. This division is always arbitrary, but often justified by the scientists in quest of explanation. Once conceptually isolated from its environment, a living system fuels itself in the last analysis on solar energy through the consumptions of wastes of the other components of the open system set apart in the description of the environment. Each component of the system which can evolve independently in the absence of constraints, must interact each other in order to maintain the viability of the system imposed by its environment.

3. or the velocity W' (t) of a time dependent connection matrix W (t) starting from the identity matrix as we shall do in the case of heavy evolution. One can also measure other features of connectionnist complexity through the sparsity of the connection matrix, i.e., the number – or the position – of entries which are equal to zero or "small". The sparser such a connectionist matrix, the less complex the system.

4. For a presentation of the ménagerie of tangent cones, we refer to chapter 4 of [6, Aubin & Frankowska].

5. as well as the "equilibrium theorems under constraints" and "optimization theorems under constraints".

6. A weaker assumption is $Im(h'(x)) + T_M(h(x)) = Y.$

References

[1] Aubin J.-P. (1981) *A dynamical, pure exchange economy with feedback pricing,* J.Economic Behavior and Organizations, 2, 95-127.

[2] Aubin J.-P. (1991) Viability Theory, Birkhäuser.

[3] Aubin J.-P. (1996) Neural Networks and Qualitative Physics: A Viability Approach, Cambridge University Press.

[4] Aubin J.-P. (1997) Dynamic Economic Theory: A Viability Approach, Springer-Verlag.

[5] Aubin J.-P. & Cellina A. (1984) Differential Inclusions, Springer-Verlag, Grundlehren der math. Wiss.

[6] Aubin J.-P. & Frankowska H. (1990) Set-Valued Analysis, Birkhäuser.

[7] Cornet B. (1976) *On planning procedures defined by multi-valued differential equations,* Système dynamiques et modèles économiques (C.N.R.S.).

[8] Cornet B. (1983) *An existence theorem of slow solutions for a class of differential inclusions.*

[9] Cornet B. (1983) *Monotone dynamic processes in a pure exchange economy.*

[10] Hildenbrand W. & Kirman A.P. (1988) Equilibrium Analysis, North-Holland.

Dynamics of
Economical Networks

An Arrow´s Theorem Perspective on Networks

Donald G. Saari
Department of Economics
Northwestern University

1. Introduction

It was obvious already in the early 1980s when I first met Åke Andersson in Vienna, as he was hurrying off to another meeting to establish new connections, that his deep commitment to the importance of networks and network dynamics extends far beyond theory. I admire and applaud his success in his pragmatic applications of networks, and I agree with his theoretical perspective. Yet, in some sense, these ideas of networks and worrying about the various connections and interactions run counter to a persistent theme of twentieth century economics.

Oh, we all know there must be interactions among the parts, we all recognize the reality of externalities, we all acknowledge the critical importance of networks for the final analysis of many problems. But, in some sense, a modern goal has been to first understand the simpler, purer models by posing and emphasizing settings that are free from the complexities of interactions, networks, and externalities. While this makes excellent sense, and I have contributed to this theme, I argue that this emphasis on the "parts" can seriously distort the resulting conclusions.

A typical goal associated with this philosophy is to find universal truths about basic economic systems. Arguably, the best example is Adam Smith's "Invisible Hand" story proposing an universal explanation for how prices change. As it is well known, the story argues that market pressures, as determined by individual greed and the heterogeneity of the marketplace, force the supply and demand to a market clearing price equilibrium. While this delightful story enjoys sufficient stature so that it even influences national policy, it crumbles under careful scrutiny – at least it does with our current understanding of how prices respond to market pressures.

Indeed, in [S1,2] I identify serious theoretical barriers which frustrate the achievement of Smith's universal explanation for prices. Instead of a global conclusion, these negative results underscore the critical need of regional approaches; they suggest the need for including interactions of the kind developed by networks. As I indicate later in this essay, a critical part of the problem is related to the "parts vs. whole" issue which is central to network analysis.

To explain my theme that a concentration on the parts (at the expense of the whole) can cause highly misleading results, I appeal to Arrow's well-known impossibility assertion from choice theory. After recalling the basic notions, I indicate why Arrow's famous conclusion is a direct consequence of his emphasis on the "parts" at the expense of the "whole".

2. Arrow's Theorem

To recall, Arrow [A] posed seemingly innocuous assumptions; requirements so basic that it is natural to believe they are satisfied by all reasonable decision procedures. Arrow then showed that these conditions have a serious, counterintuitive conclusion for society and for decision analysis – they require a dictator! His assumptions follow:

1. The *Unrestricted domain* condition is that each voter has strict transitive preferences and that no restrictions are imposed on these preferences.

Requiring transitive preferences makes perfect sense. After all, if voters are irrational, we cannot expect rational outcomes. This is why the condition of rational voters is a critical assumption for voting and choice theory as well as for economics. I return to this important point later. The second part of this assumption captures the notion that restricting individual preferences violates a normal sense of democracy. By allowing differences of opinion, it models a "heterogeneous" society.

2. A second requirement is that our procedure provides rational outcomes.

Again, "rational" means "transitive". The importance of this assumption is that without transitive outcomes, we suffer the difficulties of cycles. If, for instance, society's ranking of options is the cycle $A \succ B$, $B \succ C$, $C \succ A$, then we cannot determine which alternative achieves society's goal. In other words, without rational outcomes, rational decisions cannot be made.

3. The *Pareto Condition asserts that if all voters have the same ranking of a particular pair, then that is society's ranking of the pair.*

This "fairness condition" addresses only the obvious, extreme setting of unanimity. If everyone agrees on how a particular pair should be ranked, then it is difficult to accept a procedure which provides a contrary outcome.

4. Independence of *Irrelevant Alternatives* (IIA) is a natural extension of the Pareto. Rather than just using the extreme setting of unanimity, IIA requires all pairwise outcomes to be based, in some manner, only on each voter's relative ranking of that

particular pair. What these voters think about other (hence, irrelevant) alternatives has no effect on the relative ranking of this particular pair.

The IIA condition can be illustrated with the 1995 World Figure Skating Competition. With only Michelle Kwan left to skate, Nicole Bobek was in second place and Surya Bonaly was in third. Michelle was so far behind in the standings that it was impossible for her to place in the top three. However, her strong skating that afternoon earned her a fourth place finish. An unexpected consequence of her beautiful performance was to reverse the Bobek and Bonaly final standing! Bonaly won silver while Bobek had to settle for bronze.

Doesn't this suggest that Figure Skating uses a ridiculous procedure? Although she was delightful, what did Michelle Kwan's performance have to do with the relative merits of Nicole Bobek and Surya Bonaly? Both of them skated earlier than Michelle, and the judges publicly expressed their Bobek and Bonaly evaluations at that time. Shouldn't the judges' views of their relative merits determine the two skaters' final relative ranking? Obviously it did not. Obviously, in some mysterious manner known only to the skating federation, their assessment of Kwan flipped the Bobek-Bonaly ranking. The IIA condition is imposed to avoid these "relative ranking" difficulties.

While these conditions are natural and even expected of all elections and ranking procedures, they have flaw. Arrow [A] proved that with more than two candidates, only one undesirable procedure satisfies all of them.

Theorem (Arrow). *Once there are more than two candidates, the only procedure with transitive outcomes that satisfies Unrestricted Domain, Pareto, and IIA is a dictatorship.*

Arrow's dictator need not be benevolent. Instead, society's ranking always agrees with the dictator's preferences no matter what the other voters prefer or wish.

To appreciate how this seminal theorem contributes to our understanding of network analysis, we need to understand why the theorem is true and what it really asserts. So, before continuing, I invite the reader to spend a few minutes reflecting upon the meaning of Arrow's assertion. This is worth doing because the interpretation which follows differs radically from what has been commonly accepted for the last half century. In particular, I argue that Arrow's conclusion does not mean what we commonly thought it did. Instead, I claim that his assertion illustrates how an emphasis on the parts can cause misleading conclusions.

3. Separation of Data Loses Critical Information

Elsewhere ([S3,4]) I have published technical arguments supporting my explanation of Arrow's Theorem given here. These earlier publications, and the concomitant freedom from having to provide technical details, allow me to provide intuitive arguments which relate to network dynamics.

With reflection, it is clear that Arrow's Theorem is caused by the IIA condition. Because we know what causes the difficulty, the real issue is to understand why IIA generates such an unpleasant conclusion. As I show, the blame can be placed directly on the anti-network IIA philosophy. To explain, notice that IIA mandates that only information about a particular pair can be used when deciding society's ranking of that pair. Because IIA strictly separates information into component parts (the pairs), it explicitly forbids using "connecting" information of the kind which characterizes networks. Thus, this separation loses important information about the system. As I show, Arrow's negative conclusion merely manifests the nature of lost "connecting" information.

At least intuitively, by concentrating on pairwise rankings, we should expect information to be lost. What has not been previously understood is the critical nature of this loss. To start, the pairwise election outcomes for the three-voter Condorcet triplet defined by the rankings

$$A \succ B \succ C, \; B \succ C \succ A, \; C \succ A \succ B$$

define the cycle

$$A \succ B, B \succ C \text{ and } C \succ A$$

where each pairwise election is determined by a 2:1 vote. This cyclic outcome is precisely what we want to avoid.

A clue to the source of the difficulty comes from examining what happens when irrational voters vote. For instance, if three voters have the cyclic preferences

$$A \succ B, \; B \succ C, C \succ A,$$
$$B \succ A, \; C \succ B, C \succ A$$
$$A \succ B, \; B \succ C, C \succ A$$

then we should expect cyclic outcomes. Indeed, the first two voters, with directly opposing preferences, would cancel each others vote. Because the remaining voter would break this tie vote, this voter's cyclic preferences should define the final outcome. This is what happens with the pairwise vote; the elections define the expected $A \succ B$, $B \succ C$, and $C \succ A$ cycle where each election has a 2:1 vote.

Notice that the pairwise vote outcomes for the three rational voters (i.e., the Condorcet triplet) and the three irrational (i.e., not transitive) voters are the same. This is no coincidence; it is a direct consequence of the fact that the pairwise vote satisfies anonymity. Namely, this tallying method cannot determine who cast what vote. All the procedure knows is that for each pair, two voters prefer one ranking and the remaining voter has the opposite opinion. Consequently the pairwise vote cannot distinguish whether the ballots were cast by the rational voters (as manifested by the Condorcet triplet) or the irrational voters (as described by the second profile); for the procedure there is absolutely no difference between these profiles. Thus, the combination of anonymity and an emphasis on each pair – on the "parts" of preferences rather than the "whole" – forces the pairwise vote procedure to lose the critical assumption that the voters have transitive preferences! This is a serious indictment!

I leave it as an exercise to show that the pairwise rankings from the Condorcet triplet can be reassembled into five different profiles; four of them involve voters with irrational preferences, while the last is the Condorcet profile. It is easy to justify the cycle as the only "natural" conclusion for the four profiles with a cyclic voter. Only the Condorcet profile fails to have a natural interpretation (at this "parts" level). But because the voting procedure cannot distinguish among these profiles where 80% of them require the cycle as the natural conclusion, rather than being an anomaly, the cycle is the correct outcome.

The Condorcet triplet does have a natural conclusion, but only after we admit "connecting" information. Remember, transitivity is based on how the rankings of two pairs determine the ranking of the third – it is based on connections. By using the connecting information from transitivity, we discover that in Condorcet's triplet each candidate is ranked first, second, and last precisely once. Therefore, the connecting information shows that no candidate has an advantage over any other candidate. Consequently, the connecting information – the information dismissed by the pairwise vote – leads to the natural outcome of a $A \sim B \sim C$ tie vote. When these preferences are tallied with the Borda Count, where two and one points are assigned, respectively, to a voter's top- and second ranked candidate, the complete tie is the election outcome.

To understand what is going on, notice that the properties and characteristics of the procedure, not us, determine the procedure's natural domain. Stated in another manner, the procedure, not us, determines the class of voters for which it is intended. A convenient way to determine the domain for a procedure is to admit all possible preferences for which the procedure can be used. That is, if an outcome can be defined for a preference – no matter how weird and irrational it may be – then that preference is in the procedure's true domain. The full domain, then, includes all possible preferences admitted with this approach. Once the domain is defined, the procedure must be viewed as being designed for these types of people.

To illustrate this important notion with the pairwise vote, notice that the voter only needs to rank each pair of candidates. This is the only assumption the procedure imposes on individual preferences; it does not require the voter to sequence the pairwise rankings in any manner – in particular, the procedure does not require the voter to be transitive. So, instead of worrying about transitivity, it is correct to view the pairwise vote as favoring the needs of the unsophisticated voters who are capable only of ranking individual pairs.

For purposes of comparison, I compute the true domain for the Borda Count. Here, a voter must have a transitive ranking of the candidates for her ballot to be tallied. For instance, it is impossible to tally the ballot for the irrational voter with preferences $A \succ B$, $B \succ C$, $C \succ A$ because no candidate is first and no candidate is second. Consequently, the true domain for the Borda Count is the set of transitive preferences. By requiring individuals to have a ranking – by enforcing the transitivity connections among pairwise rankings – we find that this procedure is intended for the rational voters.

At this stage, the central point of this essay can be made. It is the nature of the procedure – not us – which determines what basic information is used, and what portion of the information is lost. It is the properties of the procedure – not us – which determine the class of people or economic environments for which it is intended. When there is a conflict between the true domain and what we had intended, expect the conflict to generate unexpected conclusions. For instance, we now know that the informational cost of using the pairwise vote is to drop the critical assumption that the voters are rational! This is troubling!

Because of its importance, let me repeat. "Transitivity of preferences" is not a natural condition for the pairwise vote; it is something we try to impose on the system. Instead of monitoring whether voters are rational, the pairwise vote is intended for highly irrational voters. While it is widely accepted that the rationality of voters plays an important role in the voting outcomes, it does not. If the procedure provides appropriate outcomes when restricted to rational voters, well, that is just a nice accident rather than the designed intent of the method.

The above description of the Condorcet triplet makes this point clear. By being agnostic, the pairwise vote cannot distinguish between the one profile with transitive voters and the four profiles with irrational voters, so it yields the outcome most appropriate for most of the indistinguishable but possible profiles. The geometric analysis introduced in [S3] shows that this assertion holds in general.

4. Arrow's Theorem and IIA

The same argument explains Arrow's IIA condition. But, guided by the above discussion, this is to be expected. Because IIA specifically requires a procedure to consider only how each voter ranks each pair, an IIA procedure only requires a voter to be able to rank each pair. Directly conflicting with the central assumption of rationality, the IIA condition strictly prohibits an IIA procedure from checking whether the preferences are transitive.

What we have learned, then, is that IIA destroys the central, critical assumption of rational voters. Rather than addressing the needs of transitive preferences, IIA admits only procedures intended to service the needs of highly unsophisticated agents. But a procedure expected to fairly service the needs of the irrational must yield cyclic and other nontransitive outcomes.

Transitivity, then, is an added domain restriction condition. This requirement should be interpreted as seeking to determine whether any of these primitive procedures provide rational outcomes when restricted to rational preferences. An immediate answer is a "dictator;" after all, this method just reports the pairwise preferences of a single voter. If this dictator has transitive preferences, then there is only one way to reassemble his pairwise rankings – they define the exact same transitive ranking. Thus, rather than being surprising, Arrow's famous "dictator" conclusion merely identifies the only procedure designed for unsophisticated voters that also yields transitive outcomes should this one voter happen to have transitive preferences. Rather than being robed with the earlier shock appeal, Arrow's conclusion now assumes trappings of the mundane and obvious.

So, instead of confronting the dilemma of a dictator or a paradox, the real message of Arrow's Theorem is that if you want rational outcomes, you cannot use procedures designed and intended for highly unsophisticated voters. Instead, we must use procedures that are intended for rational voters; we need procedures that can monitor whether voters are rational. Instead of adopting an IIA perspective which destroys networks, we need to use methods, such as the Borda Count, which monitor whether voters are rational. Remember, rationality requires the pairwise rankings to be sequenced in a particular manner; they are "networked" in a specific fashion. This suggests that acceptable resolutions of Arrow's half century mystery can be found by adding a "networking" flavor to the IIA assumption – by requiring the procedure to monitor whether voters are rational. (To see how this approach does resolve the problem in a nearly minimal fashion, see [S3,4,5].

5. Other Economic Models and Networking

The same conceptual argument explains many of the mysteries of contemporary economics. With reflection, this should not be overly surprising. After all, choice functions are elementary aggregation procedures. Almost all of economics is based on more complicated aggregation methods where, rather than just a profile, the inputs involve utility functions, initial endowments, production sets, and on and on. Rather than stabilizing the conclusion, the more options and substitutes allowed by the larger input set must be expected to create more exotic conclusions. By imposing "simplifying assumptions" that separate the information into parts (at the expense of emphasizing and requiring relationships), the resulting conclusions can be troubling and highly misleading. To model what might actually occur, we need to establish connections – we need to create some form of networking. But notice, the connections must be included in the design of procedures!

In general, the same phenomenon occurs again, and again, and again. To illustrate, a standard way to start an economic model is to impose assumptions about the environment – this describes the kinds of people, initial endowments, technology, production sets, etc., etc. that are involved. Built into basic assumptions about the environment are connections and relationships among the various factors; e.g., a worker's preferences for work and leisure and his desire for certain commodities. This approach is natural; we need to identify the group being analyzed. While conclusions from such models are insightful, we also know that when a sufficiently heterogeneous setting is considered, we can encounter disturbing, highly counterintuitive conclusions – conclusions that may, or may not occur in reality.

Actually, this weird behavior must be expected; the explanation is essentially the one I gave for Arrow's Theorem. Namely, the adopted mechanisms and procedures determine the actual class of people and environments for which they are intended – we do not! If a mechanism allows "parts" to be separated (which is standard outside of a networking perspective), then we encounter the same phenomenon whereby the procedure tries to service a far more irrational class of economic environments than we had intended. In other words, all of the carefully constructed assumptions about the environment might be for naught. Just as with the Condorcet triplet and the four profiles involving irrational voters, the procedure may not be able to distinguish between the intended "rational" environment and a large class of irrational settings which lay fair claim to an "irrational" conclusion.

The actual (rather than intended) domain of a procedure includes all environments that can be used with a procedure. For instance, suppose ballots are tallied by using the Borda Count. As already described, this procedure does monitor the transitivity of voters. But now suppose the Borda Count is applied to all triplets from four candidates. We may require the voters to be transitive, but the procedure, which now is being required to consider separate parts, cannot monitor or insist upon this rationality condition being satisfied. Instead, in this setting, the true domain includes any voter who can rank each triplet in a transitive manner – but, the triplets need not be related in any manner! This means that the true domain includes voters with cyclic preferences such as $A \succ B \succ C$, $B \succ C \succ D$, $C \succ D \succ A$, $D \succ A \succ B$. Consequently, we must expect problems and voting paradoxes; they occur. (See, for instance, [S5].) Similarly for an economic setting, it is the procedure which determines that actual connections – not our assumptions about the environment. Conversely, for those networking models where the procedures are designed to relate the parts, we may expect more realistic conclusions to occur.

As an immediate example, consider the problem of consumer surplus. As it is well known, this technique sequentially considers the excess demand of individual goods. From this information, certain predictions and assertions are made. The validity of this technique, however, has been seriously questioned through the use of examples. But, from what has been described above, we must expect this method to admit highly questionable conclusions; after all, this procedure deliberately eliminates connecting information (e.g., the interaction of all commodities). Consequently, the true domain of this procedure is highly irrational and very unpredictable. Just as with the dictator from Arrow's Theorem, the only time we can ensure a legitimate outcome is if there is only one possible way to reassemble the various parts. This suggests that conclusions from such a procedure are valid only for highly specialized preferences where the parts totally dictate the behavior of the whole. This is exactly what Chipman and Moore discovered [CM1, 2].

Another way to think about this "part vs. whole" problem is to recall the important progress made on the "free rider problem". As long as individual preferences are separated from the consequences of actions, this issue was (correctly) viewed as impossible of solution. The traditional explanation is that in a public goods setting, individuals can claim no or little interest in the public good (e.g., a public park) when it is time to make assessments. On the other hand, once the public good is made available, they can reap as much of its benefits as they wish.

For a different perspective about this problem, consider the true domain of preferences that are admitted with this separation assumption. Here, the actual class of admissible agents – when preferences are separated from consequences – includes irrational individuals who honestly have no notion that there should be a connection between what they are willing to pay and how much they use the public good. Thus, because of the built in separation, the only "fair" outcome of a procedure is to address the needs of the dominating number of irrational individuals. Just as in the resolution of Arrow's Theorem, connections – forms of networks – are mandatory to convert the true domain of the procedure into one which betters resemble what we intend to model. A solution to this free rider problem, then, is to make the assessment dependent upon other agents' messages; to "internalize externalities". This change in the procedure radically alters the "true" domain along with the resulting behavior. While the choice of words – "internalize externalities" rather than creating networks – differ, the concept remains the same.

To explain this "parts vs. whole" theme in still another manner, let me return to the price mechanism. We now know ([S1,2,5]) that the price mechanism does not behave in the nice, orderly manner as previously thought. Instead, we now know that even in simple settings of pure exchange economies of the kind often used in the class room, the resulting price dynamic can be far more complicated than any dynamics which may occur in physics, biology, or any of the other areas identified with complex behavior! Not only do we know this is the case, but there now exists an elementary technique which allows us to construct examples of simple economies exhibiting just about any desired chaotic dynamic ([S6]). So, we now have a fairly deep understanding of what can happen; it remains to understand why it occurs.

An explanation of these dynamical difficulties comes from using the above approach. Remember, the aggregate excess demand function is the sum of each agent's individual excess demand. This individual excess demand is the difference between what the agent demands and what the agent is willing to supply at each price. Each individual's excess demand is determined by the usual optimization scheme where, at a price, a budget plane is defined, and the individual preferences are then maximized subject to this constraint.

But, the aggregate excess demand procedure knows nothing about all of this rational mathematical construction! All the procedure knows is that, somehow, the demand from each agent is on the budget line. (Actually, even this is not known.) Thus, the aggregate excess demand function needs to attend to the needs of all ways such a demand could be defined – even from agents with highly irrational preferences, etc. Just as with the pairwise vote example of the Condorcet triplet and the profiles with irrational voters, once an economy becomes sufficiently heterogeneous – once Smith's conditions are met – then the individual demands become indistinguishable from the demands of highly irrational agents. From this perspective, cycles and other highly confusing dynamics must be expected!

6. Macro Versus Micro

I conclude by offering brief remarks about another one of Åke's concerns. While it is a topic often mentioned, Åke has been in the forefront of questioning, investigating, and worrying about the severe discrepancy between micro and macro results whether they occur in economics, in nature, or in any setting. In fact, one can view some of his work in Networking as an attempt to reconcile these differences. The theme offered in this essay provides an alternative explanation.

Micro analysis, of any type, requires separating connecting parts. As such, when these micro procedures are interpreted within a more general macro setting, the true domain of the procedure radically changes; no longer is it just the restricted, intended domain. Remember, the procedure – not us – determines its true domain. The procedure – not us – determines for whom it is designed. In other words, the application of several micro procedures to understand a large economy has exactly the "parts" behavior captured by Arrow's IIA conditions. Just as with Arrow's Theorem, we must expect that once the environment becomes sufficiently heterogeneous, it will become impossible to combine the parts into an aggregate, or macro, story. The problem is that the procedures are trying to service the needs of nonexistent but indistinguishable irrational agents. But, this now should be expected. Even more, just as with the resolution of Arrow's Theorem, solutions require the integration of the parts; we need to understand networks.

References

[A] Arrow, K .J., *Social Choice and Individual Values; 2nd edition,* Wiley, 1961.

[CM1]Chipman, J. and J. Moore, *Aggregate demand, real national income, and the compensation principle,* Internat. Econ. Rev 14 (1973), 153-181.

[CM2]Chipman, J. and J. Moore, *On social welfare functions and the aggregation of preferences,* Jour Econ Theory 21 (1979), 111-139.

[S1] Saari, D. G., *Doing the right thing,* Framtider International 2 (1992), 44-50.

[S2] Saari, D. G., *The mathematical complexity of simple economics,* Notices of AMS 42 (1995), 222-230.

[S3] Saari, D. G., *Basic Geometry of Voting,* Springer-Verlag, 1995.

[S4] Saari, D. G., *Resolving and connecting Arrow's and Sen's Theorems,* Social Choice & Welfare (to appear).

[S5] Saari, F. G., *A chaotic exploration of aggregation paradoxes,* SIAM Review 37 (1995), 37-52.

[S6] Saari, D. G., *The ease of generating chaotic behavior in economics,* Chaos, Solitons, and Fractals (to appear). Dept. Of Mathematics, Northwestern University, Evanston, ILL. 60208-2730.

Dynamics of International Financial Networks: Modeling, Stability Analysis, and Computation

Anna Nagurney
Department of Finance and Operations Management, School of Management
University of Massachusetts, Amherst

and
Stavros Siokos
Department of Mechanical and Industrial Engineering
University of Massachusetts, Amherst

1. Introduction

In this paper we turn to the domain of international finance and utilize for the first time in this application the methodologies of projected dynamical systems and network theory in order investigate questions of existence, stability, and computational procedures. In addition, we discover the underlying network structure of the individual sector's portfolio optimization problems which then through the dynamics presented merge into a complete network representing the equilibrium state in which markets in both the instruments and in the currencies are shown to clear.

In order to place our work in the context of the existing literature we first provide a brief historical overview of the scientific developments in financial theory, emphasizing that the history of international financial research is directly related to classical domestic finance theory. Therefore, to understand the current state of modeling in international finance, we have to keep in mind the progression of ideas in classical domestic finance theory, which is characterized by dramatic change.

The latest era in the field of finance theory begins with Modigliani and Miller (1958), with the argument that the financial structure of firms is a matter of indifference for all participants in the economy. Although this idea is simple and well accepted today, it was a controversial approach at the time. During the following year, Markowitz (1959) published a book, which is credited with being the first step in the development of portfolio theory and asset pricing theory, based on mean variance portfolio analysis. In addition, his formulation gave rise to further developments in quadratic programming. In that same decade, Debreu (1959) (see, also, Arrow (1963)) introduced the theory of the efficient allocation of resources under uncertainty.

The decades that followed the innovative work of the fifties were characterized by the continuous exploration and expansion of that work. More precisely, in the mid-sixties, Sharpe (1964), Lintner (1965), and Mossin (1966) expanded the work of Markowitz and defined the Capital Asset Pricing Model (CAPM), which, in simple words, stated that the expected rate of return of any asset can be written as the summation of a risk-free rate of interest and the asset's normalized covariance with the market times the difference between the expected rate of return on the market and the risk-free rate. Furthermore, for the first time, Grubel (1968) documented the gains from international diversified portfolios, based on the fact that exchange rates are negatively correlated with the return of a foreign asset, and, therefore, foreign investments are less risky.

In the seventies, the CAPM was tested, challenged, and extended. Extensions included multiperiod economies (cf. Merton (1973)), restrictions on borrowing (e.g., Black (1972)), and the introduction of transaction costs (cf. Milne and Smith (1980)). Even though the testing of results for the CAPM were initially satisfactory, the results were later challenged, leading to a series of theories such as the Arbitrage Pricing Theory of Ross (1976), which stated that investors can get asset prices as a linear combination of some basic contributing factors. In the same decade, Black and Scholes (1973) and Merton (1973) also established a new era in financial theory and applications with models for stock options pricing. Their model is still considered the foundation for pricing hedging financial instruments. Furthermore, a series of new models and studies (Lessard (1973) and Solnik (1974)) were conducted that proved that international portfolios outperform domestically diversified portfolios.

The decade of the eighties was the time that the model by Black and Scholes was further explored and generalized (see, e.g., Duffie and Huang (1985), Duffie (1986), Cox, Ingersoll and Ross (1985)). Moreover, the first years of the eighties witnessed a dramatic increase in the volatility of interest and currency exchange rates, leading investors to desire to hedge returns on fixed income securities against changes in the prices of financial instruments and currencies. As a result, the number of financial products available to investors had also increased and a variety of new financial products along with a combination of existing ones were introduced, providing tools for the reduction of different types of risk. Moreover, the explosive growth of computer technology, combined with a series of international trade alliances (e.g., the European Union, NAFTA, the opening of Eastern European markets), and the earlier documented gains from international diversified portfolios led investors and institutions to international trading. Because of this shift in the preferences of capital allocation, the need for international financial models and computational techniques became evident. As a result, a large portion of the literature at that time focused on the modeling and computation of international models (cf. Adler and Dumas (1983), Eun and Resnick (1988)) with interesting results.

However, the majority of existing international models (mostly optimization models) dealt only with cases that could handle a very small (mostly two) number of countries and currencies (usually as many as the countries and with ignorance of currencies such as the Eurodollar), and under somewhat unrealistic assumptions (e.g., homogeneous expectations among investors, ignorance of the issue of the estimation of the exchange risk, and the effects of capital allocation in the price of different instruments and currencies in the economy, etc.).

New mathematical tools and frameworks, which were able to capture the complexities of such large-scale problems, were introduced and applied for the modeling and computation of competitive equilibrium. Among them was the theory of variational inequalities (cf. Kinderlehrer and Stampacchia (1980) and Nagurney (1993)). In particular, the application of variational inequality theory focused on the qualitative analysis and computation of the equilibrium patterns satisfying the problem-specific equilibrium conditions. More precisely, conditions for existence and uniqueness of equilibrium patterns were found, general sensitivity results were obtained, and a variety of algorithms had been proposed and rigorously justified. To date, the theory of variational inequalities has already been successfully used in the modeling and computation of various financial applications — for a single country or currency problems — in Nagurney, Dong, and Hughes (1992), Nagurney and Dong (1996a), Nagurney (1994), and Nagurney and Dong (1996b) — for multiple countries and currencies problems — in Nagurney and Siokos (1997) — for single agent investment modeling and computation problems — in Tourin and Zariphopoulou (1994).

Nevertheless, finite-dimensional variational inequality theory focused on the equilibrium state and did not provide us with a tool for addressing the underlying dynamics. The theory of economic dynamic systems, which has been extensively studied (cf. Batten (1985), Grandmont and Malgrange (1986)) and successfully applied in a large variety of problems (see, e.g., Andersson (1989), Andersson and Andersson (1992), and Andersson, Batten, Johansson, and Nijkamp (1989)), appeared to be capable of capturing the necessary dynamic components of a financial dynamic system.

Another tool, that of dynamical programming, had been used to formulate and to solve simple (usually single country) equilibrium models for computing asset prices and capital accumulation processes. Most of the existing models consisted of a single private agent and a single government agent (living for the same amount of time, which in most of the cases is infinite). Although the main ideas of dynamic programming are simple, the details can involve sophisticated mathematical arguments. More thorough presentations of the subject can been found in Bertsekas (1976) and Chow (1981). One of the classic financial models of this ilk is the asset-pricing model of Lucas (1978), where special restrictions give the economy a recursive structure.

This model is considered as special case of the Arrow-Debreu economy. Investor preferences, endowments, and trading are specified in such a way that an unbacked (by future government surpluses) currency would not have any value at equilibrium. This model was modified as the cash-in-advance model where a valued government-issued unbacked currency is added. Agents can trade assets only with a government currency that they have previously accumulated. Moreover, it is assumed that all agents are identical in their preferences and that in the international case (Lucas (1982)) they are all endowed and taxed identically.

In this paper, we utilize the theory of projected dynamical systems (cf. Dupuis and Nagurney (1993), Zhang and Nagurney (1995), Nagurney and Zhang (1996)), for the study of general international financial equilibrium problems. Projected dynamical systems (PDS) are nonclassical in that the right-hand side, which is a projection operator, is discontinuous. These discontinuities correspond to the constraints of the specific application, which, in our case, consist of accounting balance and nonnegativity constraints. More precisely, we derive a dynamical system for the modeling and computation of international financial equilibrium. The set of stationary points of this dynamical system coincide with the set of solutions of a variational inequality and, furthermore, to a solution of a network optimization problem.

In contrast to the earlier work in international finance, our formalism can handle as many countries, sectors, instruments, and currencies, as mandated by the application in question. Indeed, the only limiting resource is the memory of the computer utilized to solve the problem using the proposed methodology.

This paper is organized as follows. In Section 2, we develop the dynamic international financial model and show that the financial adjustment process is a projected dynamical system. We also provide the underlying network associated with an indivual sector's portfolio optimization problem. We then provide some basic qualitative properties.

In Section 3, we address the stability analysis of the financial adjustment process and provide conditions that guarantee the stability and the asymptotical stability of the adjustment process.

In Section 4, we prove that the equilibrium solution corresponds to the solution of a network optimization problem. Hence, the dynamics associated with the individual networks drives the international financial system towards the complete network characterizing the structure of the problem in equilibrium.

In Section 5, we propose a discrete time method, the Euler method, for the solution of the model and provide conditions for convergence. The convergence analysis depends crucially on the stability analysis results obtained in Section 3. The algorithm decomposes the problem into network subproblems of special structure, each of which can be solved explicitly and in closed form via exact equilibration (cf. Dafermos and Sparrow (1969), Nagurney, Dong, and Hughes (1992)). We then illustrate both the model and the algorithm in Section 6 through numerical examples. We summarize our results and present our conclusions in Section 7.

2. The Dynamic International Financial Model

In this section we develop a dynamic international financial model and we study the adjustment process and the projected dynamical system that describes the dynamic behavior of different financial sectors, along with that of the instrument and currency prices in an international economy. We also illustrate the network structure of the portfolio optimization problem facing each sector in each country.

In particular, we consider an economy consisting of N countries with a typical country denoted by j. Each country, in turn, has F sectors with a typical sector denoted by f. We also consider M currencies with a typical currency denoted by i. In each currency there are K instruments with a typical instrument denoted by k.

We denote the volume of instrument k in currency i, held as an asset by sector f of country j, by $X_{i,k}^{j,f}$, and group the assets of a sector in a country into the MK-dimensional column vector $X^{j,f}$. Similarly, $Y_{i,k}^{j,f}$ denotes the volume of instrument k in currency i held as a liability by sector f of country j. The liabilities of each sector in each country are grouped into an MK-dimensional column vector $Y^{j,f}$.

Let $r_{i,k}$ represent the price of instrument k in currency i and group the instrument prices into an MK-dimensional column vector r. Finally, let e_i denote the rate of appreciation of currency i, which can be interpreted as the rate of return earned because of exchange rate fluctuations. These rates are then grouped into an M-dimensional column vector e.

The utility function for sector f of country j is denoted by $U^{j,f}(X^{j,f}, Y^{j,f}, r, e)$ and is assumed to be concave, and twice continuously differentiable, with each sector taking the prices r and e as given. We assume that the accounts of each sector of each country must balance, and denote the total financial volume held by sector f of country j by $S^{j,f}$.

The portfolio optimization problem of sector f in country j can then be expressed as:

$$\text{Maximize } U^{j,f}(X^{j,f}, Y^{j,f}, r, e) \tag{1}$$

subject to:

$$\sum_{i=1}^{M}\sum_{k=1}^{K} X_{i,k}^{j,f} = S^{j,f}, \tag{2a}$$

$$\sum_{i=1}^{M}\sum_{k=1}^{K} Y_{i,k}^{j,f} = S^{j,f}, \tag{2b}$$

$$X_{i,k}^{j,f}, \ Y_{i,k}^{j,f} \geq 0, \quad \forall i \in (1,M), \quad \forall k \in (1,K). \tag{3}$$

For each sector $f \in (1,F)$ in country $j \in (1,N)$ we let

$$\overline{X}^{j,f} = \{X^{j,f} \in R_+^{MK} : \sum_{i=1}^{M}\sum_{k=1}^{K} X_{i,k}^{j,f} = S^{j,f}\} \tag{4}$$

denote the constraint set of its assets, and, similarly, we let

$$\overline{Y}^{j,f} = \{Y^{j,f} \in R_+^{MK} : \sum_{i=1}^{M}\sum_{k=1}^{K} X_{i,k}^{j,f} = S^{j,f}\} \tag{5}$$

denote the constraint set for its liabilities. Then the feasible set for the assets and the liabilities of each sector of every country is a Cartesian product denoted by $K^{j,f}$, where $K^{j,f} \equiv \{\overline{X}^{j,f} \times \overline{Y}^{j,f}\}$.

We define the feasible set for the assets of all sectors from all countries as:

$\overline{X} \equiv \overline{X}^{1,1} \times ... \times \overline{X}^{j,f} \times ... \times \overline{X}^{N,F}$ and, similarly, for the liabilities:

$\overline{Y} \equiv \overline{Y}^{1,1} \times ... \times \overline{Y}^{j,f} \times ... \times \overline{Y}^{N,F}$, with $K \equiv \{\overline{X} \times \overline{Y}\}$.

In Figure 1, we depict the network structure of the individual sectors' optimization problems. Note that the structure of the network originally (out of equilibrium) is such that the network subproblems in assets and in liabilities are disjoint for each sector in each country. In contrast, as we shall show in Section 4, in equilibrium, the network subproblems merge, yielding structurally a synthesized and connected network that also captures market clearance for both instruments and currencies.

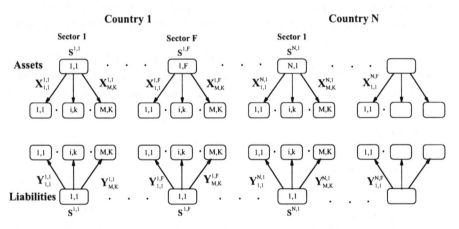

Figure 1. Network Structure of the Portfolio Optimization Problems

The optimization problem for each sector of each country can be interpreted as an effort to adjust its asset and liability pattern in a direction that maximizes its utility function, while maintaining the accounting and nonnegativity constraints. In particular, it is obvious that the ideal direction for the maximization of this utility function in the unconstrained case is given by: $(\nabla_{X^{j,f}} U^{j,f}, \nabla_{Y^{j,f}} U^{j,f})$, where:

$$\nabla_{X^{j,f}} U^{j,f} \equiv \left(\frac{\partial U^{j,f}}{\partial X_{1,1}^{j,f}}, \dots, \frac{\partial U^{j,f}}{\partial X_{i,k}^{j,f}}, \dots, \frac{\partial U^{j,f}}{\partial X_{M,K}^{j,f}} \right) \tag{6}$$

and

$$\nabla_{Y^{j,f}} U^{j,f} \equiv \left(\frac{\partial U^{j,f}}{\partial Y_{1,1}^{j,f}}, \dots, \frac{\partial U^{j,f}}{\partial Y_{i,k}^{j,f}}, \dots, \frac{\partial U^{j,f}}{\partial Y_{M,K}^{j,f}} \right), \tag{7}$$

where $\nabla_{X_{j,f}}$ denotes the gradient with respect to the variables in the vector $X^{j,f}$ with $\nabla_{Y_{j,f}}$ defined accordingly.

However, this "idealized" direction needs to be revised in order to incorporate the fact that the assets and liabilities must lie in the feasible set for each sector of each country. Consequently, the best-realizable direction of sector f of country j is the projection of the idealized direction $(\nabla_{X^{j,f}} U^{j,f}, \nabla_{Y^{j,f}} U^{j,f})$ onto the constraint set, $\kappa^{j,f}$, at its current portfolio $(X^{j,f}, Y^{j,f})$.

This best-realizable direction for a sector f in country j can be mathematically expressed as:

$$\Pi_{\kappa^{j,f}} ((X^{j,f}, Y^{j,f}), (\nabla_{X^{j,f}} U^{j,f}, \nabla_{Y^{j,f}} U^{j,f})), \tag{8}$$

where for any closed convex set $S \in R^n$, $Z \in S$ and $\upsilon \in R^n$, the projection of the vector υ at Z (with respect to S) is defined by:

$$\Pi_S(Z, \upsilon) = \lim_{\delta \to 0} \frac{P_s(Z + \delta \upsilon) - Z}{\delta} \tag{9}$$

and P_S is the norm projection defined by:

$$P_S(Z) = \arg \min_{Z' \in S} \| Z' - Z \| . \tag{10}$$

It should be pointed out that:

$$\text{If } Z \in \text{int } S \Rightarrow \Pi_S(Z, \upsilon) = \upsilon, \tag{11}$$

that is, if Z lies in the interior of the feasible set S, then the projection in the direction υ is simply υ. For additional background, see the book by Nagurney and Zhang (1996).

In view of (8), the asset and liability volumes for sector $f \in (1, F)$ of country $j \in (1, N)$ are adjusted, hence, according to the following process over time:

$$\dot{X}^{j,f} = \Pi_{\overline{X}^{j,f}} (X^{j,f}, \nabla_{X^{j,f}} U^{j,f}), \ \dot{Y}^{j,f} = \Pi_{\overline{Y}^{j,f}} (Y^{j,f}, \nabla_{Y^{j,f}} U^{j,f}). \tag{12}$$

Moreover, the instrument and currency prices also change through time, adjusting to their supply and demand fluctuations. We assume that the prices of all instruments and all currencies follow the supply-demand rule, that is, the price of an instrument or currency will increase when the demand exceeds its supply, and will decrease, otherwise.

More precisely, if we define $-\rho_{i,k}$ as the difference between supply and demand for instrument k in currency i, and $-\varepsilon_i$ as the difference between supply and demand for currency i, that is,

$$\rho_{i,k} \equiv \sum_{k=1}^{N}\sum_{f=1}^{F} -(X_{i,k}^{j,f} - Y_{i,k}^{j,f}) \tag{13a}$$

and

$$\varepsilon_i \equiv \sum_{k=1}^{K}\sum_{j=1}^{N}\sum_{f=1}^{F} -(X_{i,k}^{j,f} - Y_{i,k}^{j,f}) \tag{14}$$

and we assume that the prices of different instruments and currencies cannot be negative, the price of an instrument k in currency i will then remain zero when $r_{i,k} = 0$ and $\rho_{i,k} < 0$. Similarly, the price of a currency i will remain zero when $e_i = 0$ and $\varepsilon_i < 0$.

Hence, the instrument price $r_{i,k}$ changes according to:

$$\dot{r}_{i,k} = \begin{cases} \max(0, \rho_{i,k}) , & \text{if } r_{i,k} = 0 \\ \rho_{i,k} , & \text{if } r_{i,k} > 0 \end{cases} \tag{14}$$

which is equivalent to:

$$\dot{r}_{i,k} = \Pi_{R_+}(r_{i,k}, \rho_{i,k}) . \tag{15}$$

Similarly, the currency price e_i changes according to:

$$\dot{e}_i = \begin{cases} \max(0, \varepsilon_i), & \text{if } e_i = 0 \\ \varepsilon_i , & \text{if } e_i > 0 \end{cases} \tag{16}$$

which is equivalent to:

$$\dot{e}_i = \Pi_{R_+}(e_i, \varepsilon_i). \tag{17}$$

2.1 The Financial Adjustment Process

The financial adjustment process that derives the allocation of assets and liabilities for each sector f of country j along with the prices of each instrument and each currency can now be formulated, in view of (12), (15), and (17), as:

$$\begin{cases} \dot{X}^{j,f} = \Pi_{\overline{X}^{j,f}}(X^{j,f}, \nabla_{X^{j,f}} U^{j,f}), & \forall j \in (1,N), & \forall f \in (1,F) \\ \dot{Y}^{j,f} = \Pi_{\overline{Y}^{j,f}}(Y^{j,f}, \nabla_{Y^{j,f}} U^{j,f}), & \forall j \in (1,N), & \forall f \in (1,F) \\ \dot{r}_{i,k} = \Pi_{R^+}(r_{i,k}, \rho_{i,k}), & \forall i \in (1,M), & \forall k \in (1,K) \\ \dot{e}_i = \Pi_{R^+}(e_i, \varepsilon_i), & \forall i \in (1,M), \end{cases} \tag{18}$$

We now present a proposition which states that if S is a Cartesian product then the projection operator Π_S is decomposable over the Cartesian product. The proposition is presented without proof since it is a simple adaptation of Proposition 1 in Dong, Zhang, and Nagurney (1996).

Proposition 1

Let S_l be a closed convex set of R^{n_l}, and $Z_l \in S_l, v_l \in R^{n_l}$. Denote by ,
$S \equiv S_1 \times S_2 \times ... \times S_t \subset R^{\Sigma_{l=1}^t n_l}$, $v = (v_1, v_2, ..., v_t) \in R^{\Sigma_{l=1}^t n_l}$,
and $Z = (Z_1, Z_2, ..., Z_t) \in S$. Then we have:

$$\Pi_S(Z, v) = (\Pi_{S_1}(Z_1, v_1), ..., \Pi_{S_t}(Z_t, v_t)). \tag{19}$$

The implication of this proposition is crucial in the study of the international financial adjustment process, which, in view of the preceding proposition, can be decomposed into a series of projected dynamical systems in the space of feasible assets, liabilities, instrument and currency prices.

For simplicity, and in order to maintain a compact mathematical formulation, we introduce the following notation:

Let $B = 2MNFK + MK + M$, $\Omega \equiv \overline{X} \times \overline{Y} \times R_+^{MK} \times R_+^M$, and define the column vector $Z = (X, Y, r, e) \in \Omega$, and the column vector $F(Z)$:

$$F(Z) = \begin{bmatrix} F_1(Z) \\ ... \\ F_b(Z) \\ ... \\ F_B(Z) \end{bmatrix} \tag{20}$$

which is equal to:

$$\begin{bmatrix} -\nabla U_X(X, Y, r, e) \\ -\nabla U_Y(X, Y, r, e) \\ \sum_{j=1}^N \sum_{f=1}^F (X_{1,1}^{j,f} - Y_{1,1}^{j,f}) \\ ... \\ \sum_{j=1}^N \sum_{f=1}^F (X_{M,K}^{j,f} - Y_{M,K}^{j,f}) \\ \sum_{k=1}^K \sum_{j=1}^N \sum_{f=1}^F (X_{1,k}^{j,f} - Y_{1,k}^{j,f}) \\ ... \\ \sum_{k=1}^K \sum_{j=1}^N \sum_{f=1}^F (X_{M,k}^{j,f} - Y_{M,k}^{j,f}) \end{bmatrix}_{B \times 1} . \tag{21}$$

In view of Proposition 1, and, taking into consideration the new notation, the international financial adjustment process can be succinctly expressed mathematically as:

$$\dot{Z} = \Pi_\Omega(Z, -F(Z)) \tag{22}$$

According to Dupuis and Nagurney (1993), we have the following theorem.

Theorem 1
Equation (22) well-defines a dynamical system, in that there exists a unique solution path to (22) for any boundary condition, $Z(0) = Z_0 \in \Omega$.

It is easy to see that the dynamical system defined in (22) is a nonclassical system due to the discontinuity of the projection map Π_Ω at the boundary of Ω (see Dupuis and Nagurney (1993)). Such discontinuities arise because of the constraints underlying the application(s) in question. System (22) was first termed a "projected dynamical system" or $PDS(F, \Omega)$ by Zhang and Nagurney (1995). For additional background and applications, see the book by Nagurney and Zhang (1996).

2.2 Stationary Points and Equilibria

In this subsection we make the connection between stationary points of the projected dynamical system (22), the solutions to a variational inequality problem, and the equilibrium conditions governing the international financial model. In particular, we show the equivalence among the stationary points, the solutions to a variational inequality problem, and the equilibrium points defined by the equilibrium conditions.

Theorem 2 (Dupuis and Nagurney (1993))
The stationary points of the projected dynamical system $PDS(F, \Omega)$, that is, Z^* satisfying

$$0 = \Pi_\Omega(Z^*, -F(Z^*)) \tag{23}$$

correspond to the solutions of the variational inequality problem $VI(F, \Omega)$ defined by:
Determine $Z^* \in \Omega$, such that

$$\left\langle F(Z^*)^T, Z - Z^* \right\rangle \geq 0, \quad \forall Z \in \Omega, \tag{24}$$

where $\langle \cdot, \cdot \rangle$ denotes the inner product in R^B, since Ω is a convex polyhedron.

We now write down the variational inequality problem (24) explicitly, utilizing (20) and (21):
Determine $(X^*, Y^*, r^*, e^*) \in \Omega$, such that:

$$-\sum_{j=1}^{N}\sum_{f=1}^{F}\left[\nabla_{Y^{jf}}U^{jf}(X^{jf^*},Y^{jf^*},r^*,e^*)\right]\cdot\left[Y^{jf}-Y^{jf^*}\right]$$

$$-\sum_{j=1}^{N}\sum_{f=1}^{F}\left[\nabla_{Y^{jf}}U^{jf}(X^{jf^*},Y^{jf^*},r^*,e^*)\right]\cdot\left[Y^{jf}-Y^{jf^*}\right] \quad (25)$$

$$+\sum_{i=1}^{M}\sum_{k=1}^{K}\left[\sum_{j=1}^{N}\sum_{f=1}^{F}(X_{i,k}^{jf^*}-Y_{i,k}^{jf^*})\right]\times\left[r_{i,k}-r_{i,k}^*\right]$$

$$+\sum_{i=1}^{M}\left[\sum_{k=1}^{K}\sum_{j=1}^{N}\sum_{f=1}^{F}(X_{i,k}^{jf^*}-Y_{i,k}^{jf^*})\right]\times\left[e_i-e_i^*\right]\geq 0, \;\; \forall(X,Y,r,e)\in\Omega.$$

The Equilibrium Condition

We now describe the equilibrium conditions. As recently shown in Nagurney and Siokos (1997), the solution to variational inequality (25) satisfies the following international financial equilibrium conditions:

Portfolio Optimality

Under the assumption that the utility function for each sector for each country is concave, the necessary and sufficient conditions for an optimal portfolio for a sector $f \in (1,F)$ of country $j \in (1,N)$, $(X^{jf^*}, Y^{jf^*}) \in K^{jf}$, is that it satisfies the following inequality:

$$-\left[\nabla_{X^{jf}}U^{jf}\left(X^{jf^*},Y^{jf^*},r^*,e^*\right)\right]\cdot\left[X^{jf}-X^{jf^*}\right]$$
$$-\left[\nabla_{Y^{jf}}U^{jf}\left(X^{jf^*},Y^{jf^*},r^*,e^*\right)\right]\cdot\left[Y^{jf}-Y^{jf^*}\right]\geq 0,$$
$$\forall(X^{jf},Y^{jf})\in K^{jf}. \quad (26)$$

Furthermore, the economic system conditions for the instrument and currency prices are as follows:

Instrument Market Equilibrium Conditions

For each instrument $k \in (1,K)$ and currency $i \in (1,M)$, we must have that:

$$\sum_{j=1}^{N}\sum_{f=1}^{F}(X_{i,k}^{jf^*}-Y_{i,k}^{jf^*})\begin{cases}=0, & \text{if } r_{ik}^*>0 \\ \geq 0, & \text{if } r_{ik}^*=0\end{cases} \quad (27)$$

The system of equalities and inequalities (27) states that if the price of a financial instrument is positive, then the market must clear for that instrument and if the price is zero, then either there is an excess supply of that instrument in the economy or the market clears.

Currency Market Equilibrium Conditions

Also, for each currency $i \in (1,M)$, we must have that

$$\sum_{j=1}^{N}\sum_{f=1}^{F}\sum_{k=1}^{K}(X_{i,k}^{jf^*} - Y_{i,k}^{jf^*})\begin{cases} = 0 \ , & \text{if } e_i^* > 0 \\ \geq 0 \ , & \text{if } e_i^* = 0 \ . \end{cases} \tag{28}$$

In other words, if the exchange price of a currency is positive, then the exchange market must clear for that currency; if the exchange price is zero, then either there is an excess supply of that currency in the economy or the market clears.

We now recall the following definition from Nagurney and Siokos (1997).

Definition 1

A vector $(X,Y,r,e) \in \Omega$ *is an equilibrium point if and only if it satisfies the system of equalities and inequalities (26), (27), and (28), for all sectors* $f \in (1,F)$, *all countries* $j \in (1,N)$, *all instruments* $k \in (1,K)$, *and all currencies* $i \in (1,M)$.

Theorem 3 (Nagurney and Siokos (1997))

A vector of assets and liabilities of the sectors of the countries, and currency instrument and exchange rate prices, $(X,Y,r,e) \in \Omega$, *is an equilibrium if and only if it satisfies the variational inequality problem (25).*

Combining Theorems 2 and 3 yields the following result.

Theorem 4

Every international financial equilibrium is a stationary point of the adjustment process (22) and vice versa. In other words, $Z^* = (X^*, Y^*, r^*, e^*)$ *constitutes a financial equilibrium if and only if* $\Pi_\Omega(Z^*, -F(Z^*)) = 0$.

It is worth relating the above models to those that have appeared in the literature. In particular, if we assume that there is only a single country and, hence, a single currency, then variational inequality (25) collapses to a variational inequality for a single country model (cf. Nagurney and Dong (1996a)). Under the same assumptions, the dynamic international model described in (22) collapses to the single-country dynamic financial model developed in Dong, Zhang, and Nagurney (1996).

3. Stability Analysis

In this section we provide conditions under which a disequilibrium pattern of assets, liabilities, instrument and currency prices will reach an international financial equilibrium state under the adjustment process (22).

For consistency, and in order to simplify the study of the asymptotic behavior of the dynamic international financial model, we present the following definitions. For general stability results, see Zhang and Nagurney (1995). For additional applications, see Nagurney and Zhang (1996).

Definition 2

The international financial adjustment process is stable if each financial equilibrium, $Z^* = (X^*, Y^*, r^*, e^*)$, is a monotone attractor, that is, for any initial asset, liability, and instrument and currency price pattern, (X_0, Y_0, r_0, e_0), the distance, $\|(X(t), Y(t), r(t), e(t)) - (X^*, Y^*, r^*, e^*)\|$, is nonincreasing in time t, where $(X(t), Y(t), r(t), e(t))$ satisfies the adjustment process (22).

In other words, if an international financial adjustment process is stable, then any state of disequilibrium will remain close to an equilibrium pattern forever.

Definition 3

The international financial adjustment process is asymptotically stable if it is stable and for any initial asset, liability, and price pattern, (X_0, Y_0, r_0, e_0), the financial adjustment process starting with (X_0, Y_0, r_0, e_0) converges to some equilibrium pattern (X^*, Y^*, r^*, e^*); namely,

$$\lim_{t \to \infty} (X_0(t), Y_0(t), r_0(t), e_0(t)) \to (X^*, Y^*, r^*, e^*) \ . \qquad (29)$$

Asymptotic stability can then be interpreted as the case that any international financial disequilibrium will eventually approach equilibrium.

Theorem 5 (Stability of International Financial Adjustment Process)

If the utility function U^{jf} of each sector f from every country j is twice differentiable and concave in X^{jf} and Y^{jf}, and the utility function has the following form:

$$U^{jf}(X^{jf}, Y^{jf}, r, e) = u^{jf}(X^{jf}, Y^{jf}) + r^T(X^{jf} - Y^{jf})$$
$$+ e^T D(X^{jf} - Y^{jf}) \qquad (30)$$

where D is the $M \times MK$ dimensional matrix with a "1" in the first K elements of row 1, the next K elements of row 2, and so on, with zero elements, otherwise, then the international financial adjustment process is stable.

Proof:
Let $Z^* = (X^*, Y^*, r^*, e^*)$ be any financial equilibrium pattern of assets, liabilities, instrument and currency prices, and let $Z_0(t) = (X_0(t), Y_0(t), r_0(t), e_0(t))$ be the financial pattern at time t when the international financial adjustment process was initialized at pattern $Z_0 = (X_0, Y_0, r_0, e_0)$.

Moreover, we define $\Delta(Z_0, Z^*, t)$ as:

$$\Delta(Z_0, Z^*, t) \equiv \frac{1}{2} \left\| Z_0(t) - Z^* \right\|^2$$

and, as a result, it follows that

$$\dot{\Delta}(Z_0, Z^*, t) = \left\langle \left(Z_0(t) - Z^* \right)^T, \Pi_\Omega(Z(t), -F(Z(t))) \right\rangle. \quad (31)$$

Recalling from Dupuis and Nagurney (1993) that:

$$\Pi_S(Z(t), -F(Z(t))) = -F(Z_0(t)) + n(Z_0(t)) n * (Z_0(t)),$$

where $n * (Z(t))$ is an inward normal to Ω at $Z(t)$ and $n(Z(t))$ is a nonnegative real number, we have an upper bound for (31) given by

$$\dot{\Delta}(Z_0, Z^*, t) \leq \left\langle \left(Z_0(t) - Z^* \right)^T, -F(Z_0(t)) \right\rangle$$

$$= \sum_{j=1}^{N} \sum_{f=1}^{F} \sum_{i=1}^{M} \sum_{k=1}^{K} \left(X_0(t)_{i,k}^{j,f} - X_{i,k}^{j,f^*} \right) \left[\frac{\partial u^{j,f}(X_0(t)^{j,f}, Y_0(t)^{j,f})}{\partial X_{i,k}^{j,f}} - r_0(t)_{i,k} - e_0(t)_i \right]$$

$$= \sum_{j=1}^{N} \sum_{f=1}^{F} \sum_{i=1}^{M} \sum_{k=1}^{K} \left(Y_0(t)_{i,k}^{j,f} - Y_{i,k}^{j,f^*} \right) \left[\frac{\partial u^{j,f}(X_0(t)^{j,f}, Y_0(t)^{j,f})}{\partial Y_{i,k}^{j,f}} + r_0(t)_{i,k} + e_0(t)_i \right]$$

$$+ \sum_{i=1}^{M} \sum_{k=1}^{K} (r_0(t)_{i,k} - r_{i,k}^*) \sum_{j=1}^{N} \sum_{f=1}^{F} (X_0(t)_{i,k}^{j,f} - Y_0(t)_{i,k}^{j,f})$$

$$+ \sum_{i=1}^{M} (e_0(t)_i - e_i^*) \sum_{j=1}^{N} \sum_{f=1}^{F} \sum_{k=1}^{K} (X_0(t)_{i,k}^{j,f} - Y(t)_{i,k}^{j,f})$$

$$= \sum_{j=1}^{N} \sum_{f=1}^{F} (X_0(t)^{j,f} - X^{j,f^*}, Y_0(t)^{j,f} - Y^{j,f^*}) \nabla^2 u^{j,f}(X_0(t), Y_0(t))$$

$$(X_0(t)^{j,f} - X^{j,f^*}, Y_0(t)^{j,f} - Y^{j,f^*}) \quad . \quad (32)$$

But, since $U^{j,f}$ is concave with respect to $X^{j,f}$ and $Y^{j,f}$, the right-hand side of (32) is nonpositive. This implies that $\left\| Z_0(t) - Z^* \right\|$ is monotone nonincreasing in time t, and, by definition, Z^* is a monotone attractor. The proof is complete.

The following theorem is presented without proof since it is a simple adaptation of the proof of Theorem 5 in Dong, Zhang, and Nagurney (1996).

Theorem 6
If we assume that all sectors f in countries S have a strictly concave utility function with respect to the assets and liabilities, then all financial equilibria are comprised of unique asset and liability patterns X^* and Y^*.

The following theorem examines the asymptotical stability of the international financial process under the assumption that each sector f of country j is strictly risk averse.

Theorem 7

If we assume that all sectors f in countries j have a strictly concave utility function with respect to the asset and liability pattern, (X^{jf}, Y^{jf}), the utility functions are twice continuously differentiable, and of the form (30), then the financial adjustment process is asymptotically stable.

Proof:

As stated in Theorem 6, the international financial adjustment process starting with any initial point (X_0, Y_0, r_0, e_0) converges to a unique equilibrium asset and liability pattern (X^*, Y^*); in particular,

$$\lim_{t \to \infty}(X_0(t), Y_0(t)) \to (X^*, Y^*).$$ (33)

Moreover, since $Z_0(t)$ is stable it is also bounded and the price patterns $(r_0(t), e_0(t))$ have convergent sequences, say:

$$\lim_{k_1 \to \infty} r_0(t_{k_1}) \to r^* \text{ and } \lim_{k_2 \to \infty} e_0(t_{k_2}) \to e^*.$$ (34)

As a result, we only have to prove that (X^*, Y^*, r^*, e^*) is, indeed, a financial equilibrium pattern and that

$$\lim_{t \to \infty}(X_0(t), Y_0(t), r_0(t), e_0(t)) \to (X^*, Y^*, r^*, e^*).$$ (35)

But since (X^*, Y^*) is an equilibrium pattern we have:

$$-\rho_{i,k} = \sum_{j=1}^{N}\sum_{f=1}^{F}(X_{i,k}^{jf^*} - Y_{i,k}^{jf^*}) \geq 0, \forall i \in (1, M) \text{ and } \forall k \in (1, K)$$ (36)

and, similarly,

$$-\varepsilon_i = \sum_{k=1}^{K}\sum_{j=1}^{N}\sum_{f=1}^{F}(X_{i,k}^{jf^*} - Y_{i,k}^{jf^*}) \geq 0, \ \forall i \in (1, M).$$ (37)

But for those i,k such that $-\rho_{i,k} \geq 0$ and for those i such that $-\varepsilon_i \geq 0$, (33) implies that for a sufficiently large t:

$$\sum_{j=1}^{N}\sum_{f=1}^{F}(X_0(t)_{i,k}^{jf} - Y_0(t)_{i,k}^{jf}) > \frac{-\rho_{i,k}}{2} > 0$$ (38)

and

$$\sum_{k=1}^{K}\sum_{j=1}^{N}\sum_{f=1}^{F}(X_0(t)_{i,k}^{jf} - Y_0(t)_{i,k}^{jf}) > \frac{-\varepsilon_i}{2} > 0.$$ (39)

As a result, for those (i,k) with $-\rho_{i,k} > 0$ we have that

$$\lim_{t \to \infty} r_0(t)_{i,k} = 0 \qquad (40)$$

and, similarly, for those i with $-\varepsilon_i > 0$ the following holds:

$$\lim_{t \to \infty} e_0(t)_i = 0 \qquad (41)$$

which yields, respectively,

$$r^*_{i,k} = 0 \text{ if } -\rho_{i,k} > 0 \qquad (42)$$

and

$$e^*_{i,k} = 0 \text{ if } -\varepsilon_i > 0 . \qquad (43)$$

Recalling (36), and that the instrument prices are nonnegative, one concludes from (42) equilibrium conditions (27). Similarly, recalling (37) and the fact that the currency prices are nonnegative, one concludes from (43), equilibrium conditions (28).

Furthermore, due to Theorem 5, the financial equilibrium pattern is a monotone attractor and, therefore, in view of (33) and (34), we know that (35) must hold.

Note that the above stability results hold for utility functions of the form (30), where, explicitly,

$$U^{jf}(X^{jf}, Y^{jf}, r, e) = u^{jf}(X^{jf}, Y^{jf}) + \sum_{i=1}^{M} \sum_{k=1}^{K} r_i (X^{jf}_{i,k} - Y^{jf}_{i,k})$$

$$+ \sum_{i=1}^{M} e_i \sum_{k=1}^{K} (X^{jf}_{i,k} - Y^{jf}_{i,k}) \qquad (44)$$

The sum of the second and third terms on the right-hand side of the equality in (44) represents the value of the asset holdings minus the value of the liability holdings. An important special case of (44) is given by the quadratic utility function:

$$U^{jf}(X^{jf}, Y^{jf}, r, e) = -\begin{pmatrix} X^{jf} \\ Y^{jf} \end{pmatrix}^T Q^{jf} \begin{pmatrix} X^{jf} \\ Y^{jf} \end{pmatrix} + \sum_{i=1}^{M} \sum_{k=1}^{K} r_{i,k} (X^{jf}_{i,k} - Y^{jf}_{i,k})$$

$$+ \sum_{i=1}^{M} e_i \sum_{k=1}^{K} (X^{jf}_{i,k} - Y^{jf}_{i,k}) \qquad (45)$$

where Q^{jf} is the symmetric $2MK \times 2MK$ variance/covariance matrix representing the risk associated with the portfolio choice. The variational inequality formulation of this quadratic utility function model, along with qualitative properties and a computational procedure can be found in Nagurney and Siokos (1997). The single country version was formulated and studied in Nagurney, Dong, and Hughes (1992).

4. Network Optimization Formulation of International Financial Equilibrium

In this section we show that the international financial equilibrium coincides with the solution to a network optimization problem. Towards that end we consider utility functions of the form (30) for which stability results were obtained in the preceding section.

First, we establish, in Theorem 8, that the equilibrium solution is the solution to an optimization problem. We then, in Lemma 1, prove that, in equilibrium, the markets in instruments and in currencies must clear, and, consequently, the prices must be positive. We then utilize these results to construct the network over which the optimization takes place.

Theorem 8

The variational inequality problem of finding $(X^*, Y^*, r^*, e^*) \in \Omega$, *satisfying:*

$$\sum_{j=1}^{N}\sum_{f=1}^{F}\sum_{i=1}^{M}\sum_{k=1}^{K}\left[-\frac{\partial u^{j,f}(X^{j,f^*}, Y^{j,f^*})}{\partial X_{i,k}^{j,f}} - r_{i,k}^* - e_i^*\right] \times \left[X_{i,k}^{j,f} - X_{i,k}^{j,f^*}\right]$$

$$+\sum_{j=1}^{N}\sum_{f=1}^{F}\sum_{i=1}^{M}\sum_{k=1}^{K}\left[-\frac{\partial u^{j,f}(X^{j,f^*}, Y^{j,f^*})}{\partial Y_{i,k}^{j,f}} + r_{i,k}^* + e_i^*\right] \times \left[Y_{i,k}^{j,f} - Y_{i,k}^{j,f^*}\right]$$

$$+\sum_{i=1}^{M}\sum_{k=1}^{K}\left[\sum_{j=1}^{N}\sum_{f=1}^{F}(X_{i,k}^{j,f^*} - Y_{i,k}^{j,f^*})\right] \times \left[r_{i,k} - r_{i,k}^*\right]$$

$$+\sum_{i=1}^{M}\left[\sum_{k=1}^{K}\sum_{j=1}^{N}\sum_{f=1}^{F}(X_{i,k}^{j,f^*} - Y_{i,k}^{j,f^*})\right] \times \left[e_i - e_i^*\right] \geq 0$$

$$\forall (X, Y, r, e) \in \Omega, \tag{46}$$

is equivalent to the problem:

$$\text{Maximize } \sum_{j=1}^{N}\sum_{f=1}^{F} u^{j,f}(X^{j,f}, Y^{j,f}) \tag{47}$$

subject to:

$$\sum_{j=1}^{N}\sum_{f=1}^{F}(X_{i,k}^{j,f} - Y_{i,k}^{j,f}) \geq 0, \quad \forall i \in (1, M), \ \forall k \in (1, K) \tag{48}$$

$$\sum_{j=1}^{N}\sum_{f=1}^{F}\sum_{k=1}^{K}(X_{i,k}^{j,f} - Y_{i,k}^{j,f}) \geq 0, \quad \forall i \in (1, M), \tag{49}$$

and $(X^{j,f}, Y^{j,f}) \in \kappa^{j,f},$ $\forall j \in (1,N),$ $\forall f \in (1,F),$ (50)

where $r_{i,k}^{*} \geq 0$ corresponds to the Lagrange multipliers associated with the (i,k)-th constraint in (48), and $e_{i}^{*} \geq 0$ corresponds to the Lagrange multiplier associated with the (i)-th constraint in (49).

Proof:
See Problem 5.2 in Bertsekas and Tsitsiklis (1989).

Lemma 1
In equilibrium, the markets clear for all instruments and all currencies, i.e.,

$$\sum_{j=1}^{N}\sum_{f=1}^{F}(X_{i,k}^{j,f^{*}} - Y_{i,k}^{j,f^{*}}) = 0, \quad \forall i \in (1,M), \ \forall k \in (1,K) \quad (51)$$

and

$$\sum_{j=1}^{N}\sum_{f=1}^{F}\sum_{k=1}^{K}(X_{i,k}^{j,f^{*}} - Y_{i,k}^{j,f^{*}}) = 0, \quad \forall i \in (1,M) \quad\quad (52)$$

that is, (48) and (49) hold as strict equalities. Furthermore, the prices, $r_{i,k}^{*}$ and e_{i}^{*} for all $i \in (1,M)$ and $k \in (1,K)$ are strictly positive.

Proof:
Assume, instead, that for a particular (i,k) we have that

$$\sum_{j=1}^{N}\sum_{f=1}^{F}(X_{i,k}^{j,f^{*}} - Y_{i,k}^{j,f^{*}}) > 0 . \quad\quad (53)$$

Then, it follows that

$$\sum_{i=1}^{M}\sum_{k=1}^{K}\sum_{j=1}^{N}\sum_{f=1}^{F}(X_{i,k}^{j,f^{*}} - Y_{i,k}^{j,f^{*}}) > 0 . \quad\quad (54)$$

But, feasibility condition (2) implies that:

$$\sum_{i=1}^{M}\sum_{k=1}^{K}\sum_{j=1}^{N}\sum_{f=1}^{F}(X_{i,k}^{j,f^{*}} - Y_{i,k}^{j,f^{*}}) = 0 , \quad\quad (55)$$

which is in contradiction to (54). Hence,

$$\sum_{j=1}^{N}\sum_{f=1}^{F}(X_{i,k}^{j,f^{*}} - Y_{i,k}^{j,f^{*}}) = 0 , \quad\quad (56)$$

and this holds for all (i,k). It follows then directly that

$$\sum_{j=1}^{N}\sum_{f=1}^{F}\sum_{k=1}^{K}(X_{i,k}^{j,f^{*}} - Y_{i,k}^{j,f^{*}}) = 0 \ \ \forall i \in (1,M). \quad\quad (57)$$

Furthermore, from Lagrange multiplier theory, the Lagrange multipliers $r_{i,k}^{*}$ and e_{i}^{*} must be positive. The proof is complete.

What remains to be shown is the network structure.

Towards that end, we define the variables $R_{i,k} = r_{i,k} + e_i$ for all $i \in (1,M)$ and $k \in (1,K)$, and rewrite variational inequality (46) as:

$$\sum_{j=1}^{N}\sum_{f=1}^{F}\sum_{i=1}^{M}\sum_{k=1}^{K}\left[-\frac{\partial u^{jf}(X^{jf^*},Y^{jf^*})}{\partial X_{i,k}^{jf}} - R_{i,k}^* \right] \times \left[X_{i,k}^{jf} - X_{i,k}^{jf^*} \right]$$

$$+\sum_{j=1}^{N}\sum_{f=1}^{F}\sum_{i=1}^{M}\sum_{k=1}^{K}\left[-\frac{\partial u^{jf}(X^{jf^*},Y^{jf^*})}{\partial Y_{i,k}^{jf}} + R_{i,k}^* \right] \times \left[Y_{i,k}^{jf} - Y_{i,k}^{jf^*} \right]$$

$$+\sum_{i=1}^{M}\sum_{k=1}^{K}\sum_{j=1}^{N}\sum_{f=1}^{F}\left[X_{i,k}^{jf^*} - Y_{i,k}^{jf^*} \right] \times \left[R_{i,k} - R_{i,k}^* \right] \geq 0 ,$$

$$\forall (X^*,Y^*,R^*) \in (\overline{X},\overline{Y},R_+^{MK}) . \tag{58}$$

The following corollary is immediate from Theorem 8.

Corollary 1

The solution to the optimization problem:

$$Maximize \ \sum_{j=1}^{N}\sum_{f=1}^{F} u^{jf}(X^{jf},Y^{jf}) \tag{59}$$

subject to:

$$\sum_{j=1}^{N}\sum_{f=1}^{F}(X_{i,k}^{jf} - Y_{i,k}^{jf}) \geq 0, \ (X,Y) \in \kappa , \tag{60}$$

with Lagrange multiplier $R_{i,k}$ for constraint (i,k) in (60) coincides with the solution to problem (47)-(50), where $R_{i,k}^* \equiv (r_{i,k}^* + e_i^*)$ for all (i,k). Moreover, this solution is an equilibrium solution.

Finally, combining Lemma 1 and Corollary 1, we obtain:

Theorem 9

The international financial equilibrium pattern $(X^*,Y^*,r^*,e^*) \in \Omega$ coincides with the solution of the optimization problem:

$$Maximize \ \sum_{j=1}^{N}\sum_{f=1}^{F} u^{jf}(X^{jf},Y^{jf}) \tag{61}$$

subject to:

$$\sum_{j=1}^{N}\sum_{f=1}^{F}(X_{i,k}^{jf} - Y_{i,k}^{jf}) = 0, \ (X,Y) \in \kappa \tag{62}$$

with Lagrange multipliers $R_{i,k}^* = (r_{i,k}^* + e_i^*)$ for all (i,k) corresponding to the (i,k)-th constraint in (62).

The network structure of the problem in Theorem 9 is immediate and given in Figure 2. Note that the individual networks in Figure 1, representing the problems of the individual sectors in the countries, through the dynamics described in Section 2, now merge to yield to complete synthesized network of Figure 2. Interestingly, if one is interested in solely the computation of the international financial equilibrium pattern, then one may solve problem (61) and (62). In the subsequent section, however, we provide a discretization of the continuous time model through the Euler method, which not only yields the stationary point but also the trajectories through time.

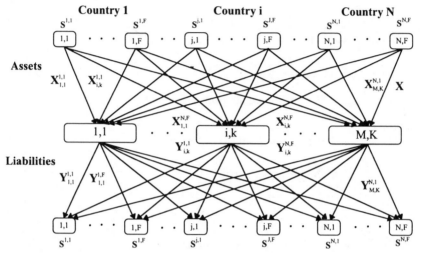

Figure 2. Network Structure of International Financial Equilibrium

5. A Discrete Time Algorithm

In this section, we present a discrete time algorithm, the Euler method, for the computation of the international financial equilibrium pattern. This algorithm may also be viewed as a discrete time approximation or discrete time adjustment process for the continuous time model given by (22).

The Euler method is a special case of the general iterative scheme introduced in Dupuis and Nagurney (1993). For completeness, and easy reference, we now state the general iterative scheme applied to the international financial model: At each iteration τ, one must compute:

$$Z^{\tau+1} = P_{\Omega}(Z^{\tau} - a_{\tau}F(Z^{\tau})), \tag{63}$$

where $\{a_{\tau}\}$ is a positive sequence, to be discussed later, along with an interpretation, P_{Ω} is the projection operator defined in (10), and F_{τ} is some suitable approximation of $F(Z)$ defined in (20) and (21).

In the case of the Euler method, we have $F_\tau = F$, and, hence, the iterative step takes the form:

$$Z^{\tau+1} = P_\Omega(Z^\tau - a_\tau F_\tau(Z^\tau)),\tag{64}$$

which is equivalent (cf. Nagurney (1993)) to the solution of the variational inequality problem:

Determine $Z^{\tau+1} \in \Omega$, satisfying:

$$\left\langle \left(Z^{\tau+1} + a_\tau F(Z^\tau) - Z^\tau\right), Z - Z^{\tau+1}\right\rangle \geq 0, \quad \forall Z \in \Omega\tag{65}$$

We now write down (65) explicitly for our problem.

Compute $(X^{\tau+1}, Y^{\tau+1}, r^{\tau+1}, e^{\tau+1}) \in \Omega$ by solving the variational inequality problem:

$$\sum_{j=1}^{N}\sum_{f=1}^{F}\sum_{i=1}^{M}\sum_{k=1}^{K}\left[X_{i,k}^{j,f^{\tau+1}} + a_\tau\left(-\frac{\partial u^{j,f}(X^{j,f^\tau}, Y^{j,f^\tau})}{\partial X_{i,k}^{j,f}} - r_{i,k}^\tau - e_i^\tau\right) - X_{i,k}^{j,f^\tau}\right]$$

$$\times\left[X_{i,k}^{j,f'} - X_{i,k}^{j,f^{\tau+1}}\right]$$

$$\sum_{j=1}^{N}\sum_{f=1}^{F}\sum_{i=1}^{M}\sum_{k=1}^{K}\left[Y_{i,k}^{j,f^{\tau+1}} + a_\tau\left(-\frac{\partial u^{j,f}(X^{j,f^\tau}, Y^{j,f^\tau})}{\partial Y_{i,k}^{j,f}} + r_{i,k}^\tau + e_i^\tau\right) - Y_{i,k}^{j,f^\tau}\right]$$

$$\times\left[Y_{i,k}^{j,f'} - Y_{i,k}^{j,f^{\tau+1}}\right]$$

$$+\sum_{i=1}^{M}\sum_{k=1}^{K}\left[e_{i,k}^{\tau+1} + a_\tau\left[\sum_{j=1}^{N}\sum_{f=1}^{F}(X_{i,k}^{j,f^\tau} - Y_{i,k}^{j,f^\tau})\right] - r_{i,k}^\tau\right]\times\left[r_{i,k} - r_{i,k}^{\tau+1}\right]$$

$$+\sum_{i=1}^{M}\left[e_i^{\tau+1} + a_\tau\left[\sum_{k=1}^{K}\sum_{j=1}^{N}\sum_{f=1}^{F}(X_{i,k}^{j,f^\tau} - Y_{i,k}^{j,f^\tau})\right] - e_i^\tau\right]\times\left[e_i' - e_i^{\tau+1}\right] \geq 0,$$

$$\forall(X', Y', r', e') \in \Omega.\tag{66}$$

Although the focus here is on the Euler method, which is the simplest scheme, we note that it is only one of many computational methods induced by the iterative scheme defined in (63) (cf. Dupuis and Nagurney (1993)). In fact, in addition to the Euler method, we can recover through (63) such well-known computational methods in dynamical systems theory as the Heun method and the Runge-Kutta method.

When the step-size sequence $\{a_\tau\}$ in the Euler method is fixed, say, $\{a_\tau\} = \rho$, for all iterations τ, then the Euler method collapses to a projection method (cf. Dafermos (1983), Bertsekas and Tsitsiklis (1989)).

In the context of the international financial problem, the projection operation entails the solution of a quadratic programming problem. Moreover, in view of the feasible set Ω, the variational inequality problem (66) decomposes into subproblems that can be solved explicitly and in closed form. In particular, we note that the feasible set consists solely of simple linear constraints and nonnegativity constraints and is a Cartesian product. In addition, the structures of the resulting subproblems are those of networks (cf. Figure 3) consisting of disjoint paths, i.e., having no links in common. Hence, each such network subproblem can be solved using an "exact" equilibration algorithm proposed by Dafermos and Sparrow (1969) (see also, e.g., Nagurney and Siokos (1996)). Interestingly, the structure of the network subproblems in Figure 3 corresponding to the sectors of the countries is precisely that illustrated in Figure 1. Here, however, in addition, we have the simple networks corresponding to the prices.

Indeed, in view of the feasible set Ω, problem (66) can be decomposed into simpler and smaller subproblems in the (X^{jf}), (Y^{jf}), $r_{i,k}$, and the e_i variables for all j,f,i,k. In particular, each asset subproblem for a sector in a country is a quadratic programming problem with a single linear constraint and nonnegativity assumption of the asset variables; the same holds for each liability subproblem for each sector in each country. Finally, each subproblem in either an instrument price or in an exchange rate consists of a single variable minimization problem subject to only a nonnegativity asssumption on the variable, which can be computed using a closed form expression.

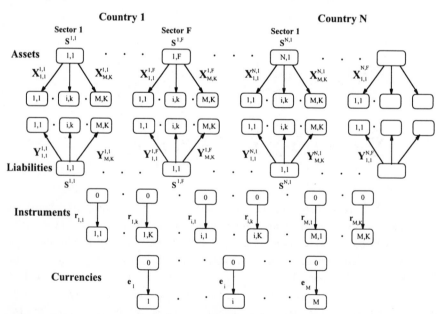

Figure 3. Structure of Algorithmic Network Subproblems

To conclude this section, the following theorem provides the convergence of the proposed Euler method in the context of the international financial problem.

Theorem 10

Suppose that the utility functions u are strictly monotone decreasing and of the form (30) (equivalently, (44)). Let $\{a_\tau\}$ be a sequence of positive real numbers that satisfies

$$\lim_{\tau \to \infty} a_\tau = 0 \tag{67}$$

and

$$\sum_{\tau=0}^{\infty} a_\tau = \infty \tag{68}$$

Then the Euler method given by

$$Z^{\tau+1} = P_\Omega(Z^\tau - a_\tau F(Z^\tau)) \tag{69}$$

where F(Z) is the vector defined in (20) and (21) converges to some international financial equilibrium pattern.

Proof:

According to Dupuis and Nagurney (1993) (Theorem 7), the sequence $\{Z^\tau\}$ generated by (69) converges to some solution to the variational inequality problem (25), equivalently, to a stationary point of the projected dynamical system (22)), provided that the following assumption is satisfied. Consequently, by taking advantage of Theorem 3, it converges to some equilibrium pattern.

Assumption

Fix an initial condition $Z^0 \in \Omega$ and define the sequence $\{Z^\tau, \tau \in N\}$ by (69).

Assume the following conditions.

1. $\sum_{\tau=0}^{\infty} a_\tau = \infty$, $a_\tau > 0$, $a_\tau \to 0$, as $\tau \to \infty$.

2. $d(F_\tau(Z), \overline{F}(Z)) \to 0$ uniformly on compact subsets of Ω as $\tau \to \infty$, where $d(Z, A) = \inf\{\|Z - y\|, y \in A\}$.

3. Define ϕ_y to be the unique solution to $\dot{Z} = \Pi_\Omega(Z, -F(Z))$ that satisfies $\phi_y(0) = y \in \Omega$. The ω-limit set $\cup_{y \in \Omega} \cap_{t \geq 0} \cup_{s \geq t} \{\phi_y(s)\}$ is contained in the set of stationary points of $\dot{Z} = \Pi_\Omega(Z, -F(Z))$.

4. The sequence $\{Z^\tau, \tau \in N\}$ is bounded.

5. The solutions to $\dot{Z} = \Pi_\Omega(Z, -F(Z))$ are stable in the sense that given any compact set Ω_1 there exists a compact set Ω_2 such that $\cup_{y \in \Omega \cap \Omega_1} \cap_{t \geq 0} \{\phi_y(t)\} \subset \Omega_2$.

Notice that condition (1) is already met by the chosen sequence according to (67) and (68). Condition (2) is also satisfied because $F(Z)$ is continuous and $F_\tau = F$. By Theorem 7, the international financial adjustment process is asymptotically stable, and, hence, every ω-limit point of the projected dynamical system (22) is an equilibrium pattern. This verifies condition (3) of that assumption. Condition (4) becomes trivial, with notice to the fact that the feasible set χ is compact. Finally, condition (5) of the assumption holds true because the financial adjustment process (22) is asymptotically stable and, therefore, it is bounded. We have completed the proof.

We now briefly give an interpretation of the sequence $\{a_\tau\}$ (cf. (67) and (68)). Note that in the discrete version of the financial adjustment process given by (69), τ may be interpreted as a stage or period. The requirement (67), in a sense, suggests that the possibility of error accumulation should be reduced to reflect learning over time. The requirement (68), on the other, suggests that the forces behind (64) should be given sufficient time to evolve. In other words, (67) and (68) together suggest that the international financial system will react more strongly at the beginning and its responsiveness will decrease later on.

6. Numerical Examples

In this section we present numerical examples to illustrate the computational procedure applied to competitive international financial equilibrium problems. The numerical examples are of increasing complexity.

The Euler method described in Section 5 was coded in FORTRAN and implemented on an IBM SP2. The CPU time for each example is reported exclusive of input/output times. The asset and liability subproblems were solved using exact equilibration and the prices were solved in closed form.

In all the examples we initialized the algorithm as follows: $X_{i,k}^{j,f} = \dfrac{S^{j,f}}{MK}$ and $Y_{i,k}^{j,f} = \dfrac{S^{j,f}}{MK}$ for all j, f; $r_{i,k} = 1$ for all i, k and $e_l = 1$ for all l. The convergence criterion was: $\left| Z_b^{\tau+1} - Z_b^\tau \right| \leq 0.001$, for all $b=1, \ldots, B$.

Example 1: 1 Country, 2 Sectors, 2 Currencies, 2 Instruments
We first considered an example consisting of a single country with two sectors trading in two currencies and two financial instruments in each currency.
The variance-covariance matrices were as follows:

For sector 1:

$$Q^{1,1} = \begin{pmatrix} 1.0 & 0.2 & 0 & 0 & -0.2 & -0.3 & 0 & 0 \\ 0.2 & 1.0 & 0 & 0 & -0.1 & -0.2 & 0 & 0 \\ 0 & 0 & 1.0 & 0 & 0 & 0 & -0.1 & 0 \\ 0 & 0 & 0 & 1.0 & 0 & 0 & -0.1 & 0 \\ -0.2 & -0.1 & 0 & 0 & 1.0 & 0.4 & 0 & 0 \\ -0.3 & -0.2 & 0 & 0 & 0.4 & 1.0 & 0 & 0 \\ 0 & 0 & -0.1 & -0.1 & 0 & 0 & 1.0 & 0 \\ 0 & 0 & 0 & 0 & 0 & 0 & 0 & 1.0 \end{pmatrix}$$

For sector 2:

$$Q^{1,2} = \begin{pmatrix} 1.0 & 0.3 & 0.2 & 0 & 0 & 0 & -0.1 & -0.2 \\ 0.3 & 1.0 & 0 & 0 & 0 & -0.15 & 0 & 0 \\ 0.2 & 0 & 1.0 & 0.5 & -0.1 & 0 & 0 & 0 \\ 0 & 0 & 0.5 & 1.0 & 0 & -0.3 & 0 & 0 \\ 0 & 0 & -0.1 & 0 & 1.0 & 0 & 0.2 & 0 \\ 0 & -0.15 & 0 & -0.3 & 0 & 1.0 & 0.3 & 0 \\ -0.1 & 0 & 0 & 0 & 0.2 & 0.3 & 1.0 & 0 \\ -0.2 & 0 & 0 & 0 & 0 & 0 & 0 & 1.0 \end{pmatrix} .$$

We set all the financial volumes $S^{j,f}$ all equal to 1.

We set the sequence $\{a_\tau\} = 1.\left\{1, \dfrac{1}{2}, \dfrac{1}{2}, \dfrac{1}{3}, \dfrac{1}{3}, \dfrac{1}{3}, \ldots\right\}$. The algorithm converged in 86 iterations and .01 CPU seconds yielding the following solution:

Instrument Prices: $r_{1,1} = .938$, $r_{1,2} = 1.054$, $r_{2,1} = 1.029$, $r_{2,2} = .979$,

Exchange Rate Prices: $e_1 = .992$, $e_2 = 1.008$,

Asset and Liability Pattern:

$$X_{1,1}^{1,1} = .251, \quad X_{1,2}^{1,1} = .264, \quad X_{2,1}^{1,1} = .255, \quad X_{2,2}^{1,1} = .230,$$
$$X_{1,1}^{1,2} = .228, \quad X_{1,2}^{1,2} = .314, \quad X_{2,1}^{1,2} = .131, \quad X_{2,2}^{1,2} = .328,$$
$$Y_{1,1}^{1,1} = .247, \quad Y_{1,2}^{1,1} = .237, \quad Y_{2,1}^{1,1} = .269, \quad Y_{2,2}^{1,1} = .246,$$
$$Y_{1,1}^{1,2} = .231, \quad Y_{1,2}^{1,2} = .338, \quad Y_{2,1}^{1,2} = .118, \quad Y_{2,2}^{1,2} = .313,$$

As proven in Lemma 1, we note that the markets cleared for each currency and, hence, the exchange rate prices were positive. Also, the markets for the instruments cleared (to within 3 decimal places), and the instrument prices were positive.

Example 2: 2 Countries, 2 Sectors, 2 Currencies, 2 Instruments

The second example consisted of two countries, two sectors in each country trading in two currencies, and two financial instruments in each currency.

The variance-covariance matrices for the sectors in the first country were as in Example 1. The variance-covariance matrices for the sectors in the second country were as follows:

For sector 1, country 2:

$$
Q^{2,1} =
\begin{pmatrix}
1.0 & 0 & 0.3 & 0 & -0.1 & -0.1 & 0 & 0 \\
0 & 1.0 & 0.1 & 0 & 0 & 0 & -0.2 & 0 \\
0.3 & 0.1 & 1.0 & 0 & -0.3 & 0 & 0 & 0 \\
0 & 0 & 0 & 1.0 & -0.4 & 0 & 0 & 0 \\
-0.1 & 0 & -0.3 & -0.4 & 1.0 & 0 & 0 & 0 \\
-0.1 & 0 & 0 & 0 & 0 & 1.0 & 0.5 & 0 \\
0 & -0.2 & 0 & 0 & 0 & 0.5 & 1.0 & 0.1 \\
0 & 0 & 0 & 0 & 0 & 0 & 0.1 & 1.0
\end{pmatrix}
$$

For sector 2, country 2:

$$
Q^{2,2} =
\begin{pmatrix}
1.0 & 0.4 & 0 & 0 & -0.2 & 0 & 0 & 0 \\
0.4 & 1.0 & 0 & 0 & 0 & -0.1 & 0 & 0 \\
0 & 0 & 1.0 & 0 & -0.3 & 0 & 0 & 0 \\
0 & 0 & 0 & 1.0 & 0 & 0 & -0.4 & 0 \\
-0.2 & 0 & -0.3 & 0 & 1.0 & 0 & 0 & 0 \\
0 & -0.1 & 0 & 0 & 0 & 1.0 & 0.3 & 0 \\
0 & 0 & 0 & -0.4 & 0 & 0.3 & 1.0 & 0.1 \\
0 & 0 & 0 & 0 & 0 & 0 & 0.1 & 1.0
\end{pmatrix}.
$$

The sector volumes for the sectors in the second country were set equal to 1.

The algorithm converged in 36 iterations and .02 CPU seconds with the sequence

$$\{a_\tau\} = 1.\left\{1, \frac{1}{2}, \frac{1}{2}, \frac{1}{3}, \frac{1}{3}, \frac{1}{3}, \dots\right\},$$ yielding the following solution:

Instrument Prices: $r_{1,1} = 1.060$, $r_{1,2} = .983$, $r_{2,1} = 1.001$, $r_{2,2} = .955$,

Exchange Rate Prices: $e_1 = 1044$, $e_2 = .956$,

Asset and Liability Pattern:

$$X^{1,1}_{1,1} = .307, \quad X^{1,1}_{1,2} = .244, \quad X^{1,1}_{2,1} = .229, \quad X^{1,1}_{2,2} = .220,$$

$$X^{1,2}_{1,1} = .297, \quad X^{1,2}_{1,2} = .279, \quad X^{1,2}_{2,1} = .154, \quad X^{1,2}_{2,2} = .270,$$

$$X^{2,1}_{1,1} = .235, \quad X^{2,1}_{1,2} = .243, \quad X^{2,1}_{2,1} = .213, \quad X^{2,1}_{2,2} = .309,$$

$$X^{2,2}_{1,1} = .242, \quad X^{2,2}_{1,2} = .180, \quad X^{2,2}_{2,1} = .288, \quad X^{2,2}_{2,2} = .290,$$

$$Y^{1,1}_{1,1} = .189, \quad Y^{1,1}_{1,2} = .264, \quad Y^{1,1}_{2,1} = .288, \quad Y^{1,1}_{2,2} = .259,$$

$$Y^{1,2}_{1,1} = .206, \quad Y^{1,2}_{1,2} = .290, \quad Y^{1,2}_{2,1} = .184, \quad Y^{12}_{2,2} = .319,$$

$$Y^{2,1}_{1,1} = .383, \quad Y^{2,1}_{1,2} = .168, \quad Y^{2,1}_{2,1} = .200, \quad Y^{2,1}_{2,2} = .249,$$

$$Y^{2,2}_{1,1} = .312, \quad Y^{2,2}_{1,2} = .178, \quad Y^{2,2}_{2,1} = .273, \quad Y^{2,2}_{2,2} = .237.$$

In this example, as expected, the markets also cleared for each currency and (approximately) cleared for each instrument.

Example 3: 3 Countries, 2 Sectors, 2 Currencies, 2 Instruments
This example consisted of three countries, with the data for the first two countries as in the preceding example. The variance/covariance matrix for the third country was given by:
For sector 1, country 3:

$$Q^{3,1} = \begin{pmatrix} 1.0 & 0.1 & 0 & 0 & 0 & 0 & 0 & -0.1 \\ 0.1 & 1.0 & 0.2 & 0 & -0.2 & 0 & 0 & 0 \\ 0 & 0.2 & 1.0 & 0 & 0 & -0.3 & 0 & 0 \\ 0 & 0 & 0 & 1.0 & 0 & 0 & 0 & 0 \\ 0 & -0.2 & 0 & 0 & 1.0 & 0.5 & 0 & 0 \\ 0 & 0 & -0.3 & 0 & 0.5 & 1.0 & 0 & 0 \\ 0 & 0 & 0 & 0 & 0 & 0 & 1.0 & 0.4 \\ -0.1 & 0 & 0 & 0 & 0 & 0 & 0.4 & 1.0 \end{pmatrix}$$

For sector 2, country 3:

$$Q^{3,2} = \begin{pmatrix} 1.0 & 0 & 0.2 & 0 & -0.1 & -0.2 & 0 & 0 \\ 0 & 1.0 & 0.3 & 0 & 0 & -0.4 & 0 & 0 \\ 0.2 & 0.3 & 1.0 & 0 & 0 & 0 & -0.1 & 0 \\ 0 & 0 & 0 & 1.0 & 0 & 0 & 0 & 0 \\ -0.1 & 0 & 0 & 0 & 1.0 & 0.4 & 0 & 0 \\ -0.2 & -0.4 & 0 & 0 & 0.4 & 1.0 & 0 & 0 \\ 0 & 0 & -0.1 & 0 & 0 & 0 & 1.0 & 0.6 \\ 0 & 0 & 0 & 0 & 0 & 0 & 0.6 & 1.0 \end{pmatrix}.$$

The algorithm was initialized as in the preceding examples with the exception that $S^{3,1} = S^{3,2} = 2$. We utilized the same $\{a_\tau\}$ sequence as in Example 2.

The algorithm converged in 76 iterations and .04 seconds of CPU time yielding the following solution:

Instrument Prices: $r_{1,1} = 1.002$, $r_{1,2} = 1.008$, $r_{2,1} = 1.047$, $r_{2,2} = .943$,

Exchange Rate Prices: $e_1 = 1.010$, $e_2 = .990$,

Asset and Liability Pattern:

$$X^{1,1}_{1,1} = .163 \quad X^{1,1}_{1,2} = .298 \quad X^{1,1}_{2,1} = .288 \quad X^{1,1}_{2,2} = .251,$$
$$X^{1,2}_{1,1} = .290 \quad X^{1,2}_{1,2} = .276 \quad X^{1,2}_{2,1} = .158 \quad X^{1,2}_{2,2} = .275,$$
$$X^{2,1}_{1,1} = .215 \quad X^{2,1}_{1,2} = .233 \quad X^{2,1}_{2,1} = .235 \quad X^{2,1}_{2,2} = .317,$$
$$X^{2,2}_{1,1} = .231 \quad X^{2,2}_{1,2} = .164 \quad X^{2,2}_{2,1} = .316 \quad X^{2,2}_{2,2} = .288,$$
$$X^{3,1}_{1,1} = .270 \quad X^{3,1}_{1,2} = .226 \quad X^{3,1}_{2,1} = .285 \quad X^{3,1}_{2,2} = .218,$$
$$X^{3,2}_{1,1} = .618 \quad X^{3,2}_{1,2} = .675 \quad X^{3,2}_{21,1} = .252 \quad X^{3,2}_{2,2} = .455,$$

$$Y^{1,1}_{1,1} = .189 \quad Y^{1,1}_{1,2} = .260 \quad Y^{1,1}_{2,1} = .281 \quad Y^{1,1}_{2,2} = .270,$$
$$Y^{1,2}_{1,1} = .215 \quad Y^{1,2}_{1,2} = .301 \quad Y^{1,2}_{2,1} = .151 \quad Y^{1,2}_{2,2} = .333,$$
$$Y^{2,1}_{1,1} = .413 \quad Y^{2,1}_{1,2} = .160 \quad Y^{2,1}_{2,1} = .175 \quad Y^{2,1}_{2,2} = .251,$$
$$Y^{2,2}_{1,1} = .337 \quad Y^{2,2}_{1,2} = .167 \quad Y^{2,2}_{2,1} = .257 \quad Y^{2,2}_{2,2} = .239,$$
$$Y^{3,1}_{1,1} = .221 \quad Y^{3,1}_{1,2} = .279 \quad Y^{3,1}_{2,1} = .209 \quad Y^{3,1}_{2,2} = .290,$$
$$Y^{3,2}_{1,1} = .391 \quad Y^{3,2}_{1,2} = .790 \quad Y^{3,2}_{2,1} = .408 \quad Y^{3,2}_{2,2} = .410.$$

In this example, the markets cleared (approximately to three decimal points) for each currency and for each financial instrument. The exchange rate prices and instrument prices were, as expected from Lemma 1, positive.

7. Summary and Conclusions

In this paper we have developed a dynamic model of competitive international financial equilibrium with multiple sectors in various countries and with different instruments in multiple currencies. We first modeled the individual behavior of a sector's portfolio optimization problem and showed the underlying network structure. We then presented the instrument and the exchange rate adjustment process and showed that the dynamical system satisfied a projected dynamical system whose set of stationary points coincides with the set of solutions to a variational inequality problem formulation of the international financial equilibrium conditions.

Stability results for the international adjustment process were then presented for the dynamic model with a class of utility functions. In particular, we established conditions on the utility functions of the sector in each country that guaranteed, respectively, stability and asymptotical stability of the adjustment process.

We then showed that, in equilibrium, the international financial equilibrium can be obtained as a solution to a network optimization problem. In fact, the network corresponding to the equilibrium solution is a synthesis of the individual sectors' networks. The adjustment process, hence, represents a movement of the individual networks through time towards the complete and synthesized equilibrium network in which markets in instruments and in currencies were shown to clear.

We then proposed a discrete time algorithm, the Euler method, which is a discretization of the continuous time model for the computation of the solution and provided convergence results. The algorithm has the notable feature that it resolves the problem into separate subproblems, respectively, for assets, liabilities, instrument, and currency prices. Moreover, these subproblems can be solved explicitly and in closed form using exact equilibration. The network structure here was also illustrated.

Finally, we illustrated the performance of the algorithm and the model on several examples of increasing size.

Acknowledgments

This research was supported by the National Science Foundation under grant DMS 902471 under the Faculty Awards for Women Program. This support is gratefully acknowledged.

This research was conducted using the resources of the Cornell Theory Center at Cornell Unviersity. The use of this facility is very much appreciated.

The authors are indebted to Åke Andersson who has never failed to push the frontiers and to raise challenging scientific questions.

References

Alder, M., and Dumas, B., 1983, "International Portfolio Choice and Corporation Finance: A Synthesis. " *The Journal of Finance*, 38, 925-984.

Andersson, Å. E., 1989, Advances in Spatial Theory and Economics. Elsevier Science, New York, New York.

Andersson, Å. E., and Andersson, S. I., 1992, Theory and Control of Dynamical Systems: Applications to Systems in Biology. World Scientific, River Edge, New Jersey.

Andersson, Å. E., Batten, D. F., Johansson, B., and Nijkamp, P., 1989, Advances in Spatial Theory and Dynamics. North-Holland, Amsterdam, The Netherlands.

Arrow, K., 1963, "The Role of Securities in the Optimal Allocation of Risk Bearing." *Review of Economic Studies*, 31, 91-96.

Batten, D., 1985, New Mathematical Advances in Economic Dynamics: Introduction. New York University Press, New York, New York.

Bertsekas, D., 1976, Dynamic Programming and Stochastic Control. Academic Press, New York, New York.

Bertsekas, D., and Tsitsiklis, J., 1989, Parallel Distributed Computation: Numerical Methods. Prentice Hall, Englewood Clifs, New Jersey.

Black, F., 1972, "Capital Market Equilibrium with Restricted Borrowing."
Journal of Business, 45, 444-455.

Black, F., and Scholes, M., 1973, "The Pricing of Options and Corporate Liabilities." *Journal of Political Economy,* 3, 637-654.

Chow, G., 1981, Econometric Analysis by Control Methods. Wiley, New York, New York.

Cox, J., Ingersoll, J., and Ross, S., 1985, "A Theory of the Term Structure of Interest Rates." *Econometrica*, 53, 363-384.

Dafermos, S., 1983, "An Iterative Scheme for Variational Inequalities." *Mathematical Programming* 26, 40-47.

Dafermos, S., and Sparrow, F. T., 1969, "The Traffic Assignment Problem for a General Network." *Journal of Research of the National Bureau of Standards,* 73B, 91-118.

Debreu, G., 1959, Theory of Value. Yale University Press, New Haven, Connecticut.

Dong, J., Zhang, D., and Nagurney, A., 1996, "A Projected Dynamical Systems Model of General Financial Equilibrium with Stability Analysis." *Mathematical and Computer Modelling*, 24, 35-44.

Duffie, D., 1986, "Stochastic Equilibria: Existence, Spanning Number, and the 'No Expected Financial Gain from Trade' Hypothesis." *Econometrica*, 54(5), 1161-1184.

Duffie, D., and Huang, C., 1985, "Implementing Arrow-Debreu Equilibria by Continuous Trading of Few Long-Lived Securities." *Econometrica*, 53, 1337-1356.

Dupuis, P., and Nagurney, A., 1993, "Dynamical Systems and Variational Inequalities." *Annals of Operations Research*, 44, 9-42.

Eun, C. and Resnick, B. 1988, "Exchange Rate Uncertainty, Forward Contracts, and International Portfolio Selection." *The Journal of Finance*, 43, 197-215.

Grandmont, J. M., and Malgrange, P., 1986, "Nonlinear Economic Dynamics: Introduction." *Journal of Economic Theory,* 40(1), 3-12.

Grubel, H., 1968, "Internationally Diversified Portfolios: Welfare Gains and Capital Flows." *American Economic Review,* 58, 1299-1314.

Kinderlehrer D., and Stampacchia G., 1980, An Introduction to Variational Inequalities and Their Applications. Academic Press, New York, New York.

Lessard, D., 1973, "World Country Industry Relationships in Equity Returns: Implications for Risk Reduction through International Diversification." *Financial Analysts Journal*, 32, 2-8.

Lintner, J., 1965, "The Valuation of Risk Assets and the Selection of Risky Investments in Stock Portfolios and Capital Budgets." *Review of Economic Statistics*, 2, 13-37.

Lucas, R. E., Jr., 1978, "Asset Prices in an Exchange Economy." *Econometrica*, 46(6), 1426-1445.

Lucas, R. E. Jr., 1982, "Interest Rates and Currency Prices in a Two-Country World." *Journal of Monetary Economics*, 10(3)}, 335-359.

Markowitz, H., 1959, Portfolio Selection: Efficient Diversification of Investments. John-Wiley and Sons Inc. New York, New York.

Merton, R., 1973, "Optimum Consumption and Portfolio Rules in a Continuous Time Model." *Journal of Economic Theory*, 3, 373-413.

Milne, F., and Smith, C., 1980, "Capital Asset Pricing with Proportional Transaction Costs." *Journal of Financial and Quantitative Analysis,* 15(2), 263-265.

Modigliani, F., and Miller, M., 1958, "The Cost of Capital, Corporate Finance and the Theory of Corporation Finance." *American Economic Review*, 48, 261-297.

Mossin, J., 1966, "Equilibrium in a Capital Asset Market."*Econometrica*, 34, 768-783.

Nagurney, A., 1993, Network Economics: A Variational Inequality Approach. Kluwer Academic Publishers, Boston, Massachusetts.

Nagurney, A., 1994, "Variational Inequalities in the Analysis and Computation of Multi-Sector, Multi-Instrument Financial Equilibria." *Journal of Economic Dynamics and Control*, 18, 161-184.

Nagurney, A., and Dong, J., 1996a, "Network Decomposition of General Financial Equilibrium with Transaction Costs." *Networks*, 28, 107-116.

Nagurney, A., and Dong, J., 1996b, "General Financial Equilibrium Modeling with Policy Interventions and Transaction Costs." *Computational Economics*, 9, 3-17.

Nagurney, A., Dong, J., and Hughes, M., 1992, "Formulation and Computation of General Financial Equilibrium." *Optimization*, 26, 339-354.

Nagurney, A., and Siokos, S., 1997, "Variational Inequalities for International General Financial Equilibrium Modeling and Computation." *Mathematical and Computer Modelling*, 25, 31-49.

Nagurney, A., and Zhang, D., 1996, Projected Dynamical Systems and Variational Inequalities with Applications, Kluwer Academic Publishers, Boston, Massachusetts.

Ross, S., 1976, "Arbitrage Theory of Capital Asset Pricing." *Journal of Economic Theory*, 13, 341-360.

Sharpe, W., 1964, "Capital Asset Prices: A Theory of Market Equilibrium under Conditions of Risk." Journal of Finance, 19(3), 425-443.

Solnik, B.H., 1974, "Equilibrium Model of the International Market." Journal of Economic Theory, 8, 500-524.

Tourin, A., and Zariphopoulou, T., 1994, "Numerical Schemes for Investment Models with Singular Transactions." Computational Economics, 7, 287-307.

Zhang, D., and Nagurney, A., 1995, "On the Stability of Projected Dynamical Systems." Journal of Optimization Theory and Applications, 85, 97-124.

Price and Nonprice Competition in Oligopoly With Scaling of Characteristics

Robert E. Kuenne
Department of Mechanical
Princeton University

In most models that include both price and quality competition, characteristics are assumed to be cardinally measurable and, therefore, capable of immediate incorporation as variables in analytical and simulation analysis. This paper deals initially with the difficult task of "scaling" characteristics which are qualities that are not at the present time directly "measurable" in the meaning of the theory of measurement or with attributes which can be measured only in a nominal 0-1 manner. The procedure in this paper moves from scaling at the product or brand level to study competition without tacit collusion to price-nonprice competition in a collusive context, and from general functional forms in closed analysis to specific forms in simulation runs.

1. Scaling by Quality Indifference Premia

We begin with the concept of the "brand" as an integral unit of analysis with the aim of characterizing n brands with scaling values to locate them in a product space. To do so, we deal with a product which consumers buy only as a single unit if they buy at all: for example, large durable goods such as refrigerators or washing machines. To obtain a scaling of each brand in consumers' preferences we choose brand 1 arbitrarily as an anchor and ask the following three questions of a sample of prospective consumers:

1) Suppose you were given a unit of brand 1 at no cost to you, but that you were forbidden to resell it. If you were offered brand j instead, how many dollars would you *have* to receive or give up (independent of your willingness or ability to do so) to make you feel as well off as you feel with brand 1?

These dollar measures I_j are *indifference premia* and will in general be negative and positive with $I_1=0$, where negative values are hypothetical payments to the individual and positive values are payments by the individual.

We then again select brand 1 arbitrarily and simply ask each consumer in the sample:

2) What is the maximum price that you would pay for a unit of brand 1?

By adding the indifference premia to the brand 1 valuation of the consumer we may then obtain the preference valuation, v_i, that the consumer gives each remaining brand. As a final question we ask the consumer:

3) Given your budget resources and the intensity of your desire for the product relative to other goods what is the maximum amount you are prepared to pay for any brand in this product group?

We symbolize this price ceiling p_c.

The valuations v_i represent prices for the brands at which the consumer is indifferent among them. In making his or her choices, however, consumers will 1) eliminate all brands whose prices, p_i, are above p_c, 2) of the remaining subset eliminate those brands whose prices are above valuations, and 3), from the remaining brands choose that brand (not necessarily unique) whose $(v_i - p_i)$ is a maximum. Formally, the consumer acts to

$$Max_i(v_i - p_i)$$
Subject to:
$$(p_i - p_c) \le 0$$
$$(p_i - v_i) \le 0 \tag{1}$$

Figure 1 illustrates the consumer's choice with an example featuring 5 brands. Their consumer's valuations, v_p and exogenously given prices, p_i are graphed. Maximum price for any member of the group, p_c, is drawn as a horizontal line. In the solution procedure, any brand whose price is above p_c is eliminated as a possibility, which removes brands 1 and 5 on Figure 1. From the remaining brand candidates, those whose valuations are below their prices are eliminated, which reduces the feasible set to brands 2 and 4. The consumer then acts to maximize his or her consumer surplus, $v_i - p_i$ which leads to the choice of brand 2 by this consumer.

For those brands whose $p_i \le p_c$, were $v_i = p_i$ for all such brands the consumer would be indifferent among the goods, and indifferent to the purchase of any one of them. In such instances the consumer would be paying exactly what his or her valuation of their utility is, within the limits of his or her budget restraint, and consumer surplus would be zero with brands' utility exactly equal to their alternative money utility.

Over a large body of consumers with constant tastes and budgets demand functions for the firms are then derivable, which we will assume are continuous:

$$x_i = x_i(p_i, p_2, ..., p_n) = x_i(P) \ , i = 1, 2, ..., n \ . \tag{2}$$

Cost functions are assumed to be

$$C_i = F_i + c_i(x_i)x_i \ , i = 1, 2, ..., n \ , \tag{3}$$

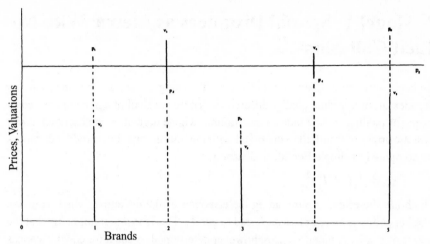

Figure 1. The Consumer's Choice Among Brands

where $c_i' > 0$, $c_i'' > 0$. Firms are price setters with profit functions

$$\Pi_i = p_i x_i(P) - C_i \ , i = 1, 2, ..., n \ .$$ (4)

To incorporate possible tacit collusion among the firms, a rivalrous consonance co-efficient matrix θ is employed, and each firm acts to maximize extended profits:

$$\Pi_i^\theta = p_i x_i(P) - C_i + \Sigma_j \theta_{ij} \Pi_j \ , i = 1, 2, ..., n, \ j \neq i \ .$$ (5)

First order conditions at a Nash equilibrium for optimal prices, P^o, may be written

$$p_i^o = \frac{x_i - (c_i' x_i + c_i) x_{i,i}' + \Sigma_j \theta_{ij} x_{j,i}' (p_j^o - c_j' x_j - c_j)}{-x_{i,i}'}$$ (6)

In (6) $c_i' = \dfrac{\delta c_i}{\delta x_i}, \ x_{i,i}' = \dfrac{\delta x_i}{\delta p_i}, \ x_{j,i}' = \dfrac{\delta x_j}{\delta p_i}, \ and \ c_j' = \dfrac{\delta c_j}{\delta x_j}.$

If we permit noninteger x_i and assume concave extended profit functions these equations can be solved efficiently by closed and iterative analysis.

Note that we may use the indifference premia to place the brands in a product space and compute the distances following some metric in that space. From the individuals' indicated indifference premia we choose some measure of central tendency as an aggregate measure of a good's distance from the origin of the product space at brand 1's location. We may choose the smallest of these values of the indifference premia as the origin of the space at 0 by subtracting its value from all original *I*-values and then normalizing them by dividing these transformed values by the range. Then we may calculate the relative distances between pairs of brand indifference premia as the differences between the normalized locations of the brands or we may opt for a different metric based on them. In the model developed in section 2 below, however, we shall use the nontransformed, nonnormalized (i.e., the original) indifference premia.

2. Model 1: Spatial Distances as Metric With No Tacit Collusion

We begin with the simplest oligopolistic model, using distances in product space as an index of interproduct quality differences. These we shall imagine are derived as means or medians of the indifference premia whose derivation is described above with the origin of 0 shifted to the brand with the minimum I_i. Interproduct distances are computed as simple Euclidean distance,

$$s_{ij} = \sqrt{(I_i - I_j)^2},$$

which has the effect of eliminating the direction of the deviation, which is to say whether good j is more or less preferred to good i. This is to assert implicitly that the degree of quality competitiveness, however determined, is equal for equal distances left or right of the origin good at 0.

Because of the costs of search activity or informational frictions, or to incorporate the variance in the indifference premia eliminated in our choice of a measure of central tendency, we assume that price competition from rival brands diminishes in a greater degree than linear in the s_{ij} by using an exponential decay factor, k. These factors can be customized by brand pairs, k_{ij}, when appropriate. Their most frequent usage is in treating advertising or other marketing expenditures aimed toward increasing the willingness consumers will have to buy brands at higher prices. Instead of reverting to derivation of new interproduct distances, it is possible to change k_{ij} or k to estimate virtual or effective changes in interbrand differences so effected. Their effectiveness is assumed to decline exponentially as brands farther away are impacted by a given firm's campaign. This is a convenient way to treat advertising because it is not a true product quality — it is external to the product — in the sense, for example, that it does not factor directly into consumer surplus as an increase in durability would do.

Demand functions therefore become:

$$x_i = a_i - b_{ii}p_i - \Sigma_{j\neq i}b_{ij}(p_i - p_j)e^{-k.s_{ij}}, \tag{7}$$

where k≥0 is a parameter, or, after defining

$$\alpha_i = b_{ii} + \Sigma_{j\neq i}b_{ij}e^{-k.s_{ij}}$$

$$\beta_i = \Sigma_{j\neq i}b_{ij}p_je^{-k.s_{ij}}.$$

we may rewrite (7) as:

$$x_i = a_i - \alpha_ip_i + \beta_i. \tag{8}$$

The b_{ii} and b_{ij} are the own- and other-price coefficients per dollar of price that would hold were Euclidean s_{ij} active in determining demand under different price regimes (i.e., if k=0). As noted above, the strength of such price effects, however, are assumed to decay more rapidly than linearly in the s_{ij} as informational frictions grow with distance and price benchmarks are less immediate to the consumer's experience. When $k=\infty$ the brands are isolated from all rival brands and each such brand becomes effectively a different product.

The form of the demand function (7) is an application of the hypothesis that consumers make their choices with the aid of benchmarks, in this case, price differentials among brands. This demand function form is a form of "benchmarking" in which consumers compare prices or quality against standards that they absorb from experience with the product or other information sources. In this paper, it is considered a more realistic approach to consumer decision making in differentiated oligopoly. Given fixed brand locations in quality space, it is assumed that consumers over time, through habit, experience or inertia, adopt one brand as a primary choice but will compare other brands' price differentials with the preferred brands in considering possible switching. In this instance of choice under brand distinctions the benchmark is this preferred brand and the variables of comparison are price differentials. Also, the notion that consumers buy a single "unit" of the product in a time period is not a great limitation on the analysis, because that unit can be defined as multiple physical units of the product where appropriate. For example, it might be five quarts of milk or three loaves of bread if the time period is one week. The assumption is made, of course, that the consumer buys only the one brand during the time period rather than an assortment of brands, but this also seems the only behavior consistent with the logic of the model.

The cost function may be written:

$$C_i = F_i + m_i x_i + n_i x_i^2 \tag{9}$$

Then, we may rewrite (6) as:

$$(6') \quad p_i^o = \frac{(a_i + \beta_i)(1 + 2\alpha_i n_i) + m_i \alpha_i}{2\alpha_i(1 + \alpha_i n_i)} + \frac{\sum_{i \neq j} \theta_{ij} b_{ji} e^{-ks_{ji}} (p_j - m_j - 2n_j x_j)}{2\alpha_i(1 + \alpha_i n_i)}$$

in which form we will solve it iteratively.

When all $\theta_{ij}=0$, $\Pi_i=p_i x_i - C_i$, and setting $\dfrac{\delta \Pi_i}{\delta p_i} = 0$ yields

1. $p_i^o = \dfrac{(a_i + \beta_i)(1 + 2n_i \alpha_i) + m_i \alpha_i}{2\alpha_i(1 + n_i \alpha_i)}$

2. $x_i^o = a_i - \alpha_i p_i^o + \beta_i$. $\tag{10}$

Of course, (10.1) is simply the condition that marginal revenue $(2p_i -(a_i+\beta_i)/\alpha_i)$ equal marginal cost $(m_i+n_i(\alpha_i+ \beta_i)-2n_i \alpha_i p_i)$.

The model has more severe restrictions in its construction that enforce more limitations on the interpretation of its comparative static propositions than exist in more conventional analyses. The first of these idiosyncrasies concerns the meaning of constancy of consumer preferences, which in the analysis of (1) above are reflected in the consumers' ceiling prices and the indifference premia. The demand functions, aggregated from (1) and its graphical depiction in Figure 1, are based upon such fixed indifference premia and their derivative scalings, interproduct distances. These aggregate functions will shift and change their slope families as product alterations, reflected in new indifference premia, are introduced. Where in conventional analyses movements along or among unchanging indifference surfaces conform to assumptions of constant preferences, the analogues of such surfaces in the model above are the indifference premia, and changes in them are changes in consumer preferences. Hence, for example, we cannot investigate the impacts of changes in the s_{ij} on α_i or β_i without assuming changes in preferences. Strictly speaking, therefore, by assuming small changes in interproduct distances we are violating the model's structure, and we must assume the changes are indeed "infinitesimally" small and be satisfied with qualitative movements wholly.

A second peculiarity in the model is that changes in the values of the b-coefficients in the demand functions must be narrowly interpreted as shifts of numbers of consumers among indifference premia that remain constant in value, which in turn implies that changes in such coefficients in one firm's function must be related to offsetting changes in other firms' functions. Under the framing assumptions of the model, with n consumers total sales over all brands must be no more than n in number. We shall again assume that changes in the b-coefficients will be very small indeed, in line with strictly construed comparative statics procedures that are frequently ignored in practice.

Consider, then, a change in s_{ij}, its consequent impacts on α_i and β_i, and movements in p_i and x_i. Movements of brand i are taken to be small enough not to change the cost function, which would be expected for finite movements in product space. When s_{ij} rises, with k given, so that competing brand j becomes less relevant in terms of effective distance for brand i and more or less competitive for remaining brands. Then

1. $\quad \alpha_i' \equiv \dfrac{\delta \alpha_i}{\delta s_{ij}} = -k \cdot b_{ij} \cdot e^{ks_{ij}}$

2. $\quad \beta_i' \equiv \dfrac{\delta \beta_i}{\delta s_{ij}} = -k \cdot b_{ij} \cdot p_j e^{ks_{ij}} = \alpha_i' p_j$

3. $\quad \dfrac{\delta p_i^o}{\delta s_{ij}} = \dfrac{2n_i \alpha_i'(a_i + \beta_i) + (1 + 2n_i \alpha_i)(\beta_i' - 2\alpha_i' p_i^o) + m_i \alpha_i'}{2\alpha_i(1 + n_i \alpha_i)}$

4. $\quad \dfrac{\delta x_i^o}{\delta s_{ij}} = -\alpha_i \dfrac{\delta p_i^o}{\delta s_{ij}} - \alpha_i' p_i^o + \beta_i' = -\alpha_i \dfrac{\delta p_i^o}{\delta s_{ij}} - \alpha_i'(p_i^o - p_j) \quad \cdot (11)$

Define:

$$\Psi = p_i^o - \frac{MC_i + (1 + 2n_i\alpha_i)p_j}{2(1 + n_i\alpha_i)}$$

where MC_i is the marginal cost of brand i. Then, from (10.1, 2) and (11.3), after some manipulation we find that:

1. $\Psi_i > 0 \rightarrow \dfrac{\delta p_i}{\delta s_{ij}} > 0$

2. $\Psi_i = 0 \rightarrow \dfrac{\delta p_i}{\delta s_{ij}} = 0$

3. $\Psi_i < 0 \rightarrow \dfrac{\delta p_i}{\delta s_{ij}} < 0$ $\hspace{2cm}$ (12)

In approximation, suppose $n_i\alpha_i \approx 0$. Then, the sign of Ψ_i is positive if $2p_i > p_j + MC_i$, so that p_i moves in the same direction as s_{ij}. We would expect, therefore, that p_i would move in a direction opposite to s_{ij} only when p_j is substantially greater than p_i.

If we assume that $\dfrac{\delta p_i}{\delta s_{ij}} > 0$, it follows from (11.4) that $\dfrac{\delta x_i}{\delta s_{ij}} < 0$ if $p_i \geq p_i$.

When $p_i < p_j$ the expression is more difficult to sign. For $\dfrac{\delta x_i}{\delta s_{ij}} < 0$ it is necessary and

sufficient that

$$-\frac{\alpha_i'}{\alpha_i}(p_i^o - p_j) < \frac{\delta p_i^o}{\delta s_{ij}} . \hspace{2cm} (13)$$

Parameter magnitudes now become important. The absolute value of the ratio on the LHS of (13) must be expected to be small in that it is the ratio of b_{ij} to the sum of all of the other-price b-coefficients, including itself. It should be expected to shrink toward zero as the number of firms rises. The difference between the prices would have to be very large in general for (13) to be violated, and therefore the expectation is that x_i will move in the opposite direction from s_{ij}.

Figure 2 depicts those expectations graphically and is based on a different manner of demonstrating their causation. Let $p_{max,1}$ be the price-axis intercept of the original demand function $D_{1,i}$. From (2) it is simply $(a_i + \beta_i)/\alpha_i$. When s_{ij} rises, differentiation of the x-axis intercept with respect to s_{ij} reveals that it will always move leftwards. On the other hand, differentiation of $p_{max,1}$ yields $-(\alpha_i'/\alpha_i)(p_{max,1} - p_j)$, which will be positive when p_j is less than p_{max}. Assuming the latter to be true, we find that $p_{max,2}$ will be above $p_{max,1}$, and connecting $p_{max,2}$ with the new, smaller x-axis intercept

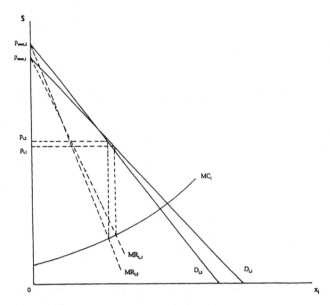

Figure 2. Price Displacement with a Rise in s_{ij}

yields a new demand function $D_{2,i}$ with steeper slope. Then $p_{i,2}$ will be above $p_{i,1}$ when profits are maximized. However, as p_j approaches $p_{max,1}$ so will $p_{max,2}$, placing $D_{2,i}$ below $D_{i,i}$ with a very steep slope, and causing p_j to fall with the rise in s_{ij}. Hence, when brand i is enjoying the sales benefits of a large $(p_j - p_i)$, a rise in s_{ij} reduces that benefit, shifts the demand curve inward, and leads brand i to compete by lowering price.

But, of course, these are only the ceteris paribus price and output movements, holding all other prices and interproduct distances constant. However, interproduct distances cannot be kept constant since, if s_{ij} is a distance for an interior brand, some brands will find brand j closer and others farther than previously, and if it is at the highest quality end-point on the brand spectrum all of its rivals will find it farther. Also, the brand which has moved will find all of its own interproduct distances have changed. Prices and outputs will rise and fall as adjustments occur throughout the industry. Because we expect prices to move in the same direction as s_{ij}, it is to be expected that interdependent effects will reinforce the impact movements. Such comparative statics become too complex for analytical solution with interesting industry sizes, and therefore we rely upon simulative theorizing to obtain insights through an example.

a. A Base Case

We assume a five-brand industry with the parameters listed in Table 1. The profit functions for each of the firms taken separately are strictly concave, and hence first-order conditions for a maximum are both necessary and sufficient for a maximum. (Initial interproduct distances are listed in Table 1 without parentheses.) By solving each successively and changing the prices of brands sequentially in a diagonalization algorithm one searches for a convergence which may indeed occur only with negative profits (and elimination) for some firms. With the base case parameters, however, all five firms remain viable with solution values listed in Table 2.

A useful index of overall price-quality competitiveness, combining the influence of demand coefficients, prices and interbrand quality distances is the measure

$$E_i = \frac{\alpha_i p_i}{\beta_i + a_i} \, .$$

The numerator of E_i rises as sales-damaging forces associated with high p_i (and through it, high costs), small interbrand distances or large demand coefficients; the denominator rises with favorable demand forces acting through high rivals' prices, large other-price coefficients, large demand function intercepts and expansive interbrand distances. Hence, low values of E_i denote overall competitiveness. They are listed for the brands in the base case in Table 2. Clearly, this set of brands falls into two subsets: the highly competitive brands 1, 2, and 3, and the disadvantaged brands 4 and 5. Profits correlate quite well with the E_i, although brand 2 is an exception.

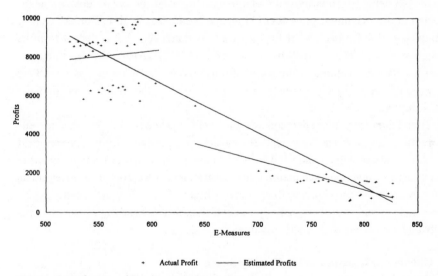

Figure 3. Relation of Profits to E-Measures

A broader investigation of this correlation is presented in Figure 3, and it is not so simply interpreted. A simple linear regression of profits on E_i for 75 observations (21 solutions with 5 brands each, with several outliers omitted) with OLS yields an adjusted r^2 of .834. This is, however, somewhat misleading. The diagram illustrates clearly the persistence of the two clusters of E_i values throughout the 21 solutions (the remaining ones of which will be given below). The steeply negative slope of the estimating equation, graphed in Figure 3 is the result of the position of the two clusters identified above. Separate estimating equations have been computed on the diagram for the clusters taken separately, and although a negative relation between profits and E-measures is found in the lower cluster, there is a slightly positive slope in the estimating equation for the upper cluster. Indeed, this equation separates two quite distinct subgroups within the upper cluster, both of which have positive profiles. The lower subset consists of the observations for brand 2 and the upper of those for brands 1 and 3. The use of the E_i measures to index competitiveness, therefore, should, in general be used cautiously. Finally, the relation between the measure and price elasticity of demand will be derived in equation (17) below.

Brands 1 and 3 bound the market interval and are farthest in product space from competitors with average distances of 5.75 and 4.25 respectively, and brand 1 is favored with the lowest variable costs and most favorable "own-price elasticity" coefficient in the industry. Its other-price elasticity coefficients are on the whole the least favorable, however. Its high price, low cost and high profit reflect these advantages, although its quantity is restrained by its other-price demand coefficient drag and its relatively low a_1. Its α_1 and β_1, the lowest in the product group, reflect both its distance advantage and other-price disadvantages.

Brand 3 benefits primarily from a large basic demand for its product (a_3), as reflected in the maximum price it could charge before losing all customers were it completely isolated from other firms' prices. As noted above, it is distant from its competitors, and its cost structure is relatively favorable. Also, other-firm demand coefficients are favorable in the cases of those rivals which are closest to it. One would not have predicted ex ante facto that it would enjoy the maximum profit in the product group.

Brand 2 enjoys a good demand and an excellent cost structure, but its competitive position might seem to be penalized by an average distance from competitors of 3 — the smallest of the brands. It does well, being the median profit earner, by virtue of its cost advantages, which permit it to benefit from its favorable price differentials with rivals and its proximity to them. Finally, brands 4 and 5 are the roundly penalized of the group, with average interproduct distances of 3.5 and high cost structures outweighing favorable demand structures.

b. Changes in Location in the Quality Space
In the real world we may distinguish five types of movements in characteristics space that interest us:

1. Limited or *marginal* moves by incumbents, in which a brand moves a small step closer to an envied rival or farther from a highly competitive brand. In general, this is the form of locational competitive moves we expect to encounter in the short or intermediate run. Firms are limited by patent restraints, technological knowledge limitations, equipment capabilities, marketing reorientation frictions, concerns about consumer reactions to changes in brand images, and so forth, in their consideration of feasible

or desirable moves in quality space. Risk-averseness will generally characterize the approach to proposed changes in the characteristics structure of a flagship brand.

2. *Quantum* changes in the location of major brands to meet major competitive challenges from closely competitive rivals or to escape them. Therefore, such moves may involve changes to positions in close proximity to such rivals or to long distances in either direction from them. Such major changes in brand images are rare, for reasons mentioned in 1. above, and in most cases are desperation moves occasioned by threats to survival.

3. *Incumbent multiplicatives* whose producers increase the number of products they sell in order to enhance sales and profits by dispersing new brands in the space. Ready-to-eat breakfast food producers are adept at this, selling new variants under their existing brand names or producing house brands for retailers, but this category also embraces truly innovative products originating within the industry.

4. *Entrant multiplicatives* which are product placements on the quality spectrum by firms previously outside the industry, including competitive variants of existing brands and innovating products from outside the industry.

5. Brand *eliminations* through competitive failures that widen gaps in the placement profile.

Table 1. Parameters for Model 1 Base Case

Parameter/Firm$_j$	1	2	3	4	5
a_{1j}	800	***	***	***	***
a_{2j}	***	700	***	***	***
a_{3j}	***	***	1100	***	***
a_{4j}	***	***	***	1000	***
a_{5j}	***	***	***	***	950
b_{1j}	15	3	5	8	3
b_{2j}	3	16	6	5	7
b_{3j}	4	7	22	6	9
b_{4j}	8	9	7	25	8
b_{5j}	7	9	11	10	26
s_{1j}	0	5 (6)	8	3	7
s_{2j}	5 (6)	0	3 (2)	2 (3)	2 (1)
s_{3j}	8 (8)	3 (2)	0	5 (5)	1 (1)
s_{4j}	3 (3)	2 (3)	5 (5)	0	4 (4)
s_{5j}	7 (7)	2 (1)	1 (1)	4 (4)	0
F_j	1000	800	1200	1500	1020
m_j	2	3	5	16	15.5
n_j	0.0005	0.00006	0.00007	0.0001	0.00012
$k_{ij} \equiv k$	0.29	0.29	0.29	0.29	0.29

Most characteristics competition in differentiated oligopoly occurs under the first category or as results of this type of competitive process. Quantum leaps pay great costs in abandoning or restructuring in a serious way sunk costs in advertising and marketing infrastructure, brand reputation and consumer recognition, and, perhaps, production efficiency accumulated over many years of learning. Incumbent multiplicatives dilute brand distinctiveness to the extent they are included as variants of the flagship brand(s), are simply ventures in copycat products, or, importantly, are essentially analytically unpredictable innovations. Entrant multiplicatives establish new brands, and may more frequently than incumbent multiplicatives be innovatory, but share the nonpredictability of point of entry into the characteristics spectrum. Finally, brand eliminations occur largely as the competitive outcomes of marginal moves causing negative profits for incumbent firms. Therefore, we confine our analysis to that of such marginal movements as we move into a rivalrous consonance environment.

Table 2. Solution Values for the Base Case of Model 1

Variables	Firms				
	1	2	3	4	5
p_i	$24.24	$19.56	$22.84	$25.02	$23.22
x_i	435	428	594	336	331
C_i	$1,963.95	$2,095.58	$4,195.66	$6,890.21	$6,165.95
Π_i	$8,571.93	$6,280.27	$9,374.08	$1,521.86	$1,524.28
α_i	19.94	25.94	33.47	37.54	43.32
β_i	$118.00	$235.53	$258.50	$275.54	$387.26
E_i	.526	.542	.563	.736	.752

As a parameter change, and in conformance to the observation above that only small changes are consistent with the logic of the model, we move brand 2 one unit to the right of brand 1 on the product line, so that s_{12} now equals 6 instead of 5. This moves brand 2 with relation to other brands, and where such interproduct distances change they are listed in bold face in parentheses in Table 1. Where unchanged, the distances are merely repeated in parentheses in the table.

The model was solved, in ceteris paribus fashion brand by brand, holding all base case prices constant but that of the firm under analysis. In Table 3, the price and output increments are the direct impacts before computing the indirect effects from changes in all other prices, and are listed in column 2 of the table. Projections of these changes are made from equations (11.3, .4) and are listed in column 3. Finally, the mutatis mutandis equilibrium was obtained by solving equations (10) sequentially, with changes from the base case values of Table 2 found in the last column. These total changes minus the direct impacts yield the price reaction impacts of the mutatis mutandis equilibrium in column 4 that temper the direct impacts of the ceteris paribus equilibrium.

The direct impacts of brand movements are in line with expectations. Brand 2 moves farther away in product space from brands 1 and 4 and closer to 3 and 5. The prices of the first two rise as their competition lessens and they exploit the shift outward of their demand functions by restricting output. The immediate impact on firms 3 and 5 is to lower their prices to meet the enhanced competition from brand 2 and to raise their outputs. As would be expected brand 2's direct price and output impacts were the largest of all the brands and reflected the necessity to compete with price reductions by brands 3 and 5. From the b_{2j} row of Table 1 it can be seen that brand 2 moved away from brands 1 and 4 with demand function coefficients of 3 and 5 to be closer to brands 3 and 5 with coefficients of 6 and 7, enhancing its ability to increase sales by decreasing its price differentials. Hence, its price declines substantially and output increases even before brands 3 and 5 react to the new situation, by lowering prices, which leads brand 2 to decrease its price further. Brand 2's E-coefficient remains the same and profits increase only slightly (see Table 4 for profit movements).

Note that the projected price and output impacts predicted the directions of ceteris paribus changes in every instance and did reasonably well pricewise and also quantitatively for brands other than brand 2, which was derived by solving equations (11.3 and .4) for each of the affected brands and summing the projected changes in Brand 2's price and output. Output projections were even more accurate percentagewise and, as in the case of the price estimates, performed relatively well as linear projections of exponentially determined variables. Also, net total changes in industry output are only .84 percent over base case output, approximating unchanged sales as formally required by the assumptions of the individual consumer model that underlies Model 1.

Table 3. Ceteris Paribus, Projected and Mutatis Mutandis Price and Output Changes

Variable	Direct Impacts	Projected Impacts	Price Reaction	Total Changes Impacts
P_1	+.48	+.10	-.36	+.12
P_2	-.33	-.89	-.04	-.37
P_3	-.30	-.36	-.04	-.34
P_4	+.27	+.21	-.05	+.22
P_5	-0.21	-.25	-.05	-.26
x_1	-5.96	-1.29	+.4.60	-1.36
x_2	+12.15	+8.53	-1.12	+11.03
x_3	+7.22	+8.21	-1.52	+5.70
x_4	-1.40	-1.90	-1.91	-3.31
x_5	+3.58	+3.88	-2.47	+1.11

We may summarize the highlights of the indirect effects, as they yield no surprising results that offer enlightening insights. These price-induced modifications to the direct effects were relatively unimportant compared with the direct impacts except for brand 5's output effects. Brand 5's high own-price sensitivity punished its sales severely as differentials between its prices and those of lower cost rivals widened. Further it is strongly sensitive to brands 2 and 3 prices, which fell both initially and after price adjustments, and this was enhanced by brand 2's enhanced proximity. In somewhat the same position as brand 5, brand 3's price fell after total adjustments, although it did benefit somewhat from brand 4's price increase, despite the distance between the two and the low b_{34} demand interaction term. Brand 4 lowered its price slightly on indirect impact account but raised its price substantially on net, benefitting from its most important rival, brand 2, moving away from it as well as its separation from the price cutting of brands 3 and 5. Also, it benefitted somewhat from brand 1's price increase. Brand 1's direct impacts permitted it to raise its price, with price cuts of rivals 2, 3 and 5 having no impact on its new solution. Its nearest and strongest competition, brand 4, was a price raiser, and its favorable cost structure and low own-price demand coefficient were other factors contributing to its rather insulated position. Finally, brand 2 suffered a bit by its movement toward its most important competitors —brands 3 and 5, both severe price cutters — and away from its least important —brands 1 and 4 — who were price raisers.

Table 4. Consumers' Surplus, Profits and Social Surplus in Model 1 Equilibria

| | Original Equilibrium | | | New Equilibrium | | |
Brands	Consumers' Surplus	Profits	Social Surplus	Consumers' Surplus	Profits	Social Surplus
1	$4,738.72	$8,571.93	$13,310.65	$3,651.54	$8,596.29	$12,247.83
2	$3.534.63	$6,280.27	$9,814.90	$3,543.87	$6,299.31	$9,843.18
3	$5,274.69	$9,374.08	$14,648.77	$5,222.35	$9,269.88	$14,492.23
4	$1,505.28	$1,521.86	$3,027.14	$1,527.44	$1,565.97	$3,093.41
5	$1,265.56	$1,524.28	$2,789.84	$1,567.35	$1,445.48	$3,012.83
Totals	$16,318.88	$27,272.42	$43,591.30	$15,512.55	$27,176.93	$42,689.48

Consumer surplus is simply calculated as $CS_i = .5x_i[((a_i + \beta_i)/\alpha_i) - p_i]$. Social surplus in the old and new mutatis mutandis equilibrium is displayed in Table 4. From the equation for consumers' surplus, it is trivial that it will rise with increases in other-firm prices (via β_i), rise with increases in x_i, and , therefore, fall with rises in p_i. Consider, however, a ceteris paribus rise in s_{ij}

$$\frac{\delta CS_i}{\delta s_{ij}} = -(\frac{CS_i \alpha_i}{x_i} + .5x_i)\frac{\delta p_i}{\delta s_{ij}} + \frac{\alpha_i'}{\alpha_i}(.5p_j x_i - CS_i) \ . \quad (14)$$

If $\delta p_i / \delta s_{ij} > 0$ then the only positive term on the RHS of (14) is $\dfrac{\alpha_i'}{\alpha_i} \cdot CS_i$, which,

given the expected low value of the ratio as developed above, seems too small to outweigh the sum of the remaining negative terms. Therefore, ceteris paribus, we expect consumers' surplus to move in a direction opposite that of distance from the firm initiating movement.

The expectations for those firms whose locations remain fixed that prices will change in the same direction as distances from firm 2, outputs will move in the opposite direction, and that consumers' surplus will move in the opposite direction are borne out in Tables 3 and 4. Note that in Table 2 the base case solution values for the prices of firms that were stationary are very close to p_2, the price of the firm whose site is changed. We should expect, therefore, that no counterintuitive movements will occur in prices, outputs or consumer surplus in the ceteris paribus solutions.

Computation of the qualitative movements of CS_i between the original and new equilibria is quite simple. From the definition of CS_i:

$$CS_i = .5x_i [\frac{a_i \beta_i}{\alpha_i} \; p_i] \; .5x_i y_i \; , \tag{15}$$

where y_i is simply the difference on the price axis of the intercept of the inverse demand function (p_{max}) and the price p_i. Letting the initial and new equilibria be denoted by $k=1,2$ respectively, we find that

$$\Delta CS_i = CS_{i2} - CS_{i1} \; \begin{matrix} > \\ = \\ < \end{matrix} \; 0 \leftrightarrow \frac{x_1}{x_2} \; \begin{matrix} \leq \\ > \end{matrix} \; \frac{y_2}{y_1} \; . \tag{16}$$

As indicated in Table 2 and the accompanying discussion, changes in the location of a brand in characteristics space will involve for each brand a shift in the demand function which registers as a new price-axis intercept, p_{max} ; a new slope of the demand function arising because of the change in the value of the decay factor involving the distance of the given brand and the moving brand; a change in the price of the stationary brand; and the implied change in the quantity of that brand. We may now broaden our ceteris paribus analysis above to include the price-output adjustments of all brands on the basis of the insights from the mutatis mutandis example. The first three of these factors determine the changes in the area of the consumer surplus triangle, and therefore, the sign of DCS_i.

It is interesting in this regard to note two opposing forces that a stationary firm experiences as another brand moves away from it. We have seen that, ceteris paribus, the slope of the inverse demand function, $1/\alpha_i$, rises because of the fall in the exponential distance term involving the moving firm, which tends toward a fall in own-price elasticity and a price increase in the face of less intense competition. The rise in the same term in the β_i term will increase the inverse demand curve intercept term enhancing its move to more own-price inelasticity and yielding an even greater likelihood of a price increase. But now we may free all other brand prices. The β_i term will also be affected by the price change of the moving firm, which has a higher likelihood of being positive. But to complicate matters, all other brands will react pricewise to the displacement of the moving firm and these prices will also register in β_i, so that the final resultant reflected in the intercept will be ambiguous. What is interesting, however, is that the initiative of the moving firm sets off advantageous and disadvantageous forces, so that what might appear to be an increase in competition force can actually lead to a fall in competitive position and profits and/or social surplus. From Table 4, in the present example, the movement of brand 2 leads to a fall in total consumer surplus and a fall in profits, so that social surplus also falls. Also, in all of the cases in which brand 2 approached rivals profits fell, and in all cases in which it moved away from rivals profits rose, which we would label the normal expectation. From the welfare standpoints of consumers' and social surplus, however, brand 2's move is not desirable.

This perception of brand competition as one of narrowly bounded movement within a bounded space with fixed total demand in the short- and intermediate-run yields a model with continuous movement and no stable equilibrium, unless one introduces relocation costs and/or minimum profit improvement constraints. Indeed, realistically, such competition among incumbents, without tacit collusion, should be the expectation: continuing marginal adjustments as firms seek profit improvements of worthwhile magnitude. The resort to advertising and other marketing costs as competition alternatives by increasing brand demand with fixed locations is readily understandable in the light of these observations. Further pursuit of Hotelling-model results with costless relocation in these conditions, absent specific industry conditions, is therefore considered realistically unfruitful, and the illustration of means of gaining some insights into the implications of firm movements using brand 2 as an arbitrary mover is deemed sufficient at this stage of the analysis. A wide choice of restrictions on movement necessary to make the analysis realistic is available to the analyst and will change with the industry under study.

c. Some Polar Displacements of Distance Discount Factors.

The implications of interbrand distances as factors in competition can be seen more clearly by study of two limiting cases, where spatial discount factors k are zero on the one hand and infinite on the other. In the first case interbrand distances derived from indifference premia are undiluted by the spatial impediments of cognition and in the second case brands are completely isolated by their densities so that price differentials play no role in the consumers' choices. Changing these k-coefficients may also be viewed as a way of increasing or decreasing interbrand distances proportionately to s_{ij}.

Table 5 displays the solutions to these two extreme cases together with the base case of Table 2 for comparison. The profit outcomes are dependent on three factors:

1) the impacts on quantities sold, 2) the implied price at which those quantities are sold, and 3) the average costs of the output. The determinants of the quantities supplied are given by the components of E_i, as explained above. The parameter α_i sums the negative impacts on sales per unit price rise, imposed both directly by b_{ii} and indirectly by the heightening in the price differential $(p_i\text{-}p_j)$ as transmitted by $b_{ij}e^{-ks_{ij}}$. The term β_i yields the positive factors acting on sales in the form of a shift in the demand function intercept caused by narrowing of the price differential through rises in the prices of other products discounted by the decay factor. If we add a_i to β_i we obtain the actual horizontal axis intercept fixing the location of the new demand function. Given the changes in p_i and the p_j, the resultant net change in output will be determined by the strength of the changes in shift and slope.

The price reactions of firms are dictated by changes in the own-price elasticities of the demand function as dictated by the changes in shift and slope parameters brought about by the change in k, where the elasticities of the new demand functions in the region of the new equilibrium are determined by the relations of (17):

$$\varepsilon_i = \frac{\delta x_i p_i}{\delta p_i x_i} = -\frac{\alpha_i p_i}{a_i + \beta_i - \alpha_i p_i}$$

$$= -\frac{E_i}{1 - \dfrac{\alpha_i p_i}{a_i + \beta_i}} = -\frac{E_i}{1 - E_i} . \tag{17}$$

Since the rises in k are effectively increases in the interbrand distances among all brands, on the basis of our previous reasoning we expect all prices to rise and outputs to fall. From Table 5 this is indeed the universal rule. All the values of α_i fall monotonically as the $b_{ij}p_i$ terms are attenuated, and the β_i fall towards 0 as the influence of the p_j diminishes. The competitiveness of all 5 brands as indexed by the E_i improve as k rises, or, minimally, remains the same, with brands 4 and 5 —the two high cost producers —benefitting most from the increasing protection from rivals' price differentials. All of these movements are those we would have predicted ex ante facto: there are no surprises.

Moreover, we would have predicted that profits would rise monotonically for all brands as k rose and the pricing power of the firms was enhanced, and for brands 1, 3, 4 and 5 our expectations would have been fulfilled. But for brand 2 we find that although profits did rise as k rose from 0 to .29, they fell as k rose further to approach ∞. This is a most interesting instance of a brand actually being injured by gaining greater independence from the pricing decisions of its competitors. Brand 2 is a lower-quality, low-cost product which prospers by offering an attractive price differential to consumers of the other brands. It follows that it must be "in view" by consumers to reap that advantage. A large part of its basic demand is generated by β_2. Hence, when spatial discount factors become too large, brand 2 is thrown back upon its own weak basic demand (a_2) to the detriment of profits. A good example of the phenomenon operating in the opposite direction is in the ready-to-eat breakfast food industry where house brands, which are low-cost, unadvertised, and of unbranded but high quality pedigree, flourish as consumer education decreases the decay factor and the cognitive distance of alternatives increases the significance of price differentials. Were marketing strategies by branded products to convince consumers that their quality has been improved and thereby further distanced themselves from house brands, the brand 2 case would be revisited.

Table 5. Solution Values for Polar Values of Spatial Discount Factors

			Firms		
Variables	1	2	3	4	5
			Spatial Discount Factor = 0		
p_i	$18.73	$16.80	$19.36	$21.11	$21.02
x_i	550	508	685	344	342
C_i	$2,251.84	$2,340.72	$4,657.25	$7,055.60	$6,341.21
Π_i	$8,055.88	$6,201.26	$8,604.99	$ 593.02	$854.94
α_i	34.00	37.00	48.00	57.00	63.00
β_i	387.17	430.06	514.37	604.77	716.47
E_i	.54	.55	.58	.79	.79
			Base Case Spatial Discount Factor = .29		
p_i	$24.24	$19.56	$22.84	$25.02	$23.22
x_i	435	428	594	336	331
C_i	$1,963.95	$2,095.58	$4,195.66	$6,890.21	$6,165.95
Π_i	$8,571.93	$6280.27	$9,374.08	$1,521.86	$1,524.28
α_i	19.94	25.94	33.47	37.54	43.33
β_i	118.00	235.53	258.50	275.54	387.26
E_i	.526	.542	.563	.736	.752
			Spatial Discount Factor = ∞		
p_i	$27.86	$23.39	$27.53	$28.03	$26.05
x_i	382	326	494	299	273
C_i	$1,837.28	$1,783.43	$3,688.31	$6,296.99	$5,254.99
Π_i	$8,808.11	$5,835.88	$9,920.37	$2,091.02	$1,848.06
α_i	15.00	16.00	22.00	25.00	26.00
β_i	0	0	0	0	0
E_i	.52	.53	.55	.70	.71

3. Model 2: Spatial Distances as Metric With the Rivalrous Consonance Mode of Tacit Collusion

Suppose now that tacit collusion pervades the relations among the five brands of section 2, modelled as rivalrous consonance relations. From (6') and (8) the expressions for optimal prices can be solved iteratively for desired values of θ_{ij}. The problem inheres in the choice of these consonant coefficients from their potential infinity of values to obtain useful insights into the nature of competition under varying degrees of tacit collusion while demonstrating the methodology. The base case of Table 1 and its solution in Table 5 is, of course, Model 1 with $\theta_{ii}=1$, $\theta_{ij}=0$. As first cases, therefore, the polar case is presented with perfect collusion when $\theta_{ii}=1, \theta_{ij}=1$, and subsequently, a midcase with $\theta_{ii}=1$, $\theta_{ij}=.5$, with the base case $k=.29$. The solutions are reproduced in Table 6 with the base case for comparison.

Prices of all firms rise and outputs fall consistently as rivalrous consonance increases tacit collusion, which is expected. The price rises reflect in large part the inclusion of marginal consonance costs which are found in the last row of each solution set. All of the firms' competitive positions (as measured by rises in E_i, which do not include marginal consonance costs) decline as cooperation rises. In the more competitive cluster of firms, the relative rise in this measure for brand 2 (11.3%) is notable and reflects its high marginal consonance cost. Taking into account the small profit rises or losses of brands 4 and 5 reduces its competitive position. Nonetheless profits do increase consistently because of the large rise in price differentials it experiences with these two high-priced brands, and indeed, it enjoys the highest percentage profit increase from zero to perfect consonance of any of the five firms.

Interestingly, the two firms that benefit least or suffer most from tacit collusion are brands 4 and 5, which are quite different in quality. From zero to perfect collusion brand 4's profits rise only 1.3 percent and brand 5's actually fall by 6.6%. Both firms do benefit modestly from midlevel collusion, but again not to the degree that the three lower quality brands do. One suspects that both of these firms would benefit maximally from tacit collusion at levels somewhat below .5.

Brand 5 is a high quality, high cost product with weak α_i own-price demand factor and whose β_i other-price demand coefficient rises slowly as competitors' prices rise sluggishly. However, the major culprit causing such profit outcomes in the case of brand 5 is the rise in consonance costs. As consonance coefficients rise the demand functions shift outward with the rise in the β_i coefficients, but (unlike the situation analyzed in Figure 2) the slopes of the functions remain the same because the α_i coefficients do not change. Hence, were the marginal cost functions unchanged, outputs, prices, revenues and profits would rise. But marginal cost functions now shift upward by large amounts in the case of brand 5, and outputs and revenues fall, with small rises or declines in profits. The marginal consonance costs are high, of

Table 6. Solution Values for Polar and Median Values of Consonance Coefficients q_{ij}

Variables			Firms		
	1	2	3	4	5
			Zero Consonance Case $\theta_{ij}=0$		
p_i	$24.24	$19.56	$22.84	$25.02	$23.22
x_i	435	428	594	336	331
C_i	$1,963.95	$2,095.58	$4,195.66	$6,890.21	$6,165.95
Π_i	$8,571.93	$6,280.27	$9,374.08	$1,521.86	$1,524.28
α_i	19.94	25.94	33.47	37.54	43.33
β_i	118.00	235.53	258.50	275.54	387.26
E_i	.526	.542	.563	.736	.752
Mar.Con.Cost	$0	$0	$0	$0	$0
			Median Case $\theta_{ij}=.5$		
p_i	$25.11	$20.99	$24.05	$26.63	$25.11
x_i	425.04	406.63	572.97	292.81	272.43
C_i	$1,940.42	$2,029.80	$4,087.81	$6,193.57	$5,251.58
Π_i	$8,731.90	$6,506.34	$9,694.66	$1,603.18	$1,589.33
α_i	19.94	25.94	33.47	37.54	43.32
β_i	125.73	251.09	278.01	292.41	410.33
E_i	.541	.572	.584	.773	.800
Mar.Con.Cost	$.68	$1.13	$.92	$1.38	$1.62
			Perfect Consonance Case With $\theta_{ij}=1$		
p_i	$26.02	$22.49	$25.33	$28.29	$27.07
x_i	414.88	384.04	550.53	248.09	211.74
C_i	$1,915.81	$1,960.98	$3,973.84	$5,475.62	$4,307.36
Π_i	$8,879.44	$6,674.79	$9,971.06	$1,542.09	$1,423.83
α_i	19.94	25.94	33.47	37.54	43.32
β_i	133.74	267.26	298.26	309.99	434.40
E_i	.556	.603	.606	.811	.847
Mar.Con.Cost	$1.39	$2.31	$1.90	$2.80	$3.30

course, because the lower quality brands' profits are rising. It is in intense competition with brand 3, whose profits rise steeply when θ_{ij} rise above .5.

The marginal consonance cost factor is also important in the case of Brand 4, but it is also a high-cost, low-quality brand intensely competitive with firms whose profits are rising. Its high basic demand from a curiously loyal band of customers, saves it from more severe profit deprivation. Its p_{max} shifts upward by a small amount between zero and perfect consonance but its marginal consonance cost shifts the cost curve up from a high base to reduce profits from their mid-level consonance value.

Brand 3 is a highest quality producer with low marginal production costs and high basic demand, with mid-level α_i own-price impacts and a low β_i other-price factor reflecting its high degree of dependence on other brands' prices. Marginal consonance costs are the second lowest in the industry. Its competitive position weakens only slightly with enhanced tacit collusion, and it is the major winner from enhanced cooperation in the industry.

Finally, brand 1's profit rises steadily as tacit collusion rises, to enjoy a modest 3.3 percent rise from zero to perfect tacit collusion. It is a low quality, low cost producer which enjoys a good deal of autonomy from its rivals. The marginal consonance cost, for example, is the smallest in the industry, its a_1 reveals the small impact of rivals' demand coefficients on its sales, and its β_1 value shows that it gains little from rises in rivals' prices.

For $k=.29$ the industry's structure reveals a dichotomy. One group of three firms is highly competitive, with monotonically rising profits as tacit collusion progresses. Brand 1 reveals a great deal of independence from rivals, brand 2 is highly dependent on its higher-pricing rivals for demand support, and brand 3 is intermediate in its dependence on rival price movements and in its marginal consonance costs. The second group, consisting of brands 4 and 5, is high-cost with large α_i values imposing punishing negative influences on demand, and only moderate offsets in β_i values. Tacit collusion benefits these firms only moderately in midranges and punishes them above those ranges. The steady rise in profits enjoyed by brands 1, 2 and 3 inflict rapidly increasing marginal consonance costs on brands 4 and 5 and reduce profits below the median collusion case and, for brand 5, below the zero collusion case as well. These cost rises forces their prices up, increase the differentials between them and group 1 prices, raising the latter group's profits, and thereby shift marginal consonance costs even higher.

As noted above, brand 4 is a lower quality product produced at high cost in very strong competition with all other firms, most notably, given the distances involved, with group 1 firms. High production costs, large adverse price differentials with these firms, and large profit gains of these firms from tacit collusion result in the poor profit experience related above. Nonetheless, it does hold on to viability rather tenaciously as collusion rises with $k=.29$. But firm 5 is something of an anomaly: it is a high quality product that somehow is in quite high competition with lower quality products as well as with brand 3, which is higher than it in the quality scale. It is, of course, that highest quality firm that inflicts most damage on it via the large profits it earns, its close distance, and the worsening price differential. But brand 5's large b_{5j}'s with such firms as 1, 2 and 4 are difficult to explain in terms of rational consumer preferences. One must wonder if it would remain viable under other rivalrous consonance regimes.

Table 7 permits us to probe more deeply into the structure and brand interrelationships in this industry. It records the solutions to combinations of four *cluster* categories:

1. the primary organizing cluster is that of *distance discounting* values, with $k = 0$, .29, and ∞ successively. These are familiar from Table 5, and can be understood as a means of stretching interbrand distances as k rises, from the pure s_{ij} values unaffected by cognitive distortion to an effective complete isolation of each firm from its rivals in product space.

2. for each distance discounting cluster brands are broken down into three additional clusters, the first of which are θ-*clusters*. These rivalrous consonance factors are set successively to 0, .5, and 1, for all firms, to obtain solutions for zero, median, and perfect tacit collusion levels. They permit study, within each k-cluster, of *generalized* rivalrous consonance, i.e., where the given θ_{ij} value is extended by all firms to all other firms.

3. for each distance discounting cluster and each θ-cluster firms are divided into E_i *clusters*, which groups the firms into the two competitive clusters discussed above, with group 1 consisting of brands 1, 2, and 3, and group 2 containing brands 4 and 5. In these solutions each group member extends a θ of 0, .5, or 1 to other group members, and of 0 to members of the other group.

4. a last cluster is that constructed by grouping brands by s_{ij} proximity, with firms 1, 2, and 4 in the first group and firms 3 and 5 in these second. For each distance discounting and θ cluster, solution to these s_{ij} *clusters* are obtained with tacit collusion ruling within the clusters only. These cluster conjunctions yield 21 independent solutions for which profit values have been listed.

Table 7. Summary of Firms' Profits Under Various Rivalrous Consonance Regimes

Solution	Π_1	Π_2	Π_3	Π_4	Π_5
		Zero Distance Discounting: $k=0$			
Zero Rivalrous Consonance:					
$\theta_{ii}=1$, $\theta_{ij}=0$, $k=0$	\$8,055.88	\$6,201.26	\$8,604.99	\$593.02	\$854.94
Median Riv. Cons.,					
$\theta_{ii}=1$, $\theta_{ij}=.5$, $k=0$	8,332.26	6,446.35	8,936.08	724.39	966.08
Perfect Riv. Con.,					
$\theta_{ii}=1$, $\theta_{ij}=1$, $k=0$	8,619.52	6,679.37	9,263.10	807.09	984.98
E_i Clusters:					
$\theta_{ii}=1$, $\theta_{ij}=.5$, $j,k=1,2,3; 4,5$, $k=0$	8,122.38	6,253.37	8,680.70	642.27	905.36
E_i Clusters:					
$\theta_{ii}=1$, $\theta_{ij}=1$, $j,k=1,2,3; 4,5$, $k=0$	8,188.89	6,298.68	8,751.88	692.17	956.21
s_{ij} Clusters:					
$\theta_{ii}=1$, $\theta_{ij}=.5$, $j,k=1,2,4; 3,5$, $k=0$	8,212.08	6,345.29	8,791.05	655.78	928.01
s_{ij} Clusters:					
$\theta_{ii}=1$, $\theta_{ij}=1$, $j,k=1,2,4; 3,5$, $k=0$	8,369.97	6,491.52	8,979.31	686.33	976.73

Table 7. Summary of Firms' Profits Under Various Rivalrous Consonance Regimes

Solution	Π_1	Π_2	Π_3	Π_4	Π_5
	Median Distance Discounting: k=.29				
Zero Rivalrous:Consonance:					
$\theta_{ii}=1, \theta_{ij}=0, k=.29$	8,571.93	6,280.27	9,374.08	1,521.86	1,524.33
Median Riv. Cons.,					
$\theta_{ii}=1, \theta_{ij}=.5, k=.29$	8,731.90	6,506.34	9,694.66	1,603.18	1,589.33
Perfect Riv. Con.,					
$\theta_{ii}=1, q_{ij}=1, k=.29$	8,879.44	6,674.79	9,971.06	1,542.09	1,423.83
E_i Clusters:					
$\theta_{ii}=1, \theta_{ij}=.5, j,k=1,2,3; 4,5, k=.29$	8,602.74	6,321.23	9,439.08	1,571.82	1,585.54
E_i Clusters:					
$\theta_{ii}=1, \theta_{ij}=1, j,k=1,2,3; 4,5, k=.29$	8,631.39	6,345.30	9,492.71	1,621.26	1,645.66
s_{ij} Clusters:					
$\theta_{ii}=1, \theta_{ij}=.5, j,k=1,2,4; 3,5, k=.29$	8,674.15	6,423.91	9,549.71	1,556.05	1,581.99
s_{ij} Clusters:					
$\theta_{ii}=1, \theta_{ij}=1, j,k=1,2,4; 3,5, k=.29$	8,767.37	6,563.70	9,719.27	1,507.75	1,578.84
	Total Distance Discounting: k= ∞				
Zero Rivalrous:Consonance:					
$\theta_{ii}=1, q_{ij}=0, k=\infty$	8,808.11	5,835.88	9,920.38	2,091.02	1,848.06
Median Riv. Cons.,					
$\theta_{ii}=1, \theta_{ij}=.5, k=\infty$	8,788.06	5,749.23	9,850.30	1,933.73	1,557.14
Perfect Riv. Con.,					
$\theta_{ii}=1, \theta_{ij}=1, k=\infty$	8,727.94	5,489.28	9,640.07	1,461.86	684.36
E_i Clusters:					
$\theta_{ii}=1, \theta_{ij}=.5, j,k=1,2,3; 4,5, k=\infty$	8,805.45	5,814.28	9,906.59	2,089.30	1,844.69
E_i Clusters:					
$\theta_{ii}=1, \theta_{ij}=1, j,k=1,2,3; 4,5, k=\infty$	8,797.46	5,749.48	9,865.24	2,084.12	1,834.56
s_{ij} Clusters:					
$\theta_{ii}=1, \theta_{ij}=.5, j,k=1,2,4; 3,5, k=\infty$	8,795.85	5,825.55	9,906.23	2,002.92	1,766.13
s_{ij} Clusters:					
$\theta_{ii}=1, \theta_{ij}=1, j,k=1,2,4; 3,5, k=\infty$	8,795.07	5,794.58	9,863.79	1,738.63	1,520.32

Brand 1's isolation from its rivals, discussed above, does not prevent it from benefiting from extensive, generalized tacit collusion. As k rises from 0 to ∞ profits rise in the θ-cluster to attain a maximum for perfect consonance when $k=.29$, then fall but in general remain higher than in the $k=0$ configuration. Thus, it generally benefits as virtual distance from all rivals rises and collusion increases. Its low price places it well to benefit from price differentials and the larger discount factors protect it from suffering large consonance costs. It benefits more from generalized collusion rather than collusion limited to its E-cluster, although profits in it rise as k rises, but clustering by distance in the s-cluster is more profitable than the E-cluster. But the s-cluster profits remain below those obtained in the generalized θ-cluster. In this industry tacit collusion benefits the lowest quality, cheapest cost firm by isolating it from negative feedbacks from rivals' policies, and, for median θ-values, permitting it to benefit from large, favorable price differentials. Although the maximum is attained for $k=.29$, profits remain high and almost identical for all seven solutions when $k = \infty$, accenting the beneficial effects from heightened virtual interbrand distances.

On the other hand, Brand 2 continues to benefit most from proximity to its rivals and, indeed, is the largest relative beneficiary of tacit collusion in the industry: its maximum profit occurs with a zero distance discount factor and perfect generalized tacit collusion. However, for the most part, its profits for the $k=.29$ category are higher than corresponding cluster values for $k=0$ and much higher than those for $k=\infty$. Except for the maximum profit value, therefore, it gains from modest enhancement of virtual distances from its rivals, but as the lowest-priced brand in the industry as k rises brand 2's β_2 depreciates and lessens its price differential advantage. In all three k-clusters the s-cluster profits are higher than the E-cluster values, indicating the greater importance to it of nearness to Brand 4 rather than Brand 3 because the former's price is higher than the latter's as well as its distance being shorter.

Brand 3 is in the happy position of being the highest quality of the five brands, with low costs and a most favorable basic demand. Moreover, the demand coefficients of other firms' price differentials inflict relatively small impacts on α_3 while its price tends to be below that of its closest neighbor, brand 5, so that it draws a large demand benefit from that differential. Its highest profit occurs in the $k=0$, perfect rivalrous consonance configuration, so that it benefits somewhat from increases in virtual interbrand distancing that increases its end-point advantage. However, as this distancing moves beyond the median k the advantages of tacit collusion pale and it finds it advantageous to revert to zero collusion. When $k=\infty$, the advantage of proximity to brand 5 in the lower k-clusters rather than brands 1 and 2 disappears (that is, the profits in the E-clusters rise to near-parity with those in the s-clusters). In the smaller k-clusters the large price differential with brand 2 relative to that with brand 5 is punishing, while when $k=\infty$ these differences effectively are eliminated and brand 3 is indifferent between the E- and s-clusters. In summary, brand 3, as a low-cost, highest quality, favorable demand-situated product, benefits to a limited extent from tacit collusion, but even more so in general from greater virtual distance from rivals. Like its polar analogue on the quality scale, brand 1, brand 3 is somewhat standoffish from its industry confreres, and opts out of increasing tacit collusion at the higher end of the spectrum.

Brand 4's experience is determined by its extremely unfavorable demand and cost conditions. Its own-price demand coefficient is extremely high, it is the highest cost brand in the industry, and it is a low quality product. In every one of the 21 solutions it suffered from the highest price of the five brands, so that it was forced to adjust to negative price differentials in every instance. The crucial variable in meliorating its plight was the distance discount factor and specifically its role in reducing α_i, the own-price demand coefficient. Although rises in k reduce β_i as well —the shift factor in the demand equation —the impact was not comparable with the relief afforded via the reduction in the slope term. For example, for the zero rivalrous consonance cluster, as k rises from 0 to .29, α_4 falls from 57 to 38, which, when multiplied by their respective p_4 yield an improvement in the reduction of sales from 1260 to 951, while β_4 falls only from 387 to 276, or by approximately one-third of the improvement brought about by the fall in α_4. For the same consonance cluster, when k is increased from .29 to ∞, all impacts of the price differentials are eliminated, and, with generalized rivalrous consonance equal to zero, brand 4 has zero marginal consonance costs. Its profits reach a maximum for this case.

In general, brand 4 does better in the E- clusters than the s-clusters, indicating that it would prefer to extend collusion to brand 5 rather than to brands 1 and 2. For $k=\infty$, its profits for zero rivalrous consonance are only negligibly different from those earned when it extends tacit collusion to brand 5 only because its consonance costs are so low when only brand 5's profits must be taken into account. Nonetheless, tacit collusion for all three of its θ-values benefits brand 4 when distance discounting is zero and except when perfect with the discount factor at .29. In each of the distance discount clusters the α_i's are fixed but the β_i's rise with the higher prices of rivals that higher values of consonance bring. In discount cluster 1 the β_4 values rise to shift the demand function with constant slope to the right, making it more inelastic and raising p_4. This effect operates partially in the second cluster, but not at all in the third, where all β's are zero. In that cluster, with infinite discounting of distance all of the firms are harmed as consonance coefficients rise above zero.

Brand 5 is a high quality product, very close in this respect to brand 3, but with average variable costs roughly triple those of its closest rival and with fixed costs only slightly below. Also, its basic demand intercept is lower and its own-price coefficient higher than brand 3's. It is, in short, highly disadvantaged when compared with its neighbor. Not surprisingly, its performance is comparable with that of brand 4, and its maximum profit about 12 percent below — the worst of the five brands. Like brand 4, it benefits from enhanced distancing from rival brands but not greatly if at all from tacit collusion.

4. A Summary and Conclusions

Several approaches were adopted in this paper that permitted consumer preferences over brands to be scaled in an operational manner, which in turn allowed the application of closed and simulative analysis of integrated competition. A formal summary of methodology and major insights derived from it may be useful.

1. The scaling method introduced yields consumer valuations in dollars, making them comparable in dimension to prices, and serving as a means of determining maximum consumer surpluses over a set of brands. Using measures of central tendency as summary aggregate indices of those preferences permitted the derivation of interbrand differences in a product space, such differences being computed by a simple Euclidean metric in the first instance, but then adjusted by an exponential decay factor. The factor k in this formulation gave us a flexible parameter to adjust advertising effects and other cognitive frictions affecting the consumer, to reflect search costs and to admit into the analysis some of the fuzziness that the failure to take into account the variance of indifference premia for each brand introduces into choice prediction. Setting $k = 0$ restores unadjusted Euclidean distances when desired.

2. Aggregate demand functions for brands are stated in terms of own-price and differentials between own- and rival-prices, instituting usage of benchmark demand theory.

3. First-order conditions are derived for optimal brand prices and outputs when firms do not engage in rivalrous consonance forms of tacit collusion (Model 1) and when they do (Model 2), and diagonalization algorithms are once more used in both cases to obtain numerical solutions to specified models.

4. Parametric displacement analysis of small changes in one firm's location are conducted and necessary and sufficient conditions derived for signing the partial derivatives of output and prices with respect to interbrand distances, s_{ij}. The induced changes are shown to be the result of changes in the intercepts and slopes of the demand functions before (in a ceteris paribus analysis) and after (in mutatis mutandis analysis) all brand price changes of firms other than the firm that changed location are taken into account. This is, however, the only experiment with firms moving in product space, i.e., with changes in the "quality" of the brand as registered in the valuations of consumers.

5. A base case is designed for simulation analysis, which features five brands with differing quality and demand and cost conditions. It is employed both to illustrate the results of the closed analysis and to derive insights where closed analysis becomes too complicated or incapable of yielding qualitative results absent numerical specifications.

6. A measure of competitiveness, E_i, for each brand i is derived, which varies directly with price elasticity of demand, to index the relative ability to compete for profit under a wide variety of environmental conditions, including the presence and

the absence of tacit collusion. It is used to divide the five firms into two subsets of differential competitiveness. Although the members of these subsets remain the same over the 21 solutions obtained , profits do not decline uniformly as the measure rose. Indeed, in the first subset, profits rise slightly with E_i , which is counter to expectations. The measure should be used with caution, therefore, as means of gauging profitability, and may be best used as a method of discerning common response subsets of brands in a variety of environments.

7. Distinctions are made among the various types of movements in product or characteristics space, and the analysis is limited to small changes on the basis that they are more realistic in quality competition. With the use of the base case in a simulation exercise, it is seen that the equations for partial derivatives of prices and quantities with respect to changes in interproduct distances give good approximations for finite changes, ceteris paribus. When mutatis mutandis changes are added, impacts frequently move in the opposite direction from the direct ceteris paribus changes, but never enough to enforce a sign change from that predicted by the ceteris paribus forecasts. Direct impacts of such movement, in the simulation, always outweigh the indirect impacts imposed by price changes of rivals.

8. Consumers' surplus is shown to move in the ceteris paribus analysis in a direction opposite to price for those firms remaining stationary in location.

9. A model with a bounded product space with continuous movement of firms will lead to no stable equilibrium without the introduction of relocation costs or minimum profit improvement constraints. Resort to advertising to affect demand may be explained by realization that brand relocations can be costly to the firm and destabilizing for the industry.

10. Displacements of the base case industry equilibrium through variations in k have two functions: they permit study of disparate brand reactions to changes in the dispersion of consumer brand decisions, or to the costs of search, or other cognitive influences such as advertising, and they permit changes in virtual distances among all brands in the space.

As k rises from 0, through the base case value of .29 to ∞, prices of all brands rise and outputs fall monotonically: increasing interbrand distance increasing firms' monopoly power, qualities held constant. Profits also rise over the whole span of k except for brand 2, which suffers a fall in profitability when k rises from .29 to ∞. This is an interesting if unsurprising result, revealing that a lower quality product with important cost advantages but low basic demand, which thrives on favorable price differentials with rivals, suffers as those rival brands increase their distances from it in quality space. Because consumers prefer higher quality to lower at constant price differentials in this example of "vertical differentiation", when such products are thrown back on their own-price demand ($\alpha_i \, b_{it}$) and price differential demand wanes ($\beta_i \rightarrow 0$), the extent of sales decline outweighs price rises and revenue losses exceed cost reductions.

11. When tacit collusion is introduced into the industry we encounter the proposition that increases in such cooperation do not necessarily benefit all firms in an industry, even when such firms are mature incumbents with significant market share. Such cooperation imposes marginal consonance costs on firms that reduce competitiveness and that can be quite significant as firms are forced to forgo own-profit possibilities to defer to rivals. The rivalrous consonance approach permits such costs to be isolated.

Firms that are likely to be benefited most by tacit collusion are low-quality, low-cost producers which rely heavily on price differentials — that is, on sheltering under high price umbrellas. As noted above, brand 2 is such a product, and it is seen to benefit consistently in Table 7 from rises in collusion for $k = 0$ and .29 both for generalized tacit cooperation and that confined to E- and s-clusters. However, when $k=\infty$, although all other firms benefit when they exploit their effective monopoly positions (brand 2 excluded on grounds explained in 10. above) brand 2 joins all firms in suffering as tacit collusion —generalized and confined to E- and s-clusters —rises. Each firm functions best as an effective monopoly because consonance costs impose penalties that drive up prices and reduce sales.

Notes

1. A version of this paper was delivered at the International Symposium on Economic Modelling, University of Oslo, July 3-5, 1996.
2. Question 2 does bring into play the consumers' valuation of money and, indirectly their income. The indifference premia also will be affected by such circumstances, much as the compensating variation of conventional demand theory.
3. This scaling approach is similar to that discussed in Jean Tirole (1993), *The Theory of Industrial Organization*, Cambridge, Mass.: MIT Press, pp. 96-97, in which he employs a "quality index", s, and a "taste parameter", θ, and defines the utility function

$$U = \begin{cases} \theta s - p, & \text{if the consumer buys a good quality } s \text{ at price } p \\ 0 & \text{if he does not buy.} \end{cases}$$

He refers to θ as a positive real number, but θs must have a dollar dimension for meaningful comparison with p. The valuation measure, v_j in (1) may be looked upon as serving the function of Tirole's θs, derived empirically, with more readily interpretable meaning than θs. Tirole, in his treatment of preferences over multiple brands, assumes a cumulative density function for consumers over θ to determine demands for brands whereas the use of the valuation approach yields these directly, albeit with the use of a measure of central tendency that ignores variance via among individuals valuations. We attempt to include this variance via the use of k parameters to alter effective interbrand distances.

4. Rivalrous consonance is an approach to the analysis of oligopoly that I have advocated for some years, and is discussed more fully in Robert E. Kuenne (1986), *Rivalrous Consonance: A Theory of General Oligopolistic Equilibrium*, Amsterdam: North Holland, and in (1992), *The Economics of Oligopolistic Competion: Price and Nonprice Rivalry*, Blackwell: Oxford. In essence it seeks to specify the power structure of an oligopolistic industry an a matrix of discount factors, θ_{ij}, which firms i apply to rivals j profit functions to make rivals' profits commensurate with their own in their maximization procedure. Hence, a consonance coefficient $\theta_{ij} = .25$ means that firm i treats a profit or loss by firm j as equivalent to \$.25 of its own profit or loss in its decision making. The larger θ_{ij} the greater the welfare of firm j assumes in firm i's extended profit maximization, by virtue of the power firm j has to inflict losses on firm i should it initiate too aggressive a pricing strategy. This extended profit function is defined in equation (5) in the body of the text.

A Schloss Laxenburg Model of Product Cycle Dynamics

Börje Johansson
Jönköping International Business School

and
Åke E. Andersson
Institute for Futures Studies, Stockholm

1. Product Cycle Theory and Location Advantage Dynamics

1.1 Introduction

In 1984, while working at IIASA in Schloss Laxenburg outside Vienna, the authors jointly produced two papers on product cycles (Andersson and Johansson, 1984a and 1984b). These papers use microeconomic models to show how product cycle assumptions generate location and relocation processes. Both papers also demonstrate how clusters of product cycles can be observed empirically in the form of aggregate specialisation patterns which describe a time-space hierarchy. In this paper we reformulate these results into a more coherent framework and emphasise new directions for this type of model formulations.

In the model framework which is formulated, knowledge intensity is a key notion. Two of the key notions are product standardisation and process routinisation. Along a product cycle path the knowledge intensity is high when a product is non-standardised and the process is non-routinised. Standardisation and routinisation means reduced knowledge intensity. The paper presents a class of models which explain this regularity. The models are used to derive interdependencies between location dynamics and product cycle development. The framework is designed to incorporate elements from models of monopolistic competition. Moreover, the notions of product and process vintages are used to classify structural properties of individual firms and the associated markets.

1.2   Product Cycle Models – An Overview

Since the middle of the Seventies empirically oriented analyses have devoted a lot of attention to how economic activities, innovations and technological development form or follow wave-like patterns in space and time. In view of this focus many researchers have exposed a revived interest in Schumpeterian hypotheses (e.g. Freeman, 1974; Kamien and Schwarz, 1982; Nelson and Winter, 1982). In regional and interregional contexts a similar interest is shared by a large number of scholars (Norton and Rees, 1979; Malecki, 1981; Markusen, 1985; Nijkamp, 1986). In this wide field one finds models of dynamic competition, innovation processes and the role of knowledge in technological change. During the 1980s and 1990s similar efforts often use evolutionary economics as a frame of reference (e.g. Batten, Casti & Johansson, 1987; Dosi et al , 1988).

The problem formulation above has strong connections with different versions of the product cycle model. In the 1960s the model was used by e.g. Vernon (1966) and Hirsch (1967) in research efforts which were stimulated by attempts to explain the Leontief paradox (1953). However, the notion of a life-cycle, along which an industry or a product group develops was suggested early by the French sociologist Tarde (1903). In aggregate forms one can recognise this idea in contributions by Kuznets (1930) and Burns (1934). Schumpeter's contributions (1926/1934, 1939) also belong to this category.

From the beginning of the 1950s we can observe detailed micro-level studies of product life-cycles in the marketing literature (Dean, 1950; Patton, 1959; Mickwitz, 1959). In the early 1960s Gaston (1961) and Gold (1964) returned to more aggregate studies of industrial development patterns. These attempts were followed by Vernon and Hirsch, both opening the analysis towards a study of spatial product cycles. They both emphasised that during the life of a product the demand for different types of knowledge, skills and other inputs changes in a systematic way (see also Freeman, 1978; Cole, 1981). This type of model framework has remained partial and emphasises the following aspects of technique and location change (Andersson and Johansson, 1984b, Johansson and Karlsson, 1987):

- During the innovation phase, the new product is developed in the most advanced regions of the world on the basis of research, experimentation and testing. The product is then produced in a small set of regions which have a location advantage in terms of R&D resources and knowledge intensity of the labour force. The new product is gradually exported from these initiating regions to other importing regions.
- In the growth stage, domestic and foreign demand expands to a point where interregional direct investment becomes feasible. In this phase, the process technology usually has become more streamlined so that it can be transferred and imitated more easily.

- When the product has matured in terms of market penetration and design of production methods, the initiating regions often lose their advantage and the production moves to regions with lower costs. Frequently this relocation also involves a decomposition of the original production process in such a way that the production of components is placed in a set of regions with an advantage to carry out routinised production, while assembly activities may stay in the original locations or in the proximity of large markets.

This stylised, descriptive model has been criticised for a set features. It is often argued that the basic model is too schematic and too ad hoc in its use of economic theory. Some claim that many products do not evolve along life-cycle paths, but are instead continuously renewed. Moreover, it has been argued that the concept of a product is elusive and ambiguous. In our reformulation of our own earlier work we suggest solutions to some of these problems. Our paper also introduces a specific formalism and an associated theoretical framework for this type of analyses. We suggest a solution to the problem of how one can delineate an individual product, and how such products can be grouped in clusters of similar but differentiated products. This approached is based on a simultaneous categorisation of product and customer groups.

The framework we put forward comprises markets in which customer-specific products are produced and delivered to a single customer or a small set of such customers. Products defined in this way may not follow the schematic development path described above. However, if such products are organised into broader categories (a product group), each category may still have a distinct life-cycle trajectory. Such life-cycle trajectories may come to rest in different types of markets, e.g. monopolistic competition with differentiated outputs as well as oligopolistic competition with standardised outputs.

The analysis concentrates on micro mechanisms which influence the dynamics of products and product groups – in space and time. This generates a series of questions. Which are the conditions that make standardisation and automation favourable? In which ways are the dynamics affected by the knowledge intensity of an urban region. In what type of regions are product cycles frequently initiated? How is relocation of production and imitation in other regions stimulated? These issues are first investigated in micro models. Thereafter the analysis is extended to aggregate trajectories of product groups and sectoral dynamics.

Attempts to empirically assess the product life cycle theory have to a large extent had the form of "statistical anecdotes", dominated by case studies, and cross-sectional approaches (Johansson and Karlsson, 1987). The collected knowledge is difficult to systematise, since each investigation tends to use its own level of product aggregation, ranging from industrial sectors to precisely defined products. Decomposing the world into OECD countries and non-OECD countries, Andersson and Johansson (1984a, 1984b) identify spatial product cycles. Such trajectories are also recorded in Batten and Johansson (1987, 1989), in a somewhat more disaggregate setting. At finer levels of disaggregation Erickson and Leihbach (1979) show for US that branch plants produce standardised products which do not require much development activities. This is especially true for establishments which are distant from their corporate headquarters. Similar results are reported for France in Aydalot (1978) and for Sweden in Wigren (1976,1984).

1.3   An Outline of the Paper

Section 2 emphasises the dynamics of knowledge intensity in the development, production and distribution activties during the life cycle of a product. The knowledge intensity is associated with two characteristics of an economic activity – the degree of product standardisation and the degree of routinisation of the production process (including production and delivery as well as all pertinent interaction activities). In section 2.2 the product life cycle is depicted by a change process in which standardisation and routinisation gradually increase. This is recorded on two separate scales – one describing a product-vintage index and the other an associated process-vintage index. New vintages are assumed to be introduced in response to a growing demand and a gradually increased price competition. The time pattern is determined by economies of scale which gradually reduce unit costs and prices, while at the same time stimulating demand. The scale phenomena are also shown to provide initiating firms and early locations a prolonged advantage in the form of lock-in effects (cf Arthur, 1983 and 1988).

Section 3 outlines the product cycle mechanisms including exit of ageing vintages and entry of new vintages. A cost function with knowledge and routine inputs is used to determine conditions such that the location advantage shifts from one region to another as the technique changes along the life cycle of a product. By specifying a production function with knowledge and routine inputs, a similar "dual" proposition about shifts in location advantage is derived.

Due to rigidities in the adjustment process, shifts in location advantage does not necessarily mean that the location of supply changes. By making use of exit and entry assumptions section 3 takes the analysis one step further and establishes conditions for link catastrophes, such that a delivery flow from region r to region s, (r,s), is replaced by a new flow (k,s). The section ends by characterising temporary equilibrium solutions which evolve through time. A proposition outlines the conditions for introducing a new technique vintage between two time periods when demand remains constant and when it increases.

Section 4 contains four subsections. The first summarises model predictions and empirical observations of where new product cycles are initiated. The second describes how regional specialisation can be described in terms of clusters of product cycles. A dynamic model is introduced to illustrate how economic activities compete for knowledge and routine resources. A third subsection suggests approaches to the aggregation of products and product groups to identify and analyse aggregate product cycles. The section ends with a discussion about observed aggregate specialisation waves in a multiregional and multicountry context. Finally, section 5 presents conclusions about future theoretical efforts and associated empirical studies. A wide set of complementary strands of research approaches are suggested. It is also argued that it is vital for further progress that one establishes a cross-communication of results between such alternative approaches.

2. Dynamics of Knowledge Intensity

2.1 Standardisation of Products and Routinisation of Production

A product is distinguished from other products by those attributes which are recognised by the customers in a market. This implies that a product's set of attributes may include a spatial identification of the (i) place of production, (ii) place of transaction, and (iii) place of consumption.

In addition to other classification principles, customer groups can be identified with regard to what use each group makes of a product. In particular, products can be described as inputs to household and firm activities. Certain products may be inputs to many different types of activities undertaken by many different customer groups, while other products are strictly specialised such that only one customer group uses it for a distinct activity. With reference to these outlined principles we suggest that consistency requires a simultaneous classification of products and customer groups. A fine specification of customer groups requires a matching classification of products and product groups. Variant forms of a specific product are then related to a corresponding customer group. Products within the same product group have only partially overlapping customer groups. The more these groups overlap, the stronger the prerequisites for substitution between the products – as a result of price changes, learning among customers etc.

Standardisation is a characteristic of how product attributes vary across customers. A non-standardised product is characterised by being supplied to few customers and often in limited amount. Ultimately every delivery is customised, i.e., every delivered product has specific attributes. One may identify several degrees of standardisation. A product may be delivered in long series (possibly over a long time period) to only one customer, a situation which may be called customised standardisation. An extended form of standardisation applies when a product is delivered in long series to many customers. If in addition several producers and sellers compete for the same customer demand, the situation is characterised by competitive standardisation.

Table 1 presents a cross-classification of the degree of routinisation of the production and the degree of product standardisation, where we recognise that products comprise both goods and services and combinations thereof. In position (I) economic activities are assumed to be knowledge and contact intensive, while in position (IV) these attributes are reduced with the intention to cut down on costs and compete with low prices. One may distinguish between two cases of standardisation. In the first case a product is standardised from the viewpoint of both the supplier and the entire market. In the second case the market is characterised by a differentiated supply of products, while each producer supplies a standardised variant. This latter case corresponds to the classical model of monopolistic competition (Chamberlin, 1933).

Table 1. Product standardisation and production routinisation

	Non-routinised production and distribution	Routinised production and distribution
Non-standard-ised product	(I) Ongoing product development; the supplier has intensive contacts with his own suppliers and customers	(II) Limited development activities; the supplier has established customer links
Standardised product	(III) Limited development activities; skill-requiring inputs, price competitive customer markets	(IV) Production and transaction activities require limited contact intensity; price competitive markets

One should observe that the concept of routinisation relates to the way Nelson and Winter (1982) treat the concept of production routines. They consider established routine activities as components of a firm's production and distribution technique, as elements of the technology in use. Consider a standardised product which is delivered in long series. It is obvious that this type of standardisation facilitates the introduction of routine and automated processes. In this way standardisation and routinisation are strongly related.

In schematic versions of product cycle paths, a young product starts in position (I) and moves gradually towards position (IV). However, one may identify several potential life cycle paths, such as (I)–(II), (I)–(III), (I)–(II)–(IV), (I)–(III)–(IV), and a direct jump (I)–(IV). The last three of these paths can be classified as ordinary product cycle trajectories.

The stylised picture of a product cycle is described by the phase diagram in figure 1. In the models introduced in the sequel we use two index-variables, θ and τ, such that $0 \leq \theta \leq 1$ and $0 \leq \tau \leq 1$. The first variable refers to standardisation, and $\theta = 0$ indicates a completely new or customer-adjusted product, while $\theta = 1$ indicates a completely standardised product. Analogously, the technique-vintage index τ takes the value $\tau = 0$ for a new, non-routinised vintage. When a vintage is mature and completely standardised, $\tau = 1$.

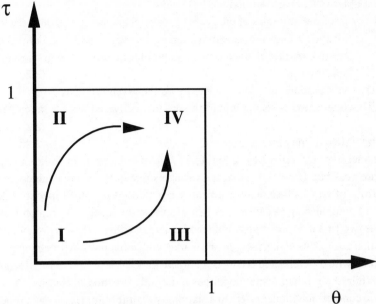

Figure1. Product cycle paths in the (τ, θ)-space

2.2   Product and Process Vintages

A sustainable hypothesis, attributed to Ricardo (1951), states for agricultural land and many other types of natural resources that in general each new vintage is associated with an increased unit cost level (or reduced productivity). In contradistinction, Heckscher (1918) observes that in manufacturing industries new vintages will be associated with improved production techniques, and that an industry generically exhibits a productivity distribution across establishments such that recent vintages are associated with higher productivity. These observations are further developed in Johansen (1972).

The above approach focuses on technology vintages or, in other words, vintages of methods applied in a production process. In the following analysis we make use of a double classification. A particular economic activity and its pertinent product (a good, service or a combination thereof) is characterised by the pair (τ, θ). Assume for a moment that this pair determines the scale of operation, with regard to a particular product in location r, so that we can specify the unit profit per output, $\pi_r(\tau, \theta)$, as follows:

$$\pi_r(\tau, \theta) = \sum_s [p_{rs}(\theta) - c_{rs}(\tau, \theta)] z_{rs}(\tau, \theta) - c_r(\tau, \theta)$$

$$z_{rs}(\tau, \theta) = x_{rs}(\tau, \theta) / \sum_s x_{rs}(\tau, \theta)$$

(2.1)

The variables in (2.1) are defined in (2.2) below

$p_{rs}(\theta)$ = price per unit sales of the given product, characterised by delivery from location r to s and standardisation of degree θ , which is defined as a relative market characteristic, comparing the product with its relevant substitutes;   (2.2a)

$c_r(\tau,\theta)$ = unit production cost associated with the given product;   (2.2b)

$c_{rs}(\tau,\theta)$ = interaction costs per unit of the product delivered from location r to s;   (2.2c)

$x_{rs}(\tau,\theta)$ = flow of the product from r to s.   (2.2d)

The model introduced in (2.1)-(2.2) refers to a particular product. Other interpretations are possible. If we instead let the variables represent average values for the product mix of an establishment or a firm, we obtain a firm profit model. Subsequently we shall also aggregate across firms in the same location or region r. With any of the two first interpretations of the model, it may either refer to a product cycle competition through its various stages or a market with monopolistic (or oligopolistic) competition with differentiated products (compare Lancaster, 1975; Dixit and Stiglitz, 1977). While the product cycle model has a definite direction of motion, the monopolistic competition model refers to a stationary equilibrium. However, a product cycle trajectory may pass through a sequence of market solutions, such that each of these solutions is characterised by monopolistic competition.

For a given point in time, let Γ denote the set of possible combinations of τ and θ such that $(\tau,\theta) \in \Gamma$. Given that $(\tau + \Delta\tau, \theta) \in \Gamma$ and $(\tau, \theta + \Delta\theta) \in \Gamma$, and that demand is sufficiently high, the conditions in formula (2.3) comprise the basic assumptions which influence the dynamics of a product cycle model:

$$\partial p_{rs}(\theta)/\partial\theta < 0 \Rightarrow p_{rs}(\theta) > p_{rs}(\theta + \Delta\theta) \qquad (2.3a)$$

$$\partial c_r(\tau,\theta)/\partial\tau < 0 \ \Rightarrow c_r(\tau,\theta) > c_r(\tau + \Delta\tau, \theta),$$

$$\partial c_{rs}(\tau,\theta)/\partial\tau < 0 \ \Rightarrow c_{rs}(\tau,\theta) > c_{rs}(\tau + \Delta\tau, \theta) \qquad (2.3b)$$

$$\frac{\partial c_r}{\partial\theta}d\theta + \frac{\partial c_r}{\partial\tau}d\tau < 0 \ , \text{ for some positive pair } (d\tau, d\theta)$$

$$\frac{\partial c_{rs}}{\partial\theta}d\theta + \frac{\partial c_{rs}}{\partial\tau}d\tau < 0 \ , \text{ for some positive pair } (d\tau, d\theta) \quad (2.3c)$$

where Δ denotes a positive change in a variable. Consider first that τ is given and cannot be changed. Then condition (2.3a) states that the price is higher the less standardised the market competition allows the product to be, where we may assume that $p_{rs}(\theta)$ describes the price-setting behaviour of firms in the market. Condition (2.3c) states that the sum of production and interaction costs diminishes as the product-vintage index increases in combination with an associated increase of the technique-vintage index. There is an obvious relation between (2.3a) and (2.3c) which is described in Remark 1.

Remark 1:
Consider a supplier in region r. Assume that this firm employs the price-setting strat-
egy $p_{rs}(\theta) = \lambda[c_r(\tau,\theta) + c_{rs}(\tau,\theta)]$ for $\lambda \geq 1$. Then condition (2.3a) follows
from (2.3c).

Remark 1 describes a mark-up price-setting behaviour. If demand is modelled
explicitly, the price-setting rule may be derived from assumptions about profit maximising behaviour. With a given price-setting rule one may specify conditions which
guarantee that there is a profit maximising $\theta = \theta^*$. We also observe that with a
given price-setting rule, the output, x_r, is determined by θ. The lower the value of
θ^*, the more strict the product attributes must be adjusted to fit a specific customer
group. Hence, low values of θ^* are generally associated with comparatively small
or few deliveries.

Next, for given competition conditions in each market s, let $\theta = \bar{\theta}$ be given such
that $p_{rs} = p_{rs}(\bar{\theta})$ for each s which also means that the corresponding volume of
sales is determined in each market. When $\bar{\theta}$ is given, (2.3b) states that it is favourable to use a process with a high technique index, τ, as long as the given demand is
sufficiently large. The demand can increase if a new and more standardised product
vintage, $\theta > \bar{\theta}$, is introduced. This conclusion indicates that, besides the technology constraint Γ, there is a relation between τ and θ via the demand. Provided
that this dependence between product and process vintages can be described by a
differentiable function and that the technology set Γ allows an interior solution, the
profit maximising combination of a firm, at a given point in time, should satisfy

$$[\partial\pi / \partial\tau]d\tau + [\partial\pi / \partial\theta]d\theta \leq 0 \qquad (2.4)$$

In ordinary product cycle models the new product is usually taken as given and the
model can then describe how the technique vintage increases over time, while the
product attributes are assumed to be approximately the same (e.g. Nelson and Norman, 1977; Andersson and Johansson, 1984). In this type of analysis, standardisation basically means that a growing number of firms are able to supply the same
product. In most product cycle models it is still understood that the product attributes
moves in the direction of standardisation as the technique vintage, τ, increases. In
this sense, all these models implicitly assume a direct relationship between
τ and θ, such that $\tau + \vartheta = 1$, where $\vartheta = 1 - \theta$. The technology simplex of
(τ, θ)-combinations is described in figure 2, in which the arrow shows the one-
directional development of technology and the gradual selection of techniques.
Moving from (I) towards (IV) in figure 2 implies further standardisation and increasing competition.

The arrow in figure 2 can be interpreted in two different ways. It may indicate how
the technology (in the form of knowledge or blueprints) becomes available over
time. In an alternative model it describes how options in this initially given technology set become applied through a gradual selection of production techniques.
With scale economies, new techniques become feasible as demand expands. In the

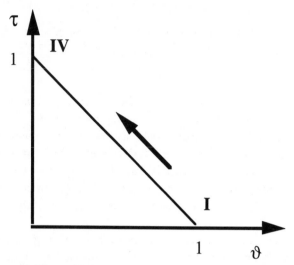

Figure 2. The orthodox product cycle assumption about technological change

sequel we shall distinguish between the change of the technology and the change of techniques put into use.

How does the technology set, Γ, develop? We assume that it evolves as a consequence of R&D investments of firms which develop new processes, in combination with R&D of firms which supply new equipment. Other firms may keep the R&D investments at a reduced level by imitating the forerunners. In summary, these assumptions imply that Γ evolves gradually as time goes by. The speed of change depends on the level of the associated R&D investments. In the following subsection we demonstrate that expansion of demand plays a crucial role in stimulating technical change.

2.3   Dynamics of Demand and Scale Phenomena

Consider the demand in region s for the product described in the preceding section, and let it be denoted by $x_{rs} = x_{rs}(p_{rs}(\theta), \theta, t)$, for given competition conditions in market s. Following the suggestions in Andersson and Johansson (1984a, 1984b), there are two major factors that ascertain that demand can continue to grow. First, for constant θ potential buyers are assumed to learn gradually about the product attributes, which implies that $\partial x_{rs} / \partial t \geq 0$, for given θ. Second, the price -setting rule $\partial p_{rs}(\theta) / \partial \theta \leq 0$ together with the condition $\partial x_{rs} / \partial p_{rs} < 0$, for given θ, provides opportunities for gradually falling market prices. Such price reductions are profit motivated if the pertinent firms over time can find new technique vintages, τ, which reduce costs sufficiently. With these conditions the following growth path becomes possible:

$$dx_{rs} = [\frac{\partial x_{rs}}{\partial p_{rs}}\frac{\partial p_{rs}}{\partial \theta} + \frac{\partial x_{rs}}{\partial \theta}]d\theta + \frac{\partial x_{rs}}{\partial t}dt \geq 0 \qquad (2.5)$$

One may observe that a larger θ does not necessarily imply that new product vintages get inferior attributes. Also in such cases the price-setting strategy of firms is assume to satisfy $\partial p_{rs}(\theta) / \partial \theta \leq 0$.

Previously we have concluded that there is an orthodox product cycle assumption which states that $\tau + \vartheta = 1$. Then it should be possible to express formula (2.5) in terms of τ instead of θ. Following this assumption, lets go back to formula (2.1) and consider a second orthodox product cycle assumption saying that there is a relation between τ and the economic scale such that

$$x_r(\tau + \Delta \tau) > x_r(\tau), \text{ and } x_{rs}(\tau + \Delta \tau) > x_{rs}(\tau)$$

Given this assumption, (2.1) can be reformulated in the following simplified way:

$$\pi_r(x_r) = \sum_s z_{rs}[p_{rs}(x_{rs}) - c_{rs}(x_{rs})] - c_r(x_r) \qquad (2.1')$$

With the simplified version in (2.1'), Remark 2 and 3 become self-evident.

Remark 2:

Assume that $\tau + \vartheta = 1$ holds. Then the conditions in (2.3b) and (2.3c) can be derived from the following assumption about scale economies in (2.1'):

(i) $dc_r / dx_r < 0$ and (ii) $dc_{rs} / dx_{rs} < 0$.

We observe that assumption (i) is assumed to hold so long as $\tau < 1$. Assumption (ii) is a natural assumption given that the capacity tensions of the relevant interaction channels are negligible. Remark 3 emphasises that scale effects on production costs can be decomposed into an effect on input prices and a second effect on input coefficients.

Remark 3:

Consider the assumptions in Remark 1, and assume that

$c_r(x_r)x_r = \sum_s \sum_i p_{sri} b_{sri} x_r$, where p_{sri} denotes the unit price of the delivery of

input i from location s to r, and b_{sri} is the coefficient describing the input of product i from region s per unit output of the given product in location r. Assumption (i) in Remark 2 can be derived from each of the following two assumptions: (iii) $\partial p_{sri} / \partial x_r < 0$, and (iv) $\partial b_{sri} / \partial x_r < 0$.

One should observe that the effect $\partial p_{sri} / \partial x_r < 0$ means that a firm's interaction with input suppliers becomes more efficient as the scale of interaction (including transportation) increases. More important, Remark 2 and 3 imply that due to scale effects a competition advantage obtains in a location in which a new economic activity is initiated. However, product cycle mechanisms as presented in section 3 imply that (in the long term) incentives to relocate become stronger as the technique vintage increases.

3. Product Cycle Mechanisms

3.1   Entry and Exit Conditions

Consider a product or a product group with a constant level of standardisation. Let $\bar{x}_r(\tau,t)$ be the production capacity in region r which applies technique τ at time t. Moreover, let τ^* be the most recent technique known in region r, and let τ^o be the oldest technique still in use in region r. Then the entire capacity, $\bar{x}_r(t)$, is the sum over existing technique (routine) vintages such that $\bar{x}_r(t)=\bar{x}_r(\tau^o,t)+...+\bar{x}_r(\tau^*,t)$. For such a distribution of vintages an exit function, $\xi(\pi_r(\tau))$, is introduced to represent the exit frequency during a time interval of length Δt.

$$\xi(\pi_r(\tau))\begin{cases} >0 \text{ if } \pi_r(\tau)+\kappa_r(\tau)\le\pi_r^o \\ =0 \text{ otherwise.} \end{cases} \tag{3.1}$$

$$0\le\xi(\pi_r(\tau))\le 1$$

where $\kappa_r(\tau)$ represents sunk costs, calculated per unit output, and where $\pi_r(\tau^o)\ge 0$ is some low gross profit level. In addition, we can assume that the exit rate is higher the lower the gross profit per output is, i.e., the lower $\pi_r(\tau)+\kappa_r(\tau)$ is. It is important to observe that exit is an ex post decision, which takes place after the technique-specific set up costs have already become committed (sunk).

The entry of new capacity is an ex ante decision and is assumed to comprise only the best practice vintage τ^* at time t. The entry of new capacity during a time interval $[t, t+\Delta t]$ is specified as follows

$$E_r(\tau,t)\begin{cases} >0 \text{ if } D_r(t)>\bar{x}_r(t) \text{ and } \pi_r(\tau^*)\ge\pi_r^* \\ =0 \text{ otherwise} \end{cases} \tag{3.2}$$

where $D_r(t)$ represents demand at time t and π_r^* the lowest profit level which is consistent with a decision to invest. Since entry is an ex ante decision the profit level π_r^* has to cover both variable and fixed costs, where the latter may become sunk as soon as the investment is made. Hence, it is natural to assume that $\pi_r^*\ge\pi_r^o+\kappa_r(\tau)$, $\tau\le\tau^*$.

Suppose that in region r at time t the only established technique is τ^o A capacity change during a time interval Δt which satisfies the following condition is then consistent with the assumptions in (3.1) and (3.2):

$$E_r(\tau^*)=f(\pi_r(\tau^*))[D_r-(1-\xi_r(\tau^o)x_r(\tau^o)]$$
$$f(\pi_r(\tau^*))>0 \text{ as } \pi_r(\tau^*)\ge\pi_r^* \tag{3.3}$$

Formula (3.3) emphasises that the entry of new techniques in a region r is influenced by three phenomena: (i) the demand directed towards the region, (ii) the region's established capacity minus removal of economically obsolete capacity (exit), and (iii) the profitability associated with the region. This implies that the introduction of new techniques is a delayed process. The delay aspects are further stressed in Remark 4.

Remark 4:
Let the exit process be governed by (3.1) and let $\theta = \overline{\theta}$ be the same for producers in region r across all markets. Assume that a new competitor (rival) from region k threatens all deliveries from r by selecting $p_{ks}(\tilde{\theta}) < p_{rs}(\overline{\theta})$, $\tilde{\theta} \geq \overline{\theta}$ in all markets s, and forcing the suppliers in r to reduce the price by Δp_{rs}. Then this does not bring about any exit of capacity in r as long as $\pi_r(\tau,\overline{\theta}) - \sum_s \Delta p_{rs} z_{rs} > \pi_r^o - \kappa_r(\tau,\overline{\theta})$, where $\pi_r(\tau,\overline{\theta})$ denotes the initial profit per unit sales before the new competition started.

Remark 4 emphasises the asymmetry between an ex ante and ex post decision. The already established capacities in r cannot avoid the associated sunk costs by removing any capacity. Hence, as long as the return covers variable costs, there is an incentive to keep the capacity. In summary, the remark shows in which way established techniques can resist fairly strong price threats from potential competitors. This form of persistence or rigidity implies that the technical improvement of a rival often must be considerable in order to force an initial deliverer to leave a market – partially or completely. At the same time, condition (3.2) shows that overall growth in demand stimulates new techniques to enter.

Consider (3.1)-(3.2) and assume that the best practice, τ^*, continues to increase until $\tau^* \approx 1$. Assume also that for all techniques $\tau < \tau^*$, exit remains positive during the sequence of future time intervals. Then we can conclude that the expected long-term outcome is a situation where all producers use the same best practice technique (cf Johansson, 1987).

3.2   Knowledge and Routine Inputs in the Cost Function

In this subsection a specific cost function is introduced. In subsections 3.3 and 3.4 this function is a major tool in the derivation of propositions about product cycle mechanisms.

The cost function is assume to incorporate the features of the cost structures specified in (2.2) and (2.3). We assume that θ remains unchanged over time or that the orthodox product cycle assumption in figure 2.2 applies, saying that $\theta = \tau$. Moreover, the trade pattern, described by z_{rs} in (2.1) and (2.1'), is assumed to remain unchanged over time. Then the following cost function can be employed to represent the cost structures introduced in (2.2)-(2.3):

$$c_r(\tau) = \kappa_r(\tau) + \rho_r a(\tau) + \omega_r b(\tau) \qquad (3.4)$$

where $\kappa_r(\tau)$ refers to set up costs (which ex post become sunk costs) associated with technique τ, $a(\tau)$ represents the input coefficient (per unit delivered output) referring to routine input resources, $b(\tau)$ represents the input coefficient referring to development resources or, in other words, input resources for knowledge activities or knowledge production. The input coefficients are assumed to be determined by the selected technique and are not dependent on the location. On the other hand, the two price variables, ρ_r and ω_r, are location specific, where ρ_r denotes the price of routine inputs and ω_r denotes the price of development, knowledge-intensive resources.

From (3.4) we may define the knowledge-intensity (in value terms) of technique τ as follows:

$$U_r(\tau) = \omega_r b(\tau) / c_r(\tau) \qquad (3.5)$$

In the next two sections we shall study how the knowledge intensity develops as the technique-vintage index increases. For these analyses we also need the following rearrangement of (3.4) into

$$c_r(\tau) = a(\tau)[\hat{\kappa}_r(\tau) + \rho_r + \omega_r \hat{b}(\tau)]$$
$$\hat{\kappa}_r(\tau) = \kappa_r(\tau) / a(\tau) \qquad (3.4')$$
$$\hat{b}(\tau) = b(\tau) / a(\tau)$$

where $\hat{b}(\tau)$ is called knowledge-routine ratio. In (3.6) we introduce a set of assumptions which ascertain that (3.4) and (3.4') are consistent with the assumptions in (2.2)-(2.3):

$$\partial \hat{\kappa}_r / \partial \tau \leq 0; \ \partial \hat{b} / \partial \tau < 0$$
$$\partial c_r / \partial \tau < 0, \text{ for } \omega_r \text{ and } \rho_r \text{ constant.} \qquad (3.6)$$

3.3   Knowledge Intensity and Shifts of Location Advantage

In this subsection the location advantage of two locations r and s are compared. Both locations are rivals with regard to a given market, for which the sales volume may grow in the course of time or due to price reductions. In the following analysis we assess two alternative locations which are supposed to serve the same market demand. In this context, region r has a location advantage over region s, given technique τ, if $c_r(\tau) < c_s(\tau)$. Location advantage is used as an ex ante concept which means that two locations are compared with regard to both set up and variable costs. Proposition 1 makes precise how the best practice must change in order to bring about a shift in location advantage. In short, the knowledge intensity must decline sufficiently much as new techniques become available, which implies that the b-coefficient must reduce faster than the a-coefficient .

Proposition 1:
Consider the cost function in (3.4') and let the assumptions in (3.6) hold. Moreover, let the two regions be characterised by $\rho_r > \rho_s$ and $\omega_r < \omega_s$ and assume that the set up costs are the same in both regions. Suppose that region r has an advantage with technique τ. Then a shift in location advantage occurs from $c_r(\tau) < c_s(\tau)$ to $c_r(\tau + \Delta\tau) > c_s(\tau + \Delta\tau)$ if and only if the knowledge-routine ratio $\hat{b}(\tau + \Delta\tau)$ is sufficiently smaller than $\hat{b}(\tau)$.

Proof:
Region r has an advantage when $c_r(\tau) - c_s(\tau) < 0$ which implies that $a(\tau)[\rho_r - \rho_s + (\omega_r - \omega_s)b(\tau)] < 0$. This may be rewritten as $0 < (\rho_r - \rho_s) / (\omega_s - \omega_r) < \hat{b}(\tau)$. From (3.4'), the knowledge-routine ratio reduces as τ grows. Obviously, $\hat{b}(\tau + \Delta\tau)$ is sufficiently smaller than $\hat{b}(\tau)$ when $\hat{b}(\tau + \Delta\tau) < (\rho_r - \rho_s) / (\omega_s - \omega_r) < \hat{b}(\tau)$.

Remark 5 contains a discussion of how much the knowledge-routine ratio must and can reduce in order to generate a shift in location advantage.

Remark 5:
Consider the assumptions in Proposition 1. A shift in location advantage from r to s is always possible during the life cycle if $\hat{b}(0) > \dfrac{\rho_r - \rho_s}{\omega_s - \omega_r} > \hat{b}(1)$ as τ changes from $\tau = 0$ to $\tau = 1$.

Remark 5 underlines that a shift in location advantage requires that the applied technology starts with a relatively high knowledge intensity and that this intensity continues to decline. It is also necessary that the relative input prices differ sufficiently between the two regions.

Proposition 1 and Remark 5 assess two different ex ante decisions as regards where to locate the production if technique τ represents the best practice, and where to locate the same production if instead $\tau + \Delta\tau$ is the best practice. From section 3.1 we know that if an establishment using technique τ is already located in region r, that establishment may, due to sunk cost effects, resist attempts to switch to the new technique, $\tau + \Delta\tau$, irrespective of where the new technique is intended be applied. That is a normal sunk cost rigidity which slows down the process of change. However, there may also be other similar effects reflecting what has been described as lock-in effects (Arthur, 1983,1988). Such an effect will occur if the set up investment, represented by the set up cost $\kappa_r(\tau)$, is useful also when a firm selects the new technology $\tau + \Delta\tau$. This aspect is stressed in Remark 6.

Remark 6:

Consider the assumptions in Proposition 1. Suppose that a given capacity using technique τ is already located in region r, and assume that when technique $\tau + \Delta\tau$ becomes available as the best practice alternative, the establishment in r will only have to consider a reduced set up cost, signified by $\Delta\kappa_r(\tau + \Delta\tau) < \kappa_r(\tau + \Delta\tau)$. This will bring about a lock-in effect which gives an advantage to the existing establishment (or firm) such that rival firms who contemplate a location in region s need more favourable input prices than what is indicated in Proposition 1 and Remark 5.

As a final exercise in this section, a Cobb-Douglas production function is introduced to describe technology options during different stages of a product life cycle. With such a specification we show that the statements in Proposition 1 remain qualitatively unchanged.

As a first step, consider the following production function which uses technique τ and is applied in region r:

$$x_r(\tau) = K_r^{k(\tau)} A_r^{\alpha(\tau)} B_r^{\beta(\tau)} \tag{3.7}$$

where K_r denotes fixed capital, A_r routine resources, and B_r knowledge resources. Moreover, the positive exponents are indexed with the technique indicator, since the product cycle assumption implies that the technology development takes the form of gradually altered exponents, such that $\beta(\tau) / \alpha(\tau)$ diminishes as τ increases. Location cost advantages will be examined by minimising the cost function in (3.8), subject to the constraint given by (3.7):

$$c_r(\tau) = \delta_r K_r + \rho_r A_r + \omega_r B_r \tag{3.8}$$

The minimisation yields the following cost expression (cf Luenberger, 1995, p 39):

$$c_r(x_r(\tau)) = \Omega[x_r(\tau)]^{1/\Omega}[\delta_r / k(\tau)]^{k(\tau)/\Omega}[\rho_r / \alpha(\tau)]^{\alpha(\tau)/\Omega}[\omega_r / \beta(\tau)]^{\beta(\tau)/\Omega} \tag{3.9}$$

where $\Omega = k(\tau) + \alpha(\tau) + \beta(\tau)$. Next we can formulate a similar expression for region s. Then we take the logarithms of each region's cost expression and subtract the cost level in region r from the cost level in region s to obtain:

$$\ln c_r(\tau) - \ln c_s(\tau) =$$
$$k(\tau)\ln(\delta_r / \delta_s) + \alpha(\tau)\ln(\rho_r / \rho_s) + \beta(\tau)\ln(\omega_r / \omega_s) \tag{3.10}$$

With the help of (3.10) we can formulate Proposition 2, which gives essentially the same result as Proposition 1. In this case the available best practice technique changes during the life cycle in such way that β / α gradually becomes smaller. Evidently, a lower β implies a lower knowledge intensity.

Proposition 2:

Assume as in Proposition 1 that $\rho_r > \rho_s$ and $\omega_r < \omega_s$. Moreover, assume that the set up costs are the same in both regions and that both regions are candidates to supply the same market. Then a shift in location advantage occurs from

$c_r(\tau) < c_s(\tau)$ to $c_r(\tau + \Delta\tau) > c_s(\tau + \Delta\tau)$ if and only if the technology set at one point in time contains a pair $[\alpha(\tau), \beta(\tau)]$ and at a later point in time a pair $[\alpha(\tau + \Delta\tau), \beta(\tau + \Delta\tau)]$ such that $\dfrac{\alpha(\tau)}{\beta(\tau)} < \dfrac{\ln(\omega_s / \omega_r)}{\ln(\rho_r / \rho_s)} < \dfrac{\alpha(\tau + \Delta\tau)}{\beta(\tau + \Delta\tau)}$

Proof:

Consider (3.10) and the assumption that the set up costs are the same in both regions. From this follows that region r has a location advantage with technique τ iff $\alpha(\tau)\ln(\rho_r / \rho_s) + \beta(\tau)\ln(\omega_r / \omega_s) < 0$. With the same reasoning region s has a location advantage with technique $\tau + \Delta\tau$ iff $\alpha(\tau + \Delta\tau)\ln(\rho_r / \rho_s) + \beta(\tau + \Delta\tau)\ln(\omega_r / \omega_s) > 0$. Combining these two conclusions yields the desired result.

One may observe that Proposition 2 like the earlier proposition formulates a condition that combines requirements about input prices in the two regions with requirements about pairs of technique parameters (a, b) and (α, β), respectively. Moreover, in both propositions the technique changes in such a way that b/a and β/α diminish.

In Remark 6 we observe a rigidity as regards the persistence of an ex post location advantage. Consider therefore the assumptions in Proposition 2. Suppose that a given capacity using technique τ is already located in region r, and assume that when technique $\tau + \Delta\tau$ becomes available as the best practice alternative, the establishment in r will only have to consider a reduced set up cost, signified by $\Delta\kappa_r(\tau + \Delta\tau) < \kappa_r(\tau + \Delta\tau)$. In a similar way as in Remark 6, this will bring about a lock-in effect which gives an advantage to the existing establishment (or firm) such that rival firms who contemplate a location in region s need much more favourable input prices than what is indicated in Proposition 2.

3.4   Link Catastrophes

The analysis in section 3.3 compares two alternative locations while assuming that they share the same potential market and that there are no essential differences in delivery costs (interaction costs). In this subsection we go back to the cost function in (2.2) and re-specify it in the following way:

$$\pi_r(\tau) = \sum_s \pi_{sr}(\tau) z_{rs}(\tau)$$
$$\pi_{sr}(\tau) = [p_{rs} - c_{rs}(\tau)] - v_r(\tau) - \kappa_r(\tau) - \kappa_{rs}(\tau) \tag{3.11}$$

The variables have essentially the same interpretation as in (2.1) and (3.4) but are specified in detail below:

$$p_{rs} = \text{price in region } s \text{ of deliveries from region } r$$
$$c_{rs} = \text{variable interaction cost per unit delivery}$$
$$z_{rs} = \text{proportion of } r's \text{ deliveries going to } s$$
$$v_r = \rho_r a(\tau) + \omega_r b(\tau) = \text{variable production cost} \quad (3.12)$$
$$\kappa_r = \text{set up cost with regard to located production}$$
$$\kappa_{rs} = \text{set up cost with regard to delivery link } r,s$$

From (3.11) one can evaluate the competitiveness of each individual link. We consider now that a firm has located the capacity $\bar{x}_r(\tau)$ in region r and established the trade links represented by the share parameters z_{rs}. We shall analyse how a rival firm in region k may reduce its selling price p_{ks} to such low levels that a link catastrophe obtains which means that $z_{rs}(\tau)$ drops to zero. In other words, region r is exporting the flow $x_{rs} = z_{rs}(\tau)\bar{x}_r(\tau)$, and region s is trying to capture this submarket. If s manages to do this we identify a link catastrophe. First the following assumptions are introduced in (3.13) with regard to Proposition 3 and Remark 7:

The set up costs κ_r and κ_k refer to the entire capacity in r and s, respectively. We assume that the submarket x_{rs} is so small that when it is added or subtracted it generates negligible changes in κ_r and κ_k. Moreover, such changes are assumed to be negligible when the technique shifts from τ to $\tau + \Delta\tau$. We also assume that the entry condition in (3.3) satisfies $\pi_{ks} \geq \pi^* = \Delta\pi^* + \pi^o$, where π^o is the profit level which defines the exit condition in (3.1). The flow x_{rs} is assumed to exit if
$$p_{rs} - c_{rs} - v_r < \pi^o. \quad (3.13)$$

Proposition 3:
Assume as in Proposition 1 that $\rho_r > \rho_s$ and $\omega_r < \omega_s$, and add the assumptions in (3.13). Moreover, let the variable link costs be approximately the same so that $c_{rs} \approx c_{ks}$ and assume that technique $\hat{\tau}$ does not allow for a link catastrophe. Such a catastrophe obtains iff $\hat{b}(\tau + \Delta\tau) = b(\tau + \Delta\tau) / a(\tau + \Delta\tau)$ is sufficiently low to satisfy $\hat{b}(\tau + \Delta\tau) < [(\rho_r - \rho_s) - \hat{\kappa}_{ks} - \Delta\hat{\pi}^*] / (\omega_s - \omega_r)$, where $\hat{\kappa}_{ks} = \kappa_{ks} / a(\tau + \Delta\tau)$, and $\Delta\hat{\pi}^* = \Delta\pi^* / a(\tau + \Delta\tau)$

Proof:
The exit and entry conditions can be combined as follows:
$$\pi_{ks} = p_{ks} - c_{ks}(\tau + \Delta\tau) - v_r(\tau + \Delta\tau) - \kappa_{ks} \geq$$
$$\pi^* > p_{rs} - c_{rs}(\tau + \Delta\tau) - v_k(\tau + \Delta\tau) + \Delta\pi^*. \text{ Since we examine a price com-}$$
petition game, it must hold that $p_{ks} = p_{rs}$. We have also assumed that the variable interaction costs are approximately the same for both supply regions. Thus, we obtain the following reduced condition: $v_r(\tau + \Delta\tau) > \kappa_{ks} + v_k(\tau + \Delta\tau) + \Delta\pi^*$,

which yields $(\rho_r - \rho_s) > \hat{\kappa}_{ks} + (\omega_s - \omega_s)\hat{b}_k + \Delta\hat{\pi}^*$ Rearranging this last expression gives the desired result.

One may observe that if $\Delta\pi^* = 0$ and $\pi^o = 0$ the link catastrophe will be governed by zero profit exit and entry conditions, which corresponds to perfect competition assumptions.

Remark 7:

Assume as in Proposition 1 that $\rho_r > \rho_s$ and $\omega_r < \omega_s$, add the assumptions in (3.13) and let $\Delta\pi^* = 0$ and $\pi^o = 0$. Moreover, let the interaction costs satisfy $c_{ks} < c_{rs}$ and assume that technique τ does not allow for a link catastrophe. Then such a catastrophe obtains if $\hat{b}(\tau + \Delta\tau) = b(\tau + \Delta\tau) / a(\tau + \Delta\tau)$ becomes sufficiently low to satisfy $v_r(\tau + \Delta\tau) = \kappa_{ks} + v_k(\tau + \Delta\tau)$.

Proof:

Remark 7 can be established in the same way as Proposition 3. The new element is that differences as regards the variable interaction or link costs are allowed to trigger the link catastrophe.

3.5 Temporary Market Equilibrium Solutions

In this subsection we use the expression in (3.11) as a starting-point for developing a complete multiregional model of capacity location and delivery flows. In a first version, all suppliers are assumed to have the same degree of standardisation, i.e., θ is given and the same for all flows. Moreover, in each supply region there is one or several capacities, and each capacity employs of given technique, τ. We assume that there is a discrete set of existing techniques and that there is an optional best practice technique denoted by τ^*. For each region we shall now refer to an overall gross profit, Π_r, from which only ex ante set up costs, $\kappa_r x_r(\tau^*) + \kappa_{rs}(\Delta x_{rs}; \overline{x}_{rs})$, should be deducted. The gross profit expression then satisfies

$$\Pi_r = \sum_s V_{rs} x_{rs} - \sum_\tau v_r(\tau) x_r(\tau) - \kappa_r x_r(\tau^*) - \kappa_{rs}(\Delta x_{rs}; \overline{x}_{rs})$$

$$V_{rs} = p_{rs} - c_{rs} \text{ and } v_r(\tau) = \rho_r a(\tau) + \omega_r b(\tau)$$

(3.14)

where for each value of $\Delta x_{rs} = (x_{rs} - \overline{x}_{rs})$, $\kappa_{rs}(\Delta x_{rs}; \overline{x}_{rs})$ is smaller the larger $\overline{x}_{rs} = z_{rs} \overline{x}_r$ is. With this specification we can assume that the partial market solution is obtained when total profits are optimised, subject to the following constraints (3.15)-(3.17):

$$\overline{x}_r(\tau) \geq x_r(\tau); \ \sigma_r(\tau)[\overline{x}_r(\tau) - x_r(\tau)] = 0, \text{ all } \tau; \text{ i.e., for each technique,}$$

the existing capacity must exceed the associated production, $\sigma_r(\tau) \geq 0$.

(3.15)

$$\sum_\tau x_r(\tau) \geq \sum_s x_{rs}; \quad \sigma_r[\sum_\tau x_r(\tau) - \sum_s x_{rs}] = 0, \text{ i.e., total capacity}$$

must excede total sales from r , $\sigma_r \geq 0$. $\qquad$ (3.16)

$$\sum_r x_{rs} \geq D_s; \quad \tilde{\sigma}_s[\sum_r x_{rs} - D_s] = 0, \text{ supply must excede demand in}$$

s, $\tilde{\sigma}_s \geq 0$. $\qquad$ (3.17)

Let us initially assume that the multiregional market is characterised by an undifferentiated product and by competitive market relations such that $p_{rs} = p_s$, for all r. Then one can describe the short-medium term competition by maximising (3.14), subject to (3.15)-(3.17). Differentiating the associated Lagrange function yields the market equilibrium conditions as follows:

$$p_s + \tilde{\sigma}_s \begin{cases} = c_{rs} + \sigma_r, \text{ as } x_{rs} > 0 \\ < c_{rs} + \sigma_r, \text{ as } x_{rs} = 0 \end{cases} \qquad (3.18)$$

$$\sigma_r \begin{cases} = \sigma_r(\tau) + p_r a(\tau) + \omega_r b(\tau), \text{ as } x_r(\tau) > 0 \\ < \sigma_r(\tau) + p_r a(\tau) + \omega_r b(\tau), \text{ as } x_r(\tau) = 0 \end{cases} \qquad (3.19)$$

where $\tilde{\sigma}_s \geq 0$ signifies how much the price in region s has to be augmented in order to avoid insufficient supply with D_s given, $\sigma_r > 0$ signifies the supply price in region r, and $\sigma_r(\tau)$ denotes the extra profit (quasi rent) associated with technique τ. One may also refer to $\sigma_r(\tau)$ as the gross profit per unit output produced with technique τ. Formula (3.18) gives the conditions for a positive delivery flow from r to s. By combining (3.18) and (3.19) one can show that in equilibrium, disadvantageous prices such that $(\rho_k - \rho_r)a(\tau) + (\omega_k - \omega_r)b(\tau) > 0$ may be compensated by advantageous interaction costs such that $c_{ks} < c_{rs}$. From (3.18) also follows that existing techniques in a region can be ordered in a sequence $\tau_0 < \tau_1 < ...$ such that $\sigma_r(\tau_0) < \sigma_r(\tau_1) < ...$, and where τ_0 represents the worst practice technique. Moreover, $\sigma_r(\tau_i) + v_r(\tau_i) = \sigma_r(\tau_m) + v_r(\tau_m)$ for two techniques i and m which both are employed. Given this we can formulate the following proposition describing introduction of new techniques (new vintages).

Proposition 4:
An equilibrium solution to (3.14)-(3.17) has the following two implications: (i) With constant demand between two periods, a new capacity $\bar{x}_r(\tau^*)$ can be introduced in the second period iff $\sigma_r(\tau^*) + v_r(\tau^*) + \kappa_r < \sigma_r(\tau_0) + v_r(\tau_0)$, given that $\bar{x}_r(\tau^*) \leq \bar{x}_r(\tau_0)$. (ii) Suppose that demand increases on links (r,s) by Δx_{rs} such that the total growth exceeds the minimum capacity level associated with technique t^*, which means that $\sum_s \Delta x_{rs} \geq \bar{x}_r(\tau^*)$. Then the new technique can be introduced in r iff $\sigma_r = \sigma_r(\tau^*) + v_r(\tau^*) + \kappa_r = p_s + \tilde{\sigma}_s - c_{rs} - \kappa_{rs}(\Delta x_{rs})$.

Proof:
Consider the solution in (3.18)-(3.19) and formulate the extended version of it which obtains as $\tau = \tau^*$. The proposition follows from this specification together with the observation that $\sigma_r(\tau_i) \geq 0$ for each technique i which is actively employed.

Next, consider the effects of agreements and contracts which imply that in the short term certain deliveries remain intact also when new rivals try to sell at lower prices. Let q_{rs} denote the share of the demand D_s which represents the established pattern of delivery. Such patterns are included in the solution if one adds the constraint (3.20) to the system given by (3.14)-(3.17). The new constraint has the following form:

$$x_{rs} \geq q_{rs}D_s; \quad \tilde{\sigma}_{rs}[x_{rs} - q_{rs}D_s] = 0. \tag{3.20}$$

With this constraint the new solution has the following link-specific price:

$$p_{rs} + \tilde{\sigma}_s = \sigma_r(\tau) + v_r(\tau) + c_{rs}; \quad p_{rs} = p_s + \tilde{\sigma}_{rs}. \tag{3.21}$$

where $\tilde{\sigma}_{rs}$ represents a rigidity-based price effect on short term deliveries between region r and s. This rigidity may be thought of as a link between sellers in region r and customers in region s, and this type of link is assumed to change only gradually when the relative competitiveness of supply regions change.

As a final exercise in this subsection we reintroduce the possibility to supply products of different varieties within a product group competing to attract the same customer groups in a market. For each market s we identify its average θ-value denoted by $\overline{\theta}_s$. Products supplied and sold on market s may be more or less standardised, where customisation increases as standardisation decreases. At the same time combinations of τ and θ must belong to the available technology set $\Gamma \ni (\tau, \theta)$. Disregarding new products and techniques we may reformulate (3.14) as follows:

$$\Pi_r = \sum_s V_{rs}(\theta)x_{rs}(\theta) - \sum_\tau v_r(\tau, \theta)x_r(\tau, \theta)$$
$$V_{rs}(\theta) = p_{rs}(\theta) - c_{rs} \text{ and } v_r(\tau, \theta) = \rho_r a(\tau, \theta) + \omega_r b(\tau, \theta) \tag{3.14'}$$

where $p_{rs}(\theta)$ represents the price established during the previous period with regard to deliveries from r to s of products with a θ-degree of standardisation. In relation to (3.14') the following constraint set is introduced:

(i) $\overline{x}_r(\tau, \theta) \geq x_r(\tau, \theta)$, all (τ, θ) with $\sigma_r(\tau, \theta)$ as Lagrange multiplier;

(ii) $\sum_\tau x_r(\tau, \theta) \geq \sum_s x_{rs}(\theta)$, all θ, with $\sigma_r(\theta)$ as multiplier;

(iii) $\sum_r x_{rs}(\theta) \geq D_s(\theta)$, with $\tilde{\sigma}_s(\theta)$, as multiplier;

(iv) $x_{rs}(\theta) \geq q_{rs}(\theta)D_s(\theta)$, with $\tilde{\sigma}_{rs}(\theta)$ as multiplier.

$$(3.22)$$

where total demand $D_s = \sum_\theta D_s(\theta)$, $D_s(\theta)$ is determined with respect to prices during the preceding period, and $q_{rs}(\theta)$ represents established contracts and agreements which influence the delivery structure during the current period. A solution which can represent a temporary equilibrium must satisfy:

$$\sigma_r(\theta) = v_r(\tau,\theta) + \sigma_r(\tau,\theta), \text{ given that the extra profit } \sigma_r(\tau,\theta) \geq 0$$

$$p_{rs}(\theta) + \tilde{\sigma}_s(\theta) + \tilde{\sigma}_{rs}(\theta) = \sigma_r(\theta) + c_{rs}, \text{ given that } x_{rs}(\theta) > 0.$$

$$(3.23)$$

For each supply region in this model, it is the link with the highest gross profit, $V_{rs} - v_r$, which determines how much the price may be raised on other links for which there is no excess supply. When more than marginal adjustments are necessary, the model exercise must also incorporate how price alterations affect the demand for each region s, where $D_s(\theta) = D_s(p_{1s}(\theta_1),...,p_{rs}(\theta_r),...)$.

If the price from the previous period, $p_{rs}(\theta)$, is an equilibrium price, such that (iii) and (iv) in (3.22) are not binding, then we have that both $\tilde{\sigma}_{rs}(\theta) = 0$ and $\sigma_s(\theta) = 0$. For $\tilde{p}_{rs} = p_{rs} + \tilde{\sigma}_s + \tilde{\sigma}_{rs}$, the price dispersion will have the following properties:

If $\sigma_r(\tau,\theta) \geq \sigma_r(\tau,\overline{\theta})$, as $\theta < \overline{\theta}$, then $\tilde{p}_{rs}(\theta) > \tilde{p}_{rs}(\overline{\theta})$, $c_r(\tau,\theta) > c_r(\tau,\overline{\theta})$.

If $x_{rs} > 0$ and $x_{ks} > 0$, $\tilde{p}_{rs}(\theta) - \tilde{p}_{rk}(\theta) = c_{rs} - c_{rk}$.

$$(3.24)$$

What can we say about the existence of the partial equilibrium described? All cost patterns can be organised in well behaved stair-cases. Hence, the short term equilibrium described exists if the supply capacities are sufficiently large, i.e., if the following holds:

$$(i) \ \sum_{r,\tau} \overline{x}_r(\tau,\theta) \geq \sum_s D_s(\theta), \text{ all } \theta;$$

$$(ii) \ \sum_\tau \overline{x}_r(\tau,\theta) \geq \sum_s q_{rs}(\theta) D_s(\theta)$$

$$(3.25)$$

4. Complex and Aggregate Product Cycles

4.1 Where Are New Product Cycles Initiated?

In which type of economic milieu are new product cycles introduced with a high frequency? Which are the corresponding location attributes? Can we classify urban regions with regard to their expected frequency of births of product cycles and early imitations of new product cycles. As has been illustrated in section 3, once the initial introduction has taken place, the model predicts a time-space hierarchy of competition-driven changes of the location pattern. In the early stages of the life cycle of a product or a whole product group there may be diminutive differences between original initiatives and associated imitation activities; in both cases it is a process of information gathering and creative experiments.

In Andersson and Johansson (1984b) one finds a set of statements describing why and how new product cycles are started with different frequency in different types of regions, as described by the following quotation: "This theory essentially states that each product undergoes a development cycle in which each new commodity enters the most highly developed regions of the world after a phase of research, laboratory testing, and implementation development. The product is then primarily produced in the region with a comparative advantage in terms of high R&D level and access to employment categories with a required profile of competence. The product is exported from this region to other regions."

It is also observed in Andersson and Johansson (1984b) that coincidences and seemingly random events may play an important role in each individual case, and that the existence of increasing returns to scale due to cumulative learning and other types of path dependency imply that the initiating place may create and maintain its location advantage in a process of dynamic reinforcement. With all respect to chance, observations tell us that innovations and other novelties are more likely to emerge in certain types of economic environments than in others. Our interest here is to identify location attributes which give a region (urban region) an advantage as a birthplace for new products and a place for early imitations. The following enumeration provides a picture of attributes which increase the region's likelihood to have a high frequency of product development:

(i) The region should have information channels which continuously brings messages to the entrepreneurs of the region about new products which are introduced across the world economy. Regions with diverse and dense import flows can be expected to have such information links. We may call this feature *import intensity*.

(ii) The firms in the region should have a high accessibility to a labour market with a rich composition of education and competence profiles. We may call this feature *knowledge intensity*.

(iii) The region should offer existing and potential firms a high accessibility to R&D competence. We may call this feature R&D *intensity*.

(iv) The region should have high accessibility to customers which are likely to be early candidates prepared to try out new products. We may call this feature *customer intensity*.

(v) The region should have a high accessibility to suppliers of specialised components and services which are especially important during the development and/or renewal phases of each product. We may call this feature *supplier intensity*.

The location attributes above include localisation economies and the wider concept agglomeration economies as discussed early by Marshall (1920). The attributes are also influenced by accessibility properties of the urban region, and by its infra-structure and quality as a built environment. These location attributes affect the supply conditions and prices on inputs that the individual firm buys, as reflected by the local prices ρ_r and ω_r in our model. Other aspects of the attributes have the nature of public goods, i.e., they influence the economic activities without causing the individual firm any direct charge, although the firm may experience higher land values and hence higher costs of premises as the location attributes become more advantageous. In Andersson (1985) regions which satisfy conditions (i)-(v) are called C-regions, stressing that their economic milieu has a rich endowment of C-resources (e.g. communication, cognitive capacity, creativity).

The location attributes in an urban region together with the composition of economic activities constitute the region's economic milieu. Let m_r denote the quality of the economic milieu as regards productivity and similar influences for which the firms are not charged directly. Then we may reformulate the profit function in (2.1') as follows:

$$\pi_r(x_r) = \sum_s z_{rs}[p_{rs}(x_{rs}, m_r) - c_{rs}(x_{rs}, m_r)] - c_r(x_r, m_r)$$

$$\partial p_{rs} / \partial m_r \geq 0; \ \partial c_{rs} / \partial m_r < 0; \ \partial c_r / \partial m_r < 0.$$

(4.1)

In particular, the inequality signs apply when attribute tensions are caused by emerging congestion phenomena. The message in (4.1) may be further stressed by referring to Remark 3 and reformulating the pertinent cost function to include the variable describing the quality of the economic milieu. This is done in formula (4.1')

$$c_r(x_r, m_r)x_r = \sum_s \sum_i p_{sri}(m_r)b_{sri}x_r(m_r)$$

$$\partial p_{sri} / \partial m_r < 0; \ \partial b_{sri} / \partial m_r < 0.$$

(4.1')

The effect on input prices has already been incorporated in the analysis in the preceding sections, in which we have noted that ρ_r and ω_r vary for different regions (with different milieu properties). The effect on input coefficients is an additional consequence of the productivity caused by the economic milieu. In this section we are especially interested in the costs of development resources and the efficiency of R&D activities as well as the interaction with suppliers and customers in the product development efforts.

In order to illustrate how the economic milieu may effect input prices we may go back to the optimisation model specified in (3.14)-(3.17). To this model we add two constraints which, for a given economic activity and a given point in time, inform us about the availability of routine and knowledge-intensive resources, given the use of such resources in all other activities in the region. Region r's activity-specific supply of routine and knowledge resources at a given point in time is denoted by A_r and B_r, respectively. The two constraints are formulated as follows:

$$A_r \geq \sum_\tau a(\tau)x_r(\tau), \text{ each } r, \text{ with } \mu_{ra} \text{ as multiplier;}$$

$$B_r \geq \sum_\tau b(\tau)x_r(\tau), \text{ each } r, \text{ with } \mu_{rb} \text{ as multiplier}$$

(4.2)

With these additional constraints, the market solution changes as described in Remark 8.

Remark 8:

Consider the optimisation model in (3.14)-(3.17) and add the constraints in (4.2). Then the Kuhn-Tucker conditions imply that the short-term market equilibrium solution has to satisfy the following condition:

$$As \ x_r(\tau) > 0, \quad \sigma_r = \sigma_r(\tau) + (\rho_r + \mu_{ra})a(\tau) + (\omega_r + \mu_{rb})b(\tau)$$

How can the condition in the remark be interpreted? Consider an economy which develops in a sequence of periods. Suppose that the local prices, ρ_r and ω_r, have been established in period t-1. Then Remark 8 informs about the necessary adjustment in period t. In an economic milieu which has a general advantage for product development, our model says that the price of knowledge resources, ω_r, should be low compared with the same price in other regions. If such a region is successful in its product development and product renewal activities, one should expect that the demand for knowledge resources continues to grow – and thereby making μ_{rb} positive. If this development goes on, the price of knowledge resources will continue to grow over time. As a consequence, the region's location advantage for product development will gradually diminish.

The above observation also implies that a location advantage as regards product development can remain in an expanding regional economy only if the critical resource, B_r, continues to grow. With regard to a particular product group (or type of economic activity) this condition can be satisfied in two ways. The total amount of knowledge-oriented activities may grow in the region. Over time this will bring about tensions with regard to the region's infrastructure and the economic milieu in general. Interaction networks will become overloaded and land values will dampen the expansion.

There is a second way to provide growth opportunities for the given type of knowledge-oriented economic activities. In this second type of dynamic process, activities with lower knowledge requirements are gradually leaving the region. The reduction of such activities will in the model have the form of relocation as long as the multiregional demand is large enough to stimulate expansion in regions with a more favourable price of routine resources.

We may summarise the observations related to the temporal solution described in Remark 8 as follows. In order to remain a place for product development and innovations, a region must continue to have prices which are comparatively high for routine resources and low for knowledge resources. A high level of ρ_r helps to push maturing activities out of the region to provide space for knowledge activities. In this context, and in the spirit of von Thünen (1866), one may observe that high land values may both represent high costs of space (associated with routine activities) and low costs of accessibility and nearness (density). A rich supply of B_r-resources helps to foster the expansion of existing knowledge-oriented production and stimulates the start of new knowledge activities. In particular, the price variable ω_r represents the costs of all frictions associated with acquiring and purchasing knowledge resources. A rich supply of B_r-resources keeps this form of friction at a low level. We shall emphasise these conclusions further in the subsequent sections.

4.2 Specialisation and the Economic Milieu in Urban Regions

Andersson and Johansson (1984b) classify metropolitan regions as the most important nodes for fostering and imitating innovations. For one thing, they are prime markets for new products. However, the following quotation also reveal the Janus face of large, often multinodal, urban regions: "The prototype metropolitan region is sometimes recognised as a system in which richness of ideas and activities provide an environment for creativity and novelty by combination. This image may be contrasted with an also common picture of an ageing metropolitan region which is characterised by large-scale production and a low frequency of innovations. A transition from the first to the second state represents a loss of long-term creativity and a gain in short-term productivity."

The analysis in sections 2 and 3 is strictly micro-oriented in the sense that the focus is on a single product or a "narrow" product group, containing goods or services which are differentiated but close substitutes. In this subsection we consider broader categories of products which may still be identified as clusters or groups. In Marshall (1920) one may find arguments why such broader product groups should be expected to be located close together with a high frequency. Marshall and followers have recognised localisation economies as the main explanation of this type of clustering together in space. Three main sources of localisation economies are then identified for a region with a relative specialisation on a particular product group or cluster of associated economic activities. The first source refers to a regional formation of a pooled market for employment categories with specialised skills and competence. The competence profile of such a labour market is adjusted to match the demand from the cluster of industries supplying products within the same product group (associated products). The second localisation source is the provision of non-traded inputs and other influences from the economic milieu in an urban region. According to this argument, the economic milieu tends to specialise in such a way that it supports the production profile in the region. In other words, the economic milieu is assumed to provide inputs of a public goods nature to the economic activities.

The third source has to do with the flow of information of importance to the cluster of firms supplying the associated products. Such information flows comprise technological spillovers, market and customer group knowledge, information about suppliers, as well as R&D-results.

A forth source of localisation economies follows by analogy from the Marshall-inspired arguments above. This fourth source can be described as a pooled market of suppliers which offer producer services and other specialised inputs to the group of firms producing the associated products.

The observations put forward above support the following assumption: individual but associated product cycles tend to develop in parallel clusters, with a base in the same urban region or in a set of interconnected regions, having a high mutual accessibility. When such a specialisation pattern obtains we shall use the term *product group specialisation* of a region. Given this assumption one may formulate models of product group specialisation, referring to aggregate product cycles. Such product groups should then be expected to follow sigmoid development paths which consist of combined individual S-shaped product life-cycle trajectories. Some of these trajectories may be in an expanding and others in a contracting phase. However, from the Marshall arguments above one can expect that the individual trajectories will develop in a partially synchronised concert, such that over time the aggregate cluster will have a distinct phase of growth which is followed by a process of decline. Observations indicate that often these aggregate paths cover several decades.

According to the model developed in this paper, product group specialisation should in general be stronger in regions which host the initiators and early imitators of a given cluster of associated products. When an individual product becomes more standardised and employs more routinised production methods, the demand for a complex economic milieu becomes weakened. However, there may still exist localisation economies also during relocation phases of each product belonging to a given product group.

As emphasised earlier in this section this switch may be severely delayed in an urban region, thereby causing maturity and obsolescence phenomena in the entire region. In an urban region which manages to retain its innovation capacity and creativity, the dynamics must satisfy the conditions illustrated in figure 3. The message

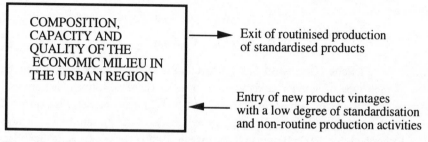

Figure 3. Self-organised renewal of the economic milieu of a vital urban region

is that when routinised production of mature and standardised products continue to stay in an urban region, they will occupy valuable space and infrastructure and obstruct the entry of new economic activities. If such a situation is prolonged, one should also expect a degradation of the economic milieu of the region.

Section 4.1 specifies five categories of location attributes which are assumed to provide advantages for non-routine and innovative activities, including product development. According to our theoretical framework, such attributes of an urban region's infrastructure and economic milieu should facilitate the adjustment of individual time-tables for the coordination of meetings between persons. They reduce the friction with regard person interaction in general. Facilities, buildings and systems in many of the region's economic zones should have a general design in order to allow new activities to replace older ones, as the latter exit from a zone.

The dynamic mechanisms of a vital urban region can ascertain the usefulness of the economic milieu and infrastructure by continuously stimulating the exit of maturing and obsolete activities to other locations, while simultaneously facilitating the entry of new activities into the same urban environment (figure 3). These arguments may be summarised by an aggregate model of urban dynamics which is presented in two steps. First the supply side is emphasised in (4.6), and then the demand side is added in (4.7).

Let I denote the group of routine-oriented, standardised activities, and let II denote non-routinised, knowledge-oriented activities. The output from activities of type I is called type I products, and the output from the second type of activities is called type II products. Moreover, let A_r and B_r represent the routine-oriented resources and the knowledge resources, respectively in region r, and let these resources comprise both market resources and other parts of the economic milieu. We also need four technical coefficients. By a_{Ir} and b_{Ir}, we denote how activities of type I use A_r and B_r resources per unit of activity (output). Activities of type II are assumed to consume from the same resources according to the coefficients a_{IIr} and b_{IIr}. With these definitions we can introduce the following model:

$$\dot{x}_{IIr} = f(x_{IIr})min \begin{cases} [A_r - a_{Ir}x_{Ir} - a_{IIr}x_{IIr}] \\ [B_r - b_{Ir}x_{Ir} - b_{IIr}x_{IIr}] \end{cases}$$

$$\dot{x}_{Ir} = f(x_{Ir})min \begin{cases} [A_r - a_{Ir}x_{Ir} - a_{IIr}x_{IIr}] \\ [B_r - b_{Ir}x_{Ir} - b_{IIr}x_{IIr}] \end{cases}$$

(4.6)

where the functions $f(x_{Ir})$ and $f(x_{IIr})$ specify the speed of adjustment, given the size of each type of activity at each point in time. These adjustment factors will express self-reinforcing adjustments if $df / dx > 0$. In many models f is expressed as $f(x_{Ir}) = \gamma_{Ir}x_{Ir}$ and $f(x_{IIr}) = \gamma_{IIr}x_{IIr}$, where γ_{Ir} and γ_{IIr} are two constants (e.g. Fisher and Pry, 1971; Batten and Johansson, 1989). The model is further characterised in Remark 9.

Remark 9:
The model in (4.6) is designed to be compatible with product cycle assumptions as expressed in (2.3) and (3.6). This requires that the following condition is satisfied $b_{IIr} / a_{IIr} > b_{Ir} / a_{Ir}$, which implies that the knowledge-routine ratio is higher in non-routine than in routine activities.

Consider an equilibrium such that both the routine and non-routine activities are positive. Any such equilibrium is unstable, and once out of equilibrium the dynamics will describe how one of the activities expand while the other declines, e.g. $\dot{x}_{IIr} > 0$ and $\dot{x}_{Ir} < 0$. The change process is characterised by competitive exclusion. And the time profile is described by one growing and one declining sigmoid curve. It is also evident from the equations in (4.6) that if x_{Ir} is not being reduced or if the pertinent process is very slow, then the entry of non-routine activities will be delayed or completely obstructed.

In order to make the analysis more precise and complete we shall add several features to the model outlined in (4.6). First, assume that there is a technique renewal process such that non-routine activities gradually are transformed to routine activities. In order to illustrate this idea we assume that there is an ongoing transformation of techniques of type II given by $\xi_{IIr} x_{IIr}$, where $0 < \xi_{IIr} < 1$ is a coefficient expressing the rate of transformation. Thus, during the time interval dt, the amount $\xi_{IIr} x_{IIr} dt$ is assumed to leave the existing capacity of type II and at the same time add to the size of x_{Ir}. Using the terminology of section 2, activities of type II refer to low $\tau - values$, while activities of type I refers to high $\tau - values$. Hence, $\xi_{IIr} x_{IIr}$ may be thought of as a shift of technique vintage.

Next, let the demand for products of type II in region s be denoted by D_{IIs} and let the corresponding demand for products of type I be denoted by D_{Is}. At each given point in time t, $\eta_{IIrs} D_{IIs}$ and $\eta_{Irs} D_{Is}$ represent the demand for products of type II and I which is directed towards region r from regions s. The coefficients η_{IIrs} and η_{Irs} reflect the competitiveness of the two activity categories in region r and may vary over time. A decrease of such coefficients implies reduced market shares for the corresponding type of product, and this may bring about an exit of the pertinent capacity.

Let us start with the resource-constrained change process in (4.6). To this we add in formula (4.7) a demand constraint for each of the two types of output, I and II, and a proportional transformation of non-routine to routine activities.

$$\dot{x}_{IIr} = f(x_{IIr})min \begin{cases} [A_r - a_{Ir}x_{Ir} - a_{IIr}x_{IIr}] \\ [B_r - b_{Ir}x_{Ir} - b_{IIr}x_{IIr}] \\ [\sum_s \eta_{IIrs}D_{IIs} - x_{IIr}] \end{cases} - \xi_{IIr}x_{IIr}$$

$$(4.7)$$

$$\dot{x}_{Ir} = f(x_{Ir})min \begin{cases} [A_r - a_{Ir}x_{Ir} - a_{IIr}x_{IIr}] \\ [B_r - b_{Ir}x_{Ir} - b_{IIr}x_{IIr}] \\ [\sum_s \eta_{Irs}D_{Is} - x_{IIr}] \end{cases} + \xi_{IIr}x_{IIr}$$

Given that the resources A_r and B_r remain unchanged over time, the output from activities of type II can, with sufficient demand, continue to grow if activities of type I continue to reduce their input consumption of the same resources. The latter is in this case equivalent with a gradually declining output of type I. One should then observe that if such a change process is constrained by the available supply of B-resources, the process will generate an excess supply of A-resources. This follows from the specification of the knowledge-routine ratios in Remark 9. Hence, for longer time periods one must also contemplate movements into and out of the region of the two categories of resources.

The technique renewal coefficient ξ_{IIr} describes how non-routine activities change technique and are transformed to routine activities. These represent product vintages which are younger than the average vintage level of type I products. Hence, if $\xi_{IIr}x_{IIr} > 0$, this should be expected to increase the competitiveness of type I activities. This may be assumed to affect the delivery coefficients η_{Irs} such that they increase or remain on a high level. If the competitiveness of type I activities improve in this way over a long time period, type II activities may be forced to decline.

In summary, the dynamic model provides a framework for an analysis of the complex dynamics of an urban region's internal structure and its interplay with its external networks. It can be used to demonstrate our earlier discussion in association with figure 4.1. The complex dynamics may generate a sequence of economic episodes such that one can observe time intervals in which type II activities dominate, while type I activities dominate in other intervals. Moreover, the model may generate waves of aggregate growth and decline episodes (e.g. Johansson and Nijkamp, 1987).

4.3 Products, Product Groups and Clusters of Products Cycles

This subsection returns to the discussion in subsection 2.1 about product and customer groups. We outline a method to identify and define a product, a product group and a customer group. Moreover, we examine and assess the conditions for aggregating across product and customer groups. Substitution processes are defined for arbitrary levels of aggregation. The subsection also presents an approach to identify and interpret aggregate waves describing growth and decline phases of product groups and industries. It is suggested that such waves can be analysed by identifying the constituting product cycles at a disaggregate level. For example, one may specify the individual product cycles which are grouped together as type *II* activities in (4.6) and (4.7).

Consider the concept of a customer group. It consists of a certain type of agents which are buyers and potential buyers of a particular set of products. This set of products is characterised by the following two things: (i) each of the products can satisfy a basic need of the customers in the customer group, and as a consequence (ii) there is a set of attributes which the products have in common, although each product has an individual composition of these attributes. Thus, the products differ by combining the attributes in different proportions. Each product may in addition have extra attributes which are not shared by the other members of the product group. Such extra attributes may be relevant for other customer groups. A typical customer group may be a set of firms in an industry, where all firms use products from the same product group as inputs in their production processes.

Let g denote a customer group and let i denote a product belonging to the product group h. The delivery of product i to customer group g at time t is denoted by $x_{ig}(t)$ and the total delivery of all products is given by

$$x_{hg} = \sum_{i \in h} x_{ig} \qquad (4.8)$$

The substitution between product i and the other members of the product group is defined as the trajectory of $z_{ig}(t) = x_{ig}(t) / [x_{hg}(t) - x_{ig}(t)]$. Generically, a substitution process generated by the type of model described in (4.6) and (4.7) follows an upward or downward sloping sigmoid curve. A second level of aggregation obtains when customer groups are added together to a grand set of customers G such that $g \in G$. In a similar way product groups h may be organised to a cluster H such that $h \in H$. In this case the substitution variable will have the form

$$z_{hG} = x_{hG} / [x_{HG} - x_{hG}]$$
$$x_{hG} = \sum_{i \in h, g \in G} x_{ig}; \; x_{HG} = \sum_{h \in H, g \in G} x_{hg} \qquad (4.9)$$

Formula (4.9) shows that if an aggregate substitution process has been constructed as suggested here, then it can also be decomposed into separate substitution dynamics for both product groups and individual products. As a consequence, aggregate substitution paths may be analysed as composed by clusters of disaggregate product cycles.

A particular aggregation of customer groups obtains if customers are grouped together depending on which region they are located in. One may for example consider a dichotomisation into domestic customers and all non-domestic ones. In particular, one may examine a product i belonging to a product group h separately for the domestic market D and for the foreign market F. First, let us define the two substitution variables:

$$z_{iD} = x_{iD} / [x_{hD} - x_{iD}] \; and \; z_{iF} = x_{iF} / [x_{hF} - x_{iF}] \qquad (4.10)$$

The typical pattern for this type of constellation is that z_{iD} initially has a higher value than z_{iF}, while the latter expands along a steeper curve as illustrated in figure 4 (Batten and Johansson, 1989).

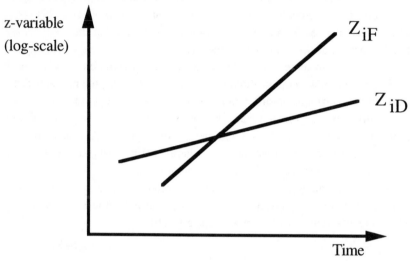

Figure 4. Expanding substitution paths on the domestic and foreign market

4.4 Aggregate Specialisation Waves

According to the theory and model exercises presented in previous sections, each spatial product cycle will generate a pattern of regional specialisation. Moreover, such patterns will change in a systematic way over time. The literature contains several alternative measures of specialisation, most of them with a considerable degree of compatibility. The availability of data in each individual study seems to determine which measure a researcher selects. In this section two different measures will be used in the discussion of specialisation waves. The first, called *location intensity*, compares a single regional area in a multiregional context. The second, called *self-sufficiency ratio*, compares for an economic area the intraregional delivery with the total delivery to the area.

The location intensity measures the regional concentration of the production of a product group. Such a measure is based on a reference value, y_r, for each region r. This reference value may e.g. be the total gross domestic product of the region. If the product group belongs to the manufacturing industry, y_r may represent total manufacturing output in the region. Let x_r denote the size of the activity referring to the product group which is investigated, and let y_r be the corresponding reference value. Then the location intensity, Ω_r, is defined as follows:

$$\Omega_r = [x_r / y_r] / [\sum_r x_r / \sum_r y_r] \qquad (4.11)$$

In Forslund and Johansson (1995) x_r and y_r are measured as the number of persons employed. Based on this measure they demonstrate specialisation waves for around 100 product groups with regard to urban regions in Sweden and Norway. In Andersson and Johansson (1984b) a related analysis is carried out to examine the location intensity of employment categories, recorded with a detailed specification. These are organised into three main categories describing the nature of each occupation: (i) knowledge, (ii) service, and (iii) goods. Table 2 shows how the location intensity of these occupation categories vary across three types of functional regions.

Table 2. Location intensity of occupational categories (Swedish regions, 1975).

Location in Regions Characterised by	Knowledge Handling Occupations	Service Handling Occupations	Goods Handling Occupations
Intraregional accessibility	**	*	*
Interregional accessibility		*	**
Land use intensity, spatial dispersion			**

One star signifies a clear correlation, and two stars a strong correlation.

The analysis illustrated by table 2 has been extended in Andersson (1986, 1988), Anderstig and Hårsman (1986). It emphasises that localisation and scale phenomena have at least two dimensions. One concerns the location of economic activities (firms) and the other concerns the location of employment categories. By combining these two dimensions we refer, once more, back to Marshall. In Anderstig and Hårsman (1993) the location of occupational categories is examined in a dynamic setting. In particular, they estimate econometrically the interurban migration of knowledge-oriented labour.

The major illustration of product cycles in Andersson and Johansson (1984b) makes use of the following definition of regional specialisation (for a specific product group):

$$S_r = x_{rr} / \sum_s x_{sr} = (x_r - e_r) / (x_r + m_r - e_r) \qquad (4.12)$$

where S_r signifies the self-sufficiency ratio of region r, x_{rr} the deliveries from suppliers in region r to customers in region r, and x_{sr} the deliveries from region s to region r. They apply this measure to countries, and then it may equivalently be defined for country r as production, x_r, minus export, e_r, divided by total consumption in country r, $(x_r + m_r - e_r)$. In particular, Andersson and Johansson (1984a and 1984b) investigates the specialisation of all OECD countries, taken as a group, in relation to the rest of the world. They define a value, S^*, which divides observations into aggregate product groups (industrial sectors) with high and low specialisation in the OECD group. In addition, the rate of change of S is examined for the period 1971-1980. With this construction all industrial sectors are classified according to the "phase diagram" in figure 5.

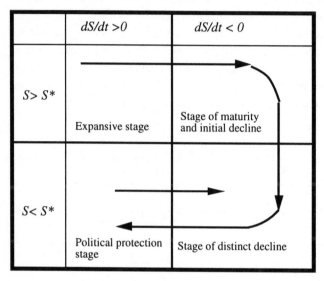

Figure 5. Aggregate product cycle trajectories; S = regional specialisation.

As a second stage in this classification analysis of industries, each industry is examined with regard to its R&D intensity, knowledge intensity, political protection etc. The empirical analyses in Andersson and Johansson (1984a and 1984b) were followed by several studies which applied the classification technique in figure 4. Batten and Johansson (1985, 1987, 1989) investigate both the OECD group and individual countries and apply a detailed classification of product groups. They also use the diagram in figure 4.2 to distinguish between regions which initiate a product cycle (or aggregate specialisation wave) and regions which at some stage imitate and follow already established product cycles. This distinction is illustrated by two parallel phase diagrams in figure 6.

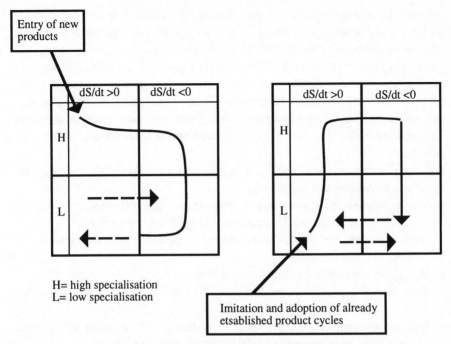

Figure 6. Product cycle trajectories in an initiating (leader) and an imitating (follower) region.

5. Conclusions

The trajectory of an ordinary product cycle portrays sales growth as exponential during a product's formative years, with an inevitable slowdown as the product mature and new competitors enter the market, followed by a decline phase which may abrupt and short but in other cases extended over long periods. This paper has selected one specific aspect of the product cycle theory by stressing the conditions that bring about location advantages for individual urban regions. For a given product and product group, the location advantage is dynamic, i.e., it is technique specific in the sense that the characteristics of the most advantageous economic milieu are different during the various phases of a product cycle. Propositions about this process are based on the assumption that a region's supply of resources and its entire economic milieu adjust on a slow time scale; the milieu functions as an arena for decision processes which determines the location, shifts of technique vintages, sales pattern etc. along the evolution of a product cycle (Andersson, 1993).

Differentials as regards location advantages comprise price differences for similar resources in different regions. Since these resources are assumed to adjust slowly, the cost differentials can be included in models adhering to the spatial price equilibrium tradition, describing temporary market solutions. With similar arguments the analyses have relied on arguments from von Thünen type models, especially as regards land values and the location of contact intensive activities.

One may expect that future theoretical efforts will include product cycle analyses in a framework of evolutionary economics. Such efforts should also comprise experimental dynamics to depict the temporal evolution of competition among alternative product variants in a product group (dynamic monopolistic competition). Both for theoretical and empirical studies on may outline a spectrum complementary approaches:

- Non-spatial models of technological progress with an emphasis on the time path of standardisation and routinisation.
- Models of product cycles related to an individual firm with a focus on technology, entrepreneurship, market strategy and organisation structure.
- Models of entry, expansion, decline and exit of product cycles with reference to individual urban regions.
- Birthplace analyses in a multiregional setting.
- Relocation and diffusion of product cycles in a multiregional (multi-urban) context.
- Location dynamics and interregional trade during various phases of a product cycle, within countries and in models of international trade.
- The slow dynamics of employment categories and location attributes in general, associated with the formation of location advantage; migration of urban resources.

References

Andersson ÅE (1985) *Kreativtitet: Storstadens framtid* (Creativity: The Future of the Metropolis), Prisma, Stockholm.

Andersson ÅE (1986) Presidential Address: The Four Logistical Revolutions, *Papers of the Regional Science Association*, 59:1-12.

Andersson ÅE (1988) *Univeristet: Regioners framtid* (University and the Future of Regions), The Stockholm Office of Regional Planning and Urban Transportation.

Andersson ÅE (1993) Capital, Knowledge, and Economic Development, in SI Andersson, ÅE Andersson and U Ottoson (eds), *Dynamical Systems – Theory and Applications*, World Scientific, Singapore.

Andersson ÅE, Anderstig, C and Hårsman, B (1990) Knowledge and Communication Infrastructure and Regional Economic Change, *Regional Science and Urban Economics*, 20:359-376.

Andersson ÅE and Johansson, B (1984a) Industrial Dynamics, Product Cycles, and Employment Structure, *WP-84-9*, IIASA, Laxenburg, Austria.

Andersson ÅE and Johansson, B (1984b) Knowledge Intensity and Product Cycles in Metropolitan Regions, *WP-84-13*, IIASA, Laxenburg, Austria.

Anderstig C and Hårsman B (1986) On Occupation Structure and Location Patterns in the Stockholm Region, Regional Science and Urban Economics, 16:97-122.

Anderstig C and Hårsman B (1993) High Technology Worker Mobility, in ÅE Andersson, DF Batten, K Kobayashi and K Yoshikawa (eds), The Cosmo-Creative Society – Logistical Networks in a Dynamic Economy, Springer-Verlag, Berlin, pp 53-66.

Aydalot P (1978) L'aménagement du Territoire en France: Une Tentative de Bilou, L'Espace Géographique, 7:245-268.

Arthur WB (1983) Competing Technologies and Lock-In by Historical Events: The Dynamics of Allocation under Increasing Returns, WP-83-90, IIASA, Laxenburg, Austria.

Arthur WB (1988) Self-Reinforcing Mechanisms in Economics, in PW Andersson, KJ Arrow and D Pines (eds), The Economy as an Evolving Complex System, Santa Fee Institute Studies in the Sciences of Complexity vol V, Addison-Wesley Publishing Company, Redwood city, California.

Batten DF (1982) On the Dynamics of Industrial Evolution, Regional Science and Urban Economics, 12:449-462.

Batten DF and B Johansson (1985) Industrial Dynamics of the Building Sector: Product Cycles, Substitution and Trade Specialisation, in F Snickars, B Johansson and TR Lakshmanan (eds), Economic Faces of the Building Sector, Document D20:1985; Swedish Council for Building Research, Stockholm.

Batten DF and Johansson B (1987) Substitution and Technological Change, in M Kallio, DP Dykstra and CS Binkley (eds), The Global Forest Sector: an Analytical Perspective, John Wiley & Sons, Chichester.

Batten DF and Johansson B (1989) Dynamics of Product Substitution, in ÅE Andersson, DF Batten, B Johansson and P Nijkamp (eds), Advances in Spatial Theory and Dynamics, North-Holland, Amsterdam.

Beckmann M & Thisse, J-F (1986) The Location of Production Activities, in P Nijkamp (ed), Handbook of Regional and Urban Economics vol I, North-Holland, Amsterdam, pp 21-95.

Burns AF (134) Production Trends in the United States, National Bureau of Economic research, New York.

Chamberlin EH (1933) The Theory of Monopolistic Competition, Harvard University Press, Cambridge Mass.

Cole S (1981) Income and Employment Effects in a Model of Innovation and Transfer of Technology, in Methods for Development Planning. Scenarios, Models and Micro Studies, The Unesco Press, Paris, pp 236-257.

Dean J (1950) Pricing Policies for New Products, Harvard Business Review, 28:48-53.

Dixit AK and Stiglitz, JE (1977) Monopolistic Competition and Optimum Product Diversity, American Economic Review, 67:297-308.

Erickson RA (1976) The Filtering-Down Process: Industrial Location in a Non-Metropolitan Area, Professional Geographer, vol XXVIII.

Fisher JC and Pry RF (1971) A Simple Substitution Model of Technological Change, Technological Forecasting and Social Change, 3:75-88.

Forslund UM and Johansson B (1995) The Mälardalen: A Leading Region in Scandinavia and Europe?, in PC Chesire and I Gordon (eds), Territorial Competition in an Integrating Europe: Local Impact and Public Policy, Aldershot, Grover.

Freeman C (1974) *The Economics of Industrial Innovation*, Penguin, Harmondsworth.

Freeman C (1978) Technology and Employment: Long Waves in Technical Change and Economic Development, Holst Memorial Lecture.

Gabszewicz JJ and Thisse, J-F (1986) Spatial Competition and the Location of Firms, in R Arnott (ed), *Location Theory*, Harwood Academic Publishers, Chur, pp 1-71.

Gaston JD (1961) Growth Patterns in Industry: a Reexamination, *Studies in Business Economics No 75*, National Industrial Board, New York.

Gold B (1964) Industry Growth Patterns: Theory and Empirical Results, *Journal of Industrial Economics*, 13:53-73.

Haken H (1983) *Advanced Synergetics*, Springer-Verlag, Berlin.

Heckscher E (1918), *Svenska produktionsproblem* (Swedish Production Problems), Bonniers, Stockholm.

Hirsch S (1967) *Location of Industry and International Competitiveness*, Oxford University press, Oxford.

Johansen L (1972) *Production Functions*, North-Holland, Amsterdam.

Johansson B (1987) Technical Vintages and Substitution Processes, in D Batten, J Casti and B Johansson (eds), *Economic Evolution and Structural Adjustment* (Lecture Notes in Economics and Mathematical Systems 293), Springer-Verlag, Berlin, pp 145-165.

Johansson B (1989) Economic Development and Networks for Spatial Interaction, *CWP-1989:28*, CERUM, University of Umeå

Johansson B (1993a), Infrastructure, Accessibility and Economic Growth, *International Journal of Transport Economics*, vol XX-2:131-156.

Johansson B (1993b) Economic Evolution and Urban Infrastructure Dynamics, in ÅE Andersson, DF Batten, K Kobayashi, K Yoshikawa (eds), *The Cosmo-Creative Society*, Springer-Verlag, Berlin.

Johansson B and Nijkamp P (1987),Analysis of Episodes in Urban Event Histories, in L van den Berg, LS Burn and LH Klaassen (eds), *Spatial Cycles*, Gower, Aldershot.

Johansson B and Karlsson C (1987) Processes of Industrial Change: Scale, Location and Type of Job, in M Fischer and P Nijkamp (eds), *Regional Labour Market Analysis*, North-Holland, Amsterdam.

Johansson B and Westin, L (1987) Technical Change, Location and Trade, *Papers of the Regional Science Association*, 62:13-25.

Kamien MI and Schwartz NL (1982) *Market Structure and Innovation*, Cambridge University press, Cambridge.

Krugman P (1979) A Model of Innovation, Technology Transfer, and the World Distribution of Income, *Journal of Political Economy*, 87:253-256.

Krugman P (1991) *Geography and Trade, Leuven* University Press, Leuven.

Kuznets S (1930) *Secular Movements in production and Prices*, Houghton Mifflin, Boston.

Lancaster K (1975), Socially Optimal Product Differentiation, *American Economic Review*, 65:567-585.

Leontief W et al (1953) *Studies in the Structure of American Economy*, Oxford University Press, London-New York.

Malecki EJ (1980) Science and Technology in the American Metropolitan System, in SD Brunn and JO Wheeler (eds), *The American Metropolitan System: Present and Future*, New York.

Malecki EJ (1981) Product Cycles and Regional Economic Change, *Technological Forecasting and Social Change*, 19:291-306.

Markusen, A (1985), *Profit Cycles, Oligopoly and Regional Development*, MIT Press, Cambridge, Massachusetts.

Marshall, A (1920) *Principles of Economics*, Macmillan, London.

Mickwitz G (1959) *Marketing and Competition*, Centraltryckeriet, Helsingfors.

Mills E & Carlino G (1989), Dynamics of County Growth, in ÅE Andersson, et al (eds) *Advances in Spatial Theory and Dynamics*, North-Holland, Amsterdam, 195-205.

Nelsson R and Norman V (1977) Technological Change and Factor Mix Over the Product Cycle: a Model of Dynamic Comparative Advantage, *Journal of Development Economics*, 4:3-24.

Nelson RR and Winter SW (1982), *An Evolutionary Theory of Economic Change*, Harvard University press, Cambridge Mass.

Nijkamp P, ed. (1986) Technological Change, Employment and Spatial Dynamics, *Lecture Notes in Economics and Mathematical Systems*, vol 270, Springer-Verlag, Berlin.

Norton R and Rees (1979) The Product Cycle and the Spatial Distribution of American Manufacturing, R*egional Studies*, 13:141-151.

Patton A (1959) Stretch your Product's Earning Years: Top Management's Stake in the Product Life Cycle, *The Management Review*, 48:9-79.

Ricardo D (1951) Principles of Political Economy, in Sraffa P (ed) *The Works and Correspondence of David Ricardo*, vol. I, Cambridge University Press, Cambridge.

Samuelson P (1952) Spatial Price Equilibrium and Linear Programming, *American Economic review*, 42:283-303.

Soete L (1981) A General Test of Technological Gap Trade Theory, *Weltwirtschafliches Archiv*, 117:638-660.

Schumpeter JA (1934) *The Theory of Economic Development,* Harvard University Press, Cambridge Mass. (First edition published in German 1912, the second 1926).

Schumpeter JA (1939) *Business Cycles: A Theoretical, Historical and Statistical Analysis of the Capitalistic Process*, McGraw-Hill, New York.

Takayama, T & Judge, GG (1971) *Spatial and Temporal Price and Allocation Models*, North-Holland, Amsterdam.

Tarde G (1903) *The Laws of Imitation*, Henry Holt, New York.

Thünen von JH (1966/1826) *Der Isolierte Staat in Beziehung auf Nationale Ökonomie und Landwirtschaft*, Gustav Fischer, Stuttgart (first edition 1826).

Vernon R (1966) International Investment and International Trade in the Product Cycle, *Quarterly Journal of Economics*, 80:190-207.

Wigren R (1976) Analys av regionala effektivitetsskillnader inom industribranscher (Analysis of Differences in Regional Efficiency within Industrial Sectors), *Memorandum 58*, Dept. of Economics, University of Gothenburg.

Wigren, R (1984) Measuring Regional Efficiency - A Method Tested on Fabricated Metal Products in Sweden 1973-75, *Regional Science and Urban Economics*, 14:363-379.

Gradient Dynamics in Webrian LocationTheory

Tönu Puu
Department of Economics
University of Umeå

Introduction

Around 1900 Wilhelm Launhardt [1] and Alfred Weber [2] invented the classical model of industrial location. Suppose that a firm produces one product, using two factors of production. The product has to be sold at a prelocated market place, and the input factors have to be bought at two likewise prelocated places where they are available. The three locations normally form the vertices of a triangle. Assuming that production requires fixed proportions of inputs to output, Weber could identify the problem of finding the best location for production with the problem of finding the location at which the sum of transportation costs is minimized.

It was shown that the model system was equivalent to the following mechanical system: Let a plate of some very smooth material (to provide for almost no friction to sliding motion) represent geographical space. Drill three holes in it, at the locations of input supplies and output demand. Next, tie three threads together in one single knot, and pass the ends through the holes. Finally, suspend three weights from these ends, proportional to transportation cost, per distance unit, and per unit of output, for the output and the inputs (using the fixed input proportions). Then the knot will be drawn to the location of minimum transportation cost, as this also minimizes the potential energy of the system. Varignon even invented a mechanical analog device with rollers fixed to the edge of a circular disk for the study of such location problems.

It was assumed that the metric for transportation cost was Euclidean, i.e. all goods could be transported along straight lines in any direction, with constant cost per unit distance. This was, of course, restrictive as both local variations due to the density and capacity of actual road networks, and possible anisotropy due to the geometric layout of the network, were disregarded.

The solution could be an inner solution in the location triangle whenever each of the weights was less than the sum of the other two. Otherwise, one of the corners would be the solution. It could be understood how production that gained in weight or bulkiness was located at the spot of delivery, like the construction of a bridge, whereas weight-losing production was located at the source of the most bulky input, like iron foundry was historically concentrated near forests with ample supply of charcoal. See [3].

It is to be noted that the prices of inputs and output did not enter the location problem at all, except as determinants of whether production would be sufficiently profitable to be established at all. This was an effect of the assumed constant proportions in production. This assumption, according to which different inputs could not be substituted for eachother, was long regarded as the most restrictive assumption in the model, in fact more restrictive than the linearity inherent in the Euclidean metric, and several attempts were made to introduce more general production technologies.

In his pathbreaking revision 1935 of the entire body of location theory Tord Palander [4] originally devoted an entire chapter to this problem, but it was removed just before printing, and the manusript to the chapter printed as a blank page was never retrieved. In 1956 Walter Isard [5] revived the problem, and in 1958 Leon Moses [6] demonstrated that only with a fixed proportion technology could the location decision be dissociated from the production decision. In particular he found that, with the inputs being perfect substitutes, the "*firm would locate at the source of one of the inputs*", quite as in the Weberian case when one weight dominates over the sum of the other two.

As shown by the author [7], such extreme substitutability is not necessary for the result. In fact location optima at both sources of inputs can coexist with an inner optimum, and, as the production and location decisions are now dependent, minor changes of input prices could cause hard bifurcations from one local optimum to another. See Fig. 2.

In the present paper we will study the process of a local gradient dynamics in such a case of three coexistent local optima. The relocation of a firm is costly, and obviously more so the longer the distance to the new location. So, when input and output prices constantly change, as they do in the real world, it would be unlikely that the firm goes on jumping between distant local optima. Rather it moves a marginal distance in the gradient direction for profits. Supposing that it is not already at one of the optima, we will see that it can keep on moving - even if the prices do not change. This movement can end at one of the optima, but it can also go on in a chaotic manner for ever.

The Weberian Location Triangle

Suppose the three prelocated points are at unit distance from the origin of the Euclidean plane, arranged as the vertices of an equilateral trangle, more specifically $(0, 1)$ for output $(-\sqrt{3}/2, -1/2)$ and $(\sqrt{3}/2, -1/2)$ for inputs. Denoting a general point by (x, y) the distances from that point to the vertices are then:

$$d = \sqrt{x^2 + (y - 1)^2} \tag{1}$$

$$d_1 = \sqrt{\left(x + \frac{\sqrt{3}}{2}\right)^2 + \left(y + \frac{1}{2}\right)^2} \tag{2}$$

$$d_2 = \sqrt{\left(x - \frac{\sqrt{3}}{2}\right)^2 + \left(y + \frac{1}{2}\right)^2} \tag{3}$$

Suppose the weights are denoted w, w_1, and w_2. The first is transportation cost for output per unit, and per unit distance, the other two are transportation costs for the inputs, per unit output as required by the fixed coefficients, and per unit distance. Total transportation cost is accordingly:

$$T = w \cdot d + w_1 \cdot d_1 + w_2 \cdot d_2 \tag{4}$$

and the conditions for an inner optimum are:

$$\frac{\partial T}{\partial x} = w \cdot \frac{x}{d} + w_1 \cdot \frac{x + \frac{\sqrt{3}}{2}}{d_1} + w_2 \cdot \frac{x - \frac{\sqrt{3}}{2}}{d_2} = 0 \tag{5}$$

$$\frac{\partial T}{\partial y} = w \cdot \frac{y - 1}{d} + w_1 \cdot \frac{y + \frac{1}{2}}{d_1} + w_2 \cdot \frac{x + \frac{1}{2}}{d_2} = 0 \tag{6}$$

Fig. 1. Isodapans for equal weights.

The distance measures are substituted from equations (1)-(3). If, for instance, all the weights are equal, then the origin is an inner optimum. Substituting $x = y = 0$ in (1)-(3) we see that all distances become unitary. As the equal weights can be divided out from (5)-(6), we see that the zero coordinates indeed reduce those equations to identities. It can also easily be proved, using the formulae, and some geometry of the triangle, that the output market or input source locations dominate provided one of the weights equals or exceeds the sum of the other two.

The situation can also be studied in terms of constant level contours in x, y-space for transportation costs, according to (4) with (1)-(3) substituted. Those curves were given the name "isodapans". In Fig. 1 we show the level curves, assuming that the weights are equal, with the best location accordingly in the centre.

Production and Location

It is now time to take production in explicit consideration. A production function defines the maximum output obtainable from each combination of inputs when they are used in the technically most advantageous combination. We will use the most frequently used standard form for it, the so called Cobb-Douglas function.

Denoting output by q and inputs by v_1, v_2 the Cobb-Douglas function reads:

$$q = A \cdot v_1^{\alpha_1} \cdot v_2^{\alpha_2} \tag{7}$$

The exponents are positive, and their sum represents the "returns to scale", i.e. the relative change of output to the relative change of inputs, provided both inputs are increased in the same proportion. There is a logarithmic linearity inherent in this production function, so, unless the returns to scale is decreasing, i.e. the sum of exponents is less than unity, problems of finding an optimal scale of operation will lack definite solutions. In regional economics increasing returns are often assumed to account for the phenomenon of spatial "agglomeration". As this is not an issue at present, we will stick to decreasing returns, i.e. $\alpha_1 + \alpha_2 < 1$. The constant A is just a multiplicative production efficiency factor. As we are free to choose a convenient volume unit for goods produced we can set its value arbitrarily, and later we will use it to get rid of some other constants.

Denoting output price (at the market place) by p, input prices (at the sources) by r_1, r_2, the transportation rate for output by t and for inputs by t_1, t_2, we can define profits:

$$\Pi = (p - t \cdot d) \cdot q - (r_1 + t_1 \cdot d_1) \cdot v_1 - (r_2 + t_2 \cdot d_2) \cdot v_2 \tag{8}$$

We can substitute for output q from (7) and then for the distances d, d_1, d_2 from (1)-(3). Prices p, p_1, p_2 and transportation rates t, t_1, t_2 being given, we see that the only remaining variables are the location coordinates x, y and the combination of inputs v_1, v_2.

The conditions for optimum production and location are obtained by equating the derivatives of (8) with respect to the remaining four variables to zero, i.e.

$$\frac{\partial \Pi}{\partial v_1} = (p - t \cdot d) \frac{\partial q}{\partial v_1} - (r_1 + t_1 \cdot d_1) = 0 \qquad (9)$$

$$\frac{\partial \Pi}{\partial v_2} = (p - t \cdot d) \frac{\partial q}{\partial v_2} - (r_2 + t_2 \cdot d_2) = 0 \qquad (10)$$

and

$$\frac{\partial \Pi}{\partial x} = -t \cdot q \cdot \frac{x}{d} - t_1 \cdot v_1 \cdot \frac{x + \frac{\sqrt{3}}{2}}{d_1} - t_2 \cdot v_2 \cdot \frac{x - \frac{\sqrt{3}}{2}}{d_2} = 0 \quad (11)$$

$$\frac{\partial \Pi}{\partial y} = -t \cdot q \cdot \frac{y-1}{d} - t_1 \cdot v_1 \cdot \frac{y + \frac{1}{2}}{d_1} - t_2 \cdot v_2 \cdot \frac{y + \frac{1}{2}}{d_2} = 0 \quad (12)$$

where, from (7), we have:

$$\frac{\partial q}{\partial v_1} = \alpha_1 \cdot \frac{q}{v_1} \qquad (13)$$

$$\frac{\partial q}{\partial v_2} = \alpha_2 \cdot \frac{q}{v_2} \qquad (14)$$

Equations (11)-(12) can be compared to equations (5)-(6). We find that the weights are no longer constant, but depend on the production decision. Equations (9)-(14) along with (7) are, however, sufficient to solve the combined production and location decision.

Using the conditions for optimal choice of production technology (9)-(10) along with (13)-(14) and (7), we can then obtain profits according to (8):

$$\Pi = \gamma \cdot (A \cdot \alpha_1^{\alpha_1} \cdot \alpha_2^{\alpha_2})^{1/\gamma} \cdot \left(\left(\frac{p - t \cdot d}{(r_1 + t_1 \cdot d_1)^{\alpha_1}} \right) \cdot \left(\frac{p - t \cdot d}{(r_2 + t_2 \cdot d_2)^{\alpha_2}} \right) \right)^{1/\gamma} \quad (15)$$

For convenience we have introduced the definition $\gamma = 1 - \alpha_1 - \alpha_2$, which is positive with decreasing returns to scale as assumed. Observe that we have not yet used the conditions (11)-(12) for optimal location choice. Thus product price, less transportation cost, and the factor prices, augmented by transportation costs, enter the profit expression (15). To calculate the optimal location in closed form for the general model would be a formidable task.

We can, however, substitute the distances according to (1)-(3), calculate profits according to (15) for different probe locations and draw a map of constant profit countours, exactly like the isodapans in Fig. 1. An example of those is displayed in Fig. 2. The shading is reversed compared to the previous Figures, because we now deal with maxima of profits instead of minima of costs.

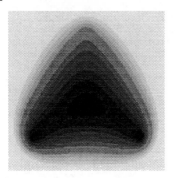

Fig. 2. Isoprofit contours in the case of three local maxima.

We also see that in the special case studied there are now three different local profit maxima. In other words the stage is set for possible chaotic motion in gradient dynamics.

Gradient Dynamics

Let us now restate equation (15), putting $\alpha_1 = \alpha_2 = 0.4$, and substituting a suitable value of A to get rid of the messy constant in that function. Thus:

$$\Pi = \frac{(p - t \cdot d)^5}{(r_1 + t_1 \cdot d_1)^2 \cdot (r_2 + t_2 \cdot d_2)^2} \qquad (16)$$

There are six parameters, three prices and three transportation rates. The distance expressions are, as always, obtained from equations (1)-(3).

The gradient is obtained as:

$$(17)$$

$$\frac{\partial \Pi}{\partial x} = \frac{\partial \Pi}{\partial d} \cdot \frac{x}{d} + \frac{\partial \Pi}{\partial d_1} \cdot \frac{x + \dfrac{\sqrt{3}}{2}}{d_1} + \frac{\partial \Pi}{\partial d_2} \cdot \frac{x - \dfrac{\sqrt{3}}{2}}{d_2}$$

$$\frac{\partial \Pi}{\partial y} = \frac{\partial \Pi}{\partial d} \cdot \frac{y - 1}{d} + \frac{\partial \Pi}{\partial d_1} \cdot \frac{y + \dfrac{1}{2}}{d_1} + \frac{\partial \Pi}{\partial d_2} \cdot \frac{y + \dfrac{1}{2}}{d_2} \qquad (18)$$

where from (16):

$$\frac{\partial \Pi}{\partial d} = -5 \cdot \Pi \cdot \frac{t}{p - t \cdot d} \tag{19}$$

and likewise:

$$\frac{\partial \Pi}{\partial d_1} = -2 \cdot \Pi \cdot \frac{t_1}{r_1 + t_1 \cdot d_1} \tag{20}$$

$$\frac{\partial \Pi}{\partial d_2} = -2 \cdot \Pi \cdot \frac{t_2}{r_2 + t_2 \cdot d_2} \tag{21}$$

and where the distances are obtained from (1)-(3). The gradient dynamics can now be formulated as the iterative system:

$$x_{t+1} = x_t + \delta \cdot \frac{\partial \Pi}{\partial x} \tag{22}$$

$$y_{t+1} = y_t + \delta \cdot \frac{\partial \Pi}{\partial y} \tag{23}$$

For drawing Fig. 2 we used $p = 8$, $r = 2$, $t = 1.25$, and $t_1 = t_2 = 1.75$, which proved sufficient to get the number of local maxima we wanted. We keep these values in the sequel and experiment with various step lengths in running the system (22)-(23) using (1)-(3) and (17)-(21). It is quite easy to obtain chaotic sequences. A $\delta = 0.0325$ was used to obtain Fig. 3.

Fig. 3. Chaotic process in the gradient dynamics.

Discussion

The shading is done according to the frequency of visiting a point. Obviously, to judge from the black central "nose" of the animal, most often the process moves in a close neighbourhood of the different local optima, though, in the chaotic case, the search algorithm is unable to actually hit any of them. Now and then, due to a very large gradient close to a sharp maximum, the process shoots far away in some arbitrary looking direction, and then starts a journey back to the central region. This journey is quite slow at first, because the gradient is small in the periphery, and it is in this way that the "hairs" of the animal arise.

Those far shots contradict our assumption of marginal adjustments. If the process can shoot that far in one step it can also jump to the global optimum location. We therefore need to reinterpret the model.

The assumption that relocation is more costly the farther the distance over which the firm moves, is itself sound. Moving people and equipment obviously are subject to distance dependent transportation costs. Moreover, information of vital importance, pertinent for instance to the acquisition of land for a new estabishment is more easily accessible in contiguous locations.

On the other hand the relocation costs must be weighed against the gains from an improved location, and if the profit gradient is large enough, then obviously a large step can be motivated despite large relocation costs.

The real problem with the action of the gradient dynamics is the interpretation of the basic location model. Let us therefore make, not the locations of input sources and the market, but rather the implicit flows of inputs and output the central piece of the model. Along the lines of these flows, prices accumulate with transportation cost. The flows need not even originate at point input sources, those can be multiple, or even dispersed over an area. Likewise neither the "market" needs be concentrated to a single point.

There are now lots of reasons why the price potentials and their orthogonal flow lines should be subject to change. If the inputs are spatially dispersed raw materials, some resources might be temporarily or permanently exhausted, if they are intermediate goods produced by other firms, those might change their location. The distribution of consumption over space, i.e. the market, changes with migration, and with changing income distribution, if the product is a consumers' good. The location of a point market also changes with relocations of other firms using the product as a semi-finished input.

All those factors can be summarized, by generalizing the three point potentials, associated with the Weberian locations, and making the potentials variable over time. By this interpretation the knowledge about the price potentials becomes extremely uncertain, and it also becomes likely that a firm moves adaptively on the basis of local information alone.

Even the case illustrated makes sense, as a simple example, of course. Point potentials also represent sources dispersed over circular disks, and change can be represented by shifts up and down of the price potentials. Even if the prices actually stay constant, there is no way a firm moving like the one in Fig.4 will find this out in finite time.

We can note that generalizing the potentials admits us to incorporate non-constant and anisotropic transportation costs, even different for inputs and output in case they are transported by different means.

Reformulation

Let us now formalize the slight reformulation indicated. We just replace the the three expressions for the local delivery prices: $p - t \cdot d$, $r_1 + t_1 \cdot d_1$, and $r_2 + t_2 \cdot d_2$ by three general potential functions: $p(x,y)$, $r_1(x,y)$, and $r_2(x,y)$. Substituting from (1)-(3) will exemplify the particular case discussed. Note the difference of meaning of the prices. Previously they were mill prices at very definite locations which would have to be corrected by transportation rates multiplied by Euclidean distances to the location of production. Now they are local net prices at the point of location considered, with no explicit specification of how the commodities are transported, who provides for transportation, or how the transportation charges are calculated.

With this reformulation profits according to (8) and the conditions for an optimal production technology (9)-(10) become:

$$\Pi = p \cdot q - v_1 \cdot r_1 - v_2 \cdot r_2 \tag{24}$$

$$\frac{\partial \Pi}{\partial v_1} = p \cdot \frac{\partial q}{\partial v_1} - r_1 = 0 \tag{25}$$

$$\frac{\partial \Pi}{\partial v_2} = p \cdot \frac{\partial q}{\partial v_2} - r_2 = 0 \tag{26}$$

whereas the marginal productrivities (13)-(14) remain unchanged.

Further, the location optimum conditions (11)-(12) are replaced by:

$$\frac{\partial \Pi}{\partial x} = q \cdot \frac{\partial p}{\partial x} - v_1 \cdot \frac{\partial r_1}{\partial x} - v_2 \cdot \frac{\partial r_1}{\partial x} = 0 \tag{27}$$

$$\frac{\partial \Pi}{\partial y} = q \cdot \frac{\partial p}{\partial y} - v_1 \cdot \frac{\partial r_1}{\partial y} - v_2 \cdot \frac{\partial r_1}{\partial y} = 0 \tag{28}$$

or just:

$$\nabla \Pi = q \cdot \nabla p - v_1 \cdot \nabla r_1 - v_2 \cdot \nabla r_2 = 0 \tag{29}$$

The profit expression (15) becomes:

$$\Pi = \gamma \cdot (A \cdot \alpha_1^{\alpha_1} \cdot \alpha_2^{\alpha_2})^{1/\gamma} \cdot \left(\left(\frac{p}{r_1^{\alpha_1}} \right) \cdot \left(\frac{p}{r_2^{\alpha_2}} \right) \right)^{1/\gamma} \tag{30}$$

Even though these formulas look simpler, this, of course, does not apply to the problems treated. Substituting the potentials implicit in (1)-(3) we have to solve exactly the same problem as before. But, now we can use any functions of the space coordinates we wish for the three prices. For more information on price potentials and gradient flows see [8].

The conditions for location optima (27)-(28) now become particularly attractive with the Cobb-Douglas function. From (13)-(14) and (25)-(26) we find that $r_i \cdot v_i = \alpha_i \cdot p \cdot q$. Substituting into (29) we get:

$$\nabla \log p = \alpha_1 \cdot \nabla \log r_1 + \alpha_2 \cdot \nabla \log r_2 \tag{31}$$

Thus any location is a local optimum where the gradient of the logarithm of product price equals the weighted average of the gradients of the logarithms of input prices, the exponents of the Cobb-Douglas function being used as weights.

As an example suppose we put:

$$p = e^{x^2 + y^2} \tag{32}$$

$$r_1 = e^{x^4 - x^2 + y^2} \tag{33}$$

$$r_2 = e^{y^4 - y^2 + x^2} \tag{34}$$

The exponentials, except fitting the Cobb-Douglas function framework, guarantee that prices are always positive. In the example product price is uniformly increasing with distance from the origin, whereas input prices are double-well potentials with minima located along the horizonal and vertical axes respectively.

We can expect that the location optima should be found in a bounded area, because the fourth order terms make input prices increase most sharply at some dis-

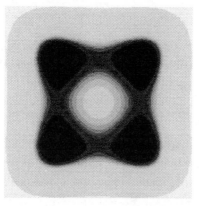

Fig. 4. Iso-profit contours for price potentials.

tance from the origin. Taking the gradients of the prices, substituting into (31), and again putting $\alpha_1 = \alpha_2 = 0.4$, we get the location condition:

$$\begin{pmatrix} 2x \\ 2y \end{pmatrix} = 0.4 \cdot \begin{pmatrix} 4x^3 \\ 4y^3 \end{pmatrix} \tag{35}$$

This system has nine solutions for the location, for $x, y = -\sqrt{5/2}, \ 0, \ \sqrt{5/2}$, among which four can be identified as profit maxima. In Fig. 4 we see the iso-profit contours according to (30) and (32)-(34). The four local maxima are in the diagonal directions, whereas close to the locations of the pairs of input wells there are saddle points. Finally there is a minimum at the origin.

The search procedure could now be reformulated in terms of gradient dynamics to any such set of potentials, and there is no longer any reason why a locating firm should know the shape of these potential functions. Local adaptive adjustment thus becomes more likely than in the original interpretation of the Weber model.

For the example given in (32)-(34) we can get the profits (30) expressed by an exponential, like the assumed prices themselves:

$$\Pi = e^{5x^2 + 5y^2 - 2x^4 - 2y^4} \tag{36}$$

The gradient vector accordingly becomes:

$$\nabla\Pi = 2 \cdot \Pi \cdot \begin{pmatrix} 5x - 4x^3 \\ 5y - 4y^3 \end{pmatrix} \tag{37}$$

and the adaptive process can be written:

$$\begin{pmatrix} x_{t+1} \\ y_{t+1} \end{pmatrix} = \begin{pmatrix} x_t \\ y_t \end{pmatrix} + \delta \cdot \nabla\Pi \tag{38}$$

The case illustrated is extremely symmetric. Running simulations we very soon detect that all movements tend towards horizontal or vertical line segments through

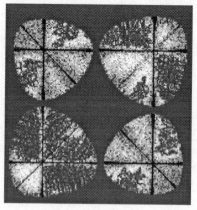

Fig.5 Four attractors in the gradient field.

the optimum location points. The process seems to be attracted very soon to the optimal location coordinate in one dimension, whereas it continues a chaotic process on a line interval in the other dimension for ever. Which direction is favoured depends on the initial conditions. In rare cases a diagonal direction can occur. Despite this fact, the process never finds the optimal location in both dimensions, given an adjustment step that leads to a chaotic regime. Chaos in the remaining dimension sets in by a period-doubling cascade of bifurcations.

This type of behaviour is non-generic, and can be destroyed by introducing some more asymmetry, for instance by adding a product term, however weak, to the squares in equation (32). Fig. 5 illustrates a search process for a step length $\delta = 0.00075$ and a slight product term with the coeeficient 0.0035 introduced. What we see are four different co-existent attractors, with each its basin. The shading is again done by the frequency of visiting the point, and we see that the horizontal, vertical, and diagonal directions are still favoured in the adjustment process.

In Fig. 6. we let the step length increase to $\delta = 0.00095$. As a result the four attractors merge to one single attractor which has the entire plane as just one basin.

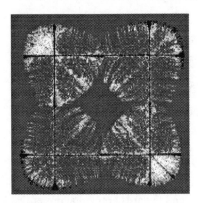

Fig.6. One single merged attractor.

Conclusion

As a final remark we should add that the re-interpretation of the Launhardt-Weber model in terms of potentials favoures an outlook where a spatially dispersed economy with many different agents is considered, i.e. the outlook of a competitive economy. In comparison to this, the original model has a more monopolistic flavour. Now the actions of many consumers and producers are responsible for the trade flows and price potentials. The firm whose movements have been considered contributes itself to these potentials, but its marginal contributions are themselves atomistic, and the potentials may thus be taken as given, though imperfectly known.

References

[1] W. Launhardt, "Die Bestimmung des zweckmässigsten Standorts einer Gewerblichen Anlage", *Zeitschrift des VereinsDeutscher Ingenieure 26:105-115* (1882)

[2] A. Weber, *Über den Standort der Industrien.* Tübingen (1909)

[3] M. J. Beckmann, *Location Theory.* Random House (1968)

[4] T. F. Palander, *Beiträge zur Standortstheorie.* Uppsala (1935)

[5] W. Isard, *Location and Space-Economy.* The M.I.T. Press (1956)

[6] L. N. Moses, "Location and the theory of production", *Quarterly Journal of Economics 72:259-272* (1958)

[7] M. J. Beckmann and T. Puu, *Spatial Structures.* Springer-Verlag (1990)

[8] M. J. Beckmann and T. Puu, *Spatial Economics.* North-Holland (1985)

Creativity - Learning and Knowledge in Networks

Human Contacts in Knowledge Society: An Analytical Perspective

Kiyoshi Kobayashi
Department of Civil Engineering
Graduate School of Engineering
Kyoto University

and
Kei Fukuyama
Department of Social Systems Engineering
Tottori University

1. Introduction

Supported by the highly advanced technologies of information/communication and transportation, the new *knowledge society* has been emerging. Newly developed transportation and communication technologies are diffused in the society, resulting in the rapid increase in the flexibility and the degree of freedom of human communication behaviours. The technological innovation of communications in the society does not only mean more rapid and efficient transmission of information and knowledge, it also expands the possibility of interactions of various types of activities in the spatially distant areas. The increased opportunity of communications in the society, greatly affecting the communication behaviours, brings about the structural evolution of social systems themselves.

The *agora* in ancient Athens was a meeting facility. People in Athens had gathered in the agora and had exchanged goods and ideas. In many cities in Europe, the plazas had functioned as meeting places/facilities. *Fin-de-Siécle Vienna* has had the cultural meeting facility called *Café*. These meeting facilities had played important roles in formation of urban structures and the growth of creativity in the cities. In modern cities, huge amount of ideas and knowledge have been accumulating. The smooth and easy transmission of ideas facilitates the agglomeration externality of the cities. The meeting facilities have significant meaning in the society, acting as the media to fulfill efficiently the expanding demand for the knowledge exchange in the spatially-limited urban systems.

In most communication behaviours, the decisions made by travel agents cannot be independent of the decision and/or intention of other agents. One's decision on his/her communication behaviour is more or less affected by the others' will. Especially, in face-to-face communication (referred to as 'meetings' hereafter), the agreement with the other party to have a meeting is the prerequisite for the meeting. In various aspects of daily life, mankind organizes meetings. People who make business trips, such as those for negotiation, collecting money and preconcert, and association trips with friends and loved ones, aim at the meetings themselves. Many institutional conditions and conventions which rule daily trips such as going to school, hospital, work and shopping, have stemmed from the need for meetings. In addition, there is a tremendous amount of meetings at home within the family, and the relationships among the family members affect in many ways the occurrence of their trip/traffic behaviours.

In this chapter, we focus on the meeting as a key communication behaviour in the knowledge society, and research perspectives for infrastructure arrangement and communication policies in the knowledge societies. In section 2, the social roles of meeting partner or opponent are investigated. In section 3, the fact that the communication network consists of two networks, human network and physical network, is clarified. Sections 4 and 5 give simple models to explain the effects of human interactions and information, respectively, on the transportation (physical) network. In section 6, research issues for effective consolidation and refinement of transportation infrastructure in knowledge society are mentioned.

2. Rediscovering the Existence of the Meeting 'Partner'

The major tenet of this chapter is to rediscover the 'partner' or 'opponent' in human communications, and to provide a new perspective on human contacts modelling. Let us first investigate the social role of meetings, by briefly referring to the historical facts, that the discovery of the "opponent" in human communications had brought about the *time revolution* in Europe in the Middle Ages.

The promotion of commerce in Europe in the Middle Ages had allowed the present day time system to see the light – the so-called *time revolution*. Since the acceptance of Christianity by the emperor *Constantinus*, the "Church time system" had governed the human behaviour in the Middle Ages. The Church time system was originally introduced to gather monasteries and church councilor members and to organize choirs. The noontime was divided into three. Though the Church time system is a very irregular time clock system, it was rather rational for farmers' work patterns.

matines	midnight
laudes	around 3:00
prime	around 3:00
tierce	around 3:00
sixte	around 3:00
none	around 3:00
vêpres	around 3:00
complies	around 3:00

Figure 1. The Church time system.

The time system is very conservative and rarely changes. In medieval Europe the machinery clock with which the public can realize time had not yet been innovated and only the bell ring by the church for sacred affairs was available for them. Farmers and workers had depended their "time to start working", "time to have lunch", "time to go back to work" and "time to finish work", on the bell ring by church.

In the thirteenth century, the time system underwent a dramatic change caused by the emergence of the "time of merchants". The merchants needed to have meetings with customers and trade opponents all the time. For them, the Church time system with only three hour units was extremely inconvenient. It was incompatible with the merchants' rhythm. Churches had been gradually changing its time system based on the requirement by the merchants and workers in the town. J. Le Goff states that "during the period between tenth and thirteenth century, only time system in noon have evolved. Time on *nine* was originally around three o'clock p.m. in current time system, but had moved backwards gradually, and stopped around the noon (this explains why the word 'noon' come from 'nine'). The *noon*, that brings people working in the city about time to have a rest under the holy time told by the bell" (Le Goff, 1980). After all, the time system that divides labour times into two, morning and afternoon, appeared; this is the emergence of the time evolution.

The time revolution provides a reasonable material for understanding meetings. The Church time system is the system for people to have holy communications with

Table 1. The Types of Meetings.

	High Frequency	Low Frequency
Direct	Meeting with high frequency in high density space (daily contacts; friend intercourse)	Meeting with multi-persons at limited time and space (business meeting; preconcert)
Indirect	Information to be consumed very frequently in short time; quick acquisition of goods (stock market, etc.)	Transmission of highly insepa-rable and conservative knowledge and service (conferences)

"God". The morning-afternoon halving time system had appeared in order to communicate not with "God" but "Partners". This is not a system made up artificially by the man of power at the time. It is rather self-organized through the repetition of meetings by mankind. The essence of meeting is the discovery of the existence of the "opponent". Meeting never happens by one individual. The necessity of agreement with opponents had formed the halving of working time, commercial conventions, work system, holidays and others over the long time.

3. Roles of Meetings

3.1 Meeting Types

T. Palander, an economist, focuses on the role of negotiation in the modern cities and points out the importance of communication costs in location decision in the city activities beyond the scope of the pioneering works by Thünen and other earlier works in location theory (Palander, 1935). As he reveals, one of the conspicuous characteristics of the modern city activities is the frequent and wide-range communications with activities of other cities. The amount of knowledge produced and concentrated in the big cities is increasing remarkably.

While the types of meetings vary, they can be categorized as given in Table 1 by focusing on the spatial and temporal characteristics. Human beings exchange their scientific ideas (such as knowledge and information), psychological services (such as friendship and affection), and goods via meetings. If the thing to be exchanged differs, the frequency and characteristic also differ. The one-to-one communication is the elemental form of meeting and the way to exchange ideas in the most condensed manner. Especially, when exchanging psychological services such as friendship and affection, this kind of communication form takes places. Similarly, in the case of scientific ideas, one-to-one communication allows people to exchange ideas with high concentration. The one-to-one meeting is, however, not always the most effective way of knowledge exchange. The progress of the society towards the knowledge one and activation of research and development lead to the increase in the variety of scientific ideas and also the frequency of knowledge exchange. If all the knowledge exchanges are done by one-to-one contacts, then the total number of exchanges should increase exponentially. If the large number of people gather at the same places such as conferences and conventions, the efficiency of knowledge exchange would be improved extensively. The scale of meeting increases, the probability that one can get to know the persons who hold the same knowledge that he/she seeks increases. Once came to know the person, one can have personal meetings and communications with him/her afterwards.

Table 2. The Comparison of Network Characteristics

	Transportation Network	Human Network
Node	Origin; Destination	Individual
Link	Roads, Railways	Meeting; Communication
Input	Trip Demand	Idea; Friendship, etc
Output	Realized Trips	Evolution of Ideas; Deepening of Friendships, etc.
Observable Variable	Transportation Trip	The Number of Meeting State
Variable(s)	Transportation Conditions	Function of Ideas, Friendship and others
Objective Variable(s)	Travel Time; Costs	Exchange of Knowledge and Ideas
Medium	Transportation Methods	Discussion
Activity Reachability	Long Distance	Short Distance

3.2 Duality of Communication Networks

The traffic and communication behaviours expand on two kind of networks: "the human network" and "the physical network". Except for traffic behaviours conducted independently from the other decision makers, these two networks play important roles in the realization of traffic and communications. Table 2 clarifies the characteristic differences of the transportation network and the human network both of which play substantial roles in the face-to-face communication aspect (Beckmann, 1994):

The human network is the network on which the scientific ideas, psychological services and other ideas flow. One remarkable characteristic of the human network is that the individual human functions as a node which accumulates ideas, and the meeting as a link at which ideas are exchanged. In the human network, "whether to meet or not", "where to meet and when", and others are determined. Here, the "agreement by all participants" is a fundamental principle, and the decision making by multi persons takes part in the formation of a meeting. The meetings would vary in their decision making processes and manners; sometimes a meeting is organized compulsorily by a specific leader in power such as many formal business meetings, and sometimes a meeting is determined in its contents by the intention of the person who has the lowest incentive to meet, such as voluntary meeting and man-woman intimacy. It is, however, clear that the decisions by the multitudinous persons is concerned in the formation of meetings.

The physical network basically consists of the "telecommunication network" and the "transportation network". In the telecommunication network, the network itself functions as the medium so as to realize the meetings, such as the television meeting and facsimile. In this case, the monetary and temporal resources to be consumed by the communicating agents are relatively low. Though the parties concerned communicate with each other while implicitly taking for granted the formation of communication, it does not require mutual agreement by parties very much. This type, with fewer requirements and conditions for pre-agreement on communication and meeting, is utilized as an efficient method to organize meetings that often require a tremendous amount of energy. The transportation network is always different from the human network in characteristics. A node and a link in the communication network do not always correspond to a node and a link in the human network. Individuals, that are nodes in the human network, move on the links in the transportation network. Many meetings use meeting facilities (nodes) such as hotels, convention halls, and cafés in town. The household, a most important link in the human networks, is also a node in the transportation network. The consultation during the transit in a mean of transportation and the unexpected chat with unknowns represent the cases where the link in the transportation network is utilized as a link in the human network.

One remarkable characteristic of the transportation network is the point that decision making on usage of the network is entrusted to the discretion of the persons who make trips in the network. The transportation network is the place which many trip makers use simultaneously, and each individual decision is affected by the "results" from the decision making by others. Therefore, the point in concern here is the "results" of the decision making by many and unspecific persons, and no agreement among the participating individuals is formed. In the transportation network without any agreement, individuals have to make decisions under uncertain conditions. The decisions under uncertainty may not lead to the socially preferable, efficient and effective state. The transportation and traffic information provided by public sectors may induce individual behaviours, resulting in the more desirable decisions as a whole.

3.3 Issues in Traffic Behaviour Modelling

Research on traffic behaviours stands on the viewpoint of methodological individualism in which the behaviour of trip makers is modelled separately from others' behaviours. The methodological individualism, the expression Schumpeter first used, means the scientific approach to clarify and understand a phenomenon in social or economic system by reducing it into independent individual behaviours and then aggregating them. While recognizing the operational handifulness and usefulness of this paradigm, he criticizes it as having the essential difficulty to understand the social phenomena caused by the interactions of composing individuals (Schumpeter, 1908). So far, traffic behaviour modellings have neglected the interactions of individual decision making on mutual adjustment in time schedule and on (same) meeting/gathering location and time, under the paradigm of the methodological individualism. To model the meeting behaviour, the mutual effects among individuals must be considered explicitly.

The decision making on the human network and the transportation network are closely connected. Meetings in the human network appear in the transportation network. The meeting properties such as "where", "when", "who" and "how" are determined by the potential meeting participants collectively. While there may be exceptional cases where the results of individual decisions on the transportation network affect the decision making processes regarding meetings, such as those on the accessibility of the location, the contents of meetings are essentially determined by the human network. Therefore, the decision on meeting has the Stackelberg-game like structure with the decision making in the human network as upper problem and the decision making in the transportation network as a lower problem. The development of behavioural models for meeting formation will be one of the most important and promising research topic.

Due to the transition of the society to become knowledge-oriented, many social and economic fundamental structures are reorganizing. For example, the standardization of the five-day-week system and the progress of research and development change the structure of and decisions in the human network. This may lead to the change of individual traffic behaviour. The person trip survey designed based on the fundamental paradigm of the methodological individualism had contributed to the clarification of the individual traffic behaviour. It is, however, a survey to investigate traffic behaviour in the transportation network, and therefore brings about highly limited and fragmental information on the meeting behaviour on the human network. There is still a long way before one can fully grasp the whole picture of meeting, which is one of the most fundamental communication form in the modern knowledge society.

4. Meeting and Traffic Behaviour

4.1 Necessity of Meeting Behaviour Modelling

In the last two decades, a huge amount of research has been accumulated on traffic behaviour. The discrete choice modelling based on the random utility theory has accelerated the development of more flexible and various traffic behaviour models (refer to, for example, Finney (1971), McFadden (1974), Domencich and McFadden (1975), Daganzo (1979) and Ben-Akiva and Lerman (1987). The traffic behaviour models based on the random utility theory maintains consistency with the utility maximization theory, and therefore enables us to evaluate effects of various traffic/

transportation policies without causing any conflict and inconsistency with economic theory. The analyses of economic effects in traffic policies had not been studied until recently. The reason for this must relate to the fact that the main focus of the traffic behavioural analyses had been on the modelling of the behaviour itself.

When somebody decides his/her meeting behaviour, the decisions by others intervene. At extreme cases, one may have to join a meeting by someone's strong intention. When an individual's decision is intervened by others' intentions, the resulting equilibrium of the traffic behaviour is unlikely to be the Pareto optimum. This kind of inefficiency causes the problems of over-supply and/or under-supply of trips. For example, the frequency of interactions with friends would tend to be less than the socially optimum level. On the other hand, many business trips governed by constitutional and/or mandatory conditions may be realized with more than optimal frequency. Among those, some trips may be substituted by telecommunication. Surprisingly, this kind of externalities in traffic behaviour has not been discussed in the literature on traffic/transportation behaviour. Within the framework of the methodological individualism, there has been no room for argument on this kind of trip externality.

Once recognizing the existence of the economic externality based on the interactions among individuals, one can notice many research topics left to be solved. The traffic demand management (TDM) policies in the comprehensive traffic planning in modern society should also be looked at again, by viewing it as the measure to correct the externality on trip generation. The problem of externalities caused by traffic has persistently been in focus, regarding the social costs by the vehicle traffic and the congestion taxes. The externality, however, exists in phenomena other than traffic congestion. A new era has emerged, requiring the systematization of traffic management policies (which had been sometimes proposed ad hoc way), as measures to correct and internalize the externalities.

4.2 Modelling of Meeting Behaviour

A meeting is organized through the negotiation process by the participants. In the formation of a convention, a private intercourse, or any other meeting, there is always a proposer/originator who calls for the potential meeting participants. With the agreement on the meeting, the size, place, time and other details of the meeting are then determined. While some are simultaneously determined, and adjustment and feedback based on each individual private affair may take place in the midst of them, the meeting agreement formation processes can be, in general, expressed as the sequential decision making process consisting of 1) decisions on the purpose of the meeting and on whether to have it, and 2) decision on the details of the meeting. Further investigation and research are necessary to clarify the process of meeting formations.

An agreement mechanism on the meeting is expressed as a simple model. Consider a human network consisting of the two individuals, i and j. The utility of the meeting for the individual i is defined by

$$U_i = V_i + e_i \tag{1}$$

where V_i is the fixed utility term to be determined by the individual attributes, the characteristics of the meeting, and others, and e_i is the probability variable representing the unobservable characteristics. Assume that the decision that whether an individual will join the meeting depends on whether the utility to be obtained by participating in the meeting is greater than the least utility level. The probability that the individual i agrees with the participation in the meeting is then given as

$$P_i = \text{Prob}\{V_i + e_i \geq \overline{V}_i + \overline{\varepsilon}_i\} \tag{2}$$

where $\overline{V}_i$: the fixed least utility level, and $\overline{\varepsilon}_i$: the probability error term. Thus, the probability that the individuals i and j agree with holding the meeting is expressed as

$$P_{ij} = \text{Prob}\{V_i + e_i \geq \overline{V}_i + \overline{\varepsilon}_i, V_j + e_j \geq \overline{V}_j + \overline{\varepsilon}_j\}. \tag{3}$$

This model expresses the formation mechanism of a meeting as the agreement by the multi persons, and therefore can be called as the *random matching model*. By using this random matching model, the realization of the face-to-face communication can be expressed. Also, it can be used to model terminal trips such as kiss-and-ride, and traffic behaviours involving simultaneous decisions by multi persons such as pick-up and give-a-ride traffic.

By employing the random matching model, problems existing in the modern knowledge society can be expressed. Assume that due to the intensification of the society, the opportunity cost of the meeting has become higher, resulting in the increase in the level of the least utility levels of $\overline{V}_i$ and $\overline{V}_j$. The increase in the least utility level leads to the decrease in the frequency of meeting occurrence. Moreover, notice that the probability of the meeting occurrence P_{ij} should be always smaller than the individual probability of the meeting agreement P_i and P_j. If the probability of the meeting agreement for each individual is mutually independent and takes value of 0.1, for example, the probability of the meeting occurrence will take a considerably small value of 0.01. Therefore, there holds the law of contradiction; while meetings play very important roles in development and intensification of society, economy and culture, the increase in the time value brings about increased difficulty of the formation of meetings. The knowledge society is the society where human communications are highly required but also are highly difficult to self-organize. The technological innovations of the transportation and communication, and the consolidation of the traffic and urban infrastructures are valuable means to support the realization of meetings, coping with the increase in the time value. Economic development is realized only by achieving both the realization of frequent meetings and the increase in the time value in the society.

Consider the over-supply and under-supply problems of traffic trips, by expanding the basic model given above. Let the marginal utility of the individual i when meeting the individual j n times, $MU_i(n)$, be expressed as

$$MU_i(n) = \frac{\partial V_i(n)}{\partial n} + \varepsilon_i \qquad (4)$$

Whether one agrees with the nth meeting is determined by whether the marginal utility of the meeting is greater or smaller than the shadow price of time λ_i. The shadow price is the probability variable that varies with the individual in concern. The probability that the individuals i and j agree with the nth meeting is then expressed as

$$\Phi(n) = \mathrm{Prob}\{MU_i(n) \geq \lambda_i,\ MU_j \geq \lambda_j\} \qquad (5)$$

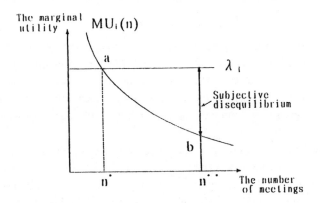

a) Person i

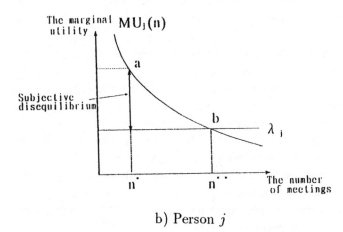

b) Person j

Figure 2. The Agreement Mechanism of the Meeting.

Figure 2 shows the relationship between the number of meeting and the utilities for i and j obtained from the meetings. Assume that the marginal utility functions are identical for both individuals and only the shadow prices to be hold are different between them. The individual i is very busy and his or her shadow price is very high. Contrarily, the individual j has time to spare and his or her shadow price is relatively low. For a meeting to be organized only by their intentions (spontaneously), the marginal utilities of both individuals should be greater than their own shadow prices at the same time. At point a in Figure 2 a), the marginal utility of the individual i coincides with his or her shadow price. For the individual j, however, the marginal utility is still bigger than his or her shadow price. The individual i consumes meetings optimally, while for the individual j this amount of meetings is regarded as under-consuming. The number of spontaneous meetings, n^*, is governed by the individual with higher shadow price, and therefore, in the society as a whole the meetings are under-consumed. On the other hand, assume that the individual j can set the number of meeting to be n^{**} as in Figure 2 b), by some binding power. In this case, the marginal utility of the individual i takes a value smaller than his or her shadow price. As a result, this kind of compulsory meeting tend to be societally over-consumed.

In the modern society, there is a quite remarkable tendency that spontaneous or voluntary meeting such as friendships and private intercourses are under-consumed. The energy and the opportunity cost for individuals in different environments to use and consume in order to meet at the same time and at the same place by adjusting their schedules can never be small. On the other hand, meetings with some compulsory power including many business and institutional meetings are likely over-held. The number of meetings is not determined to be societally optimal, but would rather involve individual disequilibrium. Moreover, the over-consumption and under-consumption of the meetings cause over-supply and under-supply, respectively. It is therefore necessary to strive to induce the societally desirable level of communication by the introduction of new institutional innovations and by communication technologies such as traffic policies and television meetings. Unfortunately, the research with this point of view has not yet been started.

4.3 Meeting Accessibility

The decision on the location to hold the meeting also determines the major parts of the traffic behaviour of the participants. Consider the case in which an individual participates in a meeting to be held at a hotel. The only reason why the destination of his or her trip results in the hotel was because the meeting was taking place at that hotel. The hotel might be chosen as a place to meet because it is located at a convenient location to access for all participants. The attractiveness of the meeting location is determined as a result of the selection behaviour in the human network. One important characteristic of the meeting location is that it is determined endogenously depending on the patterns of the land utilization and the transportation systems in the city.

Assume that the environment of the communication in the city is expressed by the meeting accessibility (Kobayashi *et al.* (1993)). Let us assume that the meeting attractiveness of a site (zone) j, ATT_j, is determined by the capacity of the meeting facility in concern, Q_j, and the meeting size, D_j.

$$ATT_j = \alpha_1 Q_j^{\alpha_2} D_j^{\alpha_3} \tag{6}$$

where α_1, α_2 and α_3 are parameters. The accessibility from the zone i to the zone j is expressed as

$$ACC_{ij} = \exp(-\beta d_{ij}) \, ATT_j \tag{7}$$

where d_{ij} is the distance between zones i and j, and β is the parameter representing the resistance to the distance. The number of participants from zone k in the meeting held in zone j can be written as the gravity model given as follows:

$$G_{kj} = \alpha_1 N_k^\gamma f_{kj} Q_j^{\alpha_2} D_j^{\alpha_3} \tag{8}$$

where $f_{kj} = \exp(-\beta d_{ij})$, N_k is the number of knowledge workers, and $\gamma(>0)$, the parameter. By summing up the equation (8) on k's, the meeting size in the zone j, D_j, is given as follows:

$$D_j = \left\{ \alpha_1 Q_j^{\alpha_2} \left(\sum_k N_k^\gamma f_{kj} \right) \right\}^\varepsilon \tag{9}$$

where $\varepsilon = 1/(1 - \alpha_3)$. By substituting equations (6) and (9) into equation (7), the accessibility to the meetings to be held in all sites in the city from zone i is given as follows.

$$ACC_i = \sum_j \left\{ \Phi_{ij} Q_j^\delta \left(\sum_k N_k^\gamma f_{kj} \right)^\sigma \right\} \tag{10}$$

where $\Phi_{ij} = \alpha_1^\varepsilon f_{ij}$, $\delta = \alpha_2 \alpha_3 \varepsilon$, and $\sigma = \alpha_3 \varepsilon$. Equation (10) can be called, then, the meeting accessibility. Notice here that the meeting attractiveness is determined depending on the decision on the human network: "how many individuals participate in the meeting".

Let us look at a numerical example of the meeting accessibility. The streets in the city of Venice in Italy have a unique structure. The access to the city by vehicle, a modern transportation method, is impossible, and except for the surface transportation, walking is the only choice of method available. Venice has very few streets that are wide enough to be called "streets", and others are paths running through the binding buildings, constituting the captivating city of labyrinth. Walking on the street, one may suddenly face a open space called *"Campo"*. In the city of Venice, many kinds of Campo exists and are utilized by residents as the places to leisure, relax, and exchange information. It is interesting to observe that only very few Campos are used busily and highly crowded, while the others are hardly ever crowded except for some use by the neighbours. This kind of phenomenon on usage pattern of Campo can be analysed by the meeting accessibility. Assume that the residents are evenly distributed on the island. Let the capacity of the meeting facility be represented by the space size of the Campo. The parameter setting is given in Figure 3. As shown in Figure 3, the meeting size of only a few Campos are relatively large. It is not necessarily true that for Campos, bigger space means higher demand. When the parameter γ, representing the elasticity of the meeting size to demand, increases, the meetings tend to concentrate on fewer Campos. In the numerical example given above, something like "which Campo should be the center of the meeting" is assumed. It would instead be determined endogenously as a result of individual choices and behaviours under the specific characteristics of the street/path network in Venice. The meeting accessibility is a system with very strong non-linearity. If a new bridge is constructed over the great canal, the whole structure of the meeting accessibility in Venice changes. This may lead to the abandonment of the center Campo and the rapid growth of another Campo as a new center.

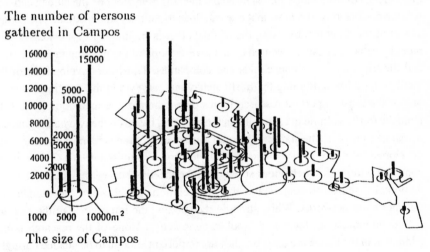

Figure 3. The Meeting ccessibility of Campos.

4.4 Human Contacts Modelling Focused on the "Meeting": Perspectives

The critical concerns on the modelling of the meeting behaviour are presented above, by briefly introducing the ideas and models that the authors have developed. To systematize these ideas into a new methodology for human contacts modelling, there are many issues to be addressed in the future. First, it is necessary to develop a model that can explicitly consider both the meeting formation and the traffic behaviour at the same time. In order to analyze policies, various types of (fragmental) modelling of the meeting behaviour have to be accumulated. For example, the further research and modelling of the random matching modelling presented above may be beneficial. Second, the development of the technique for aggregation of meeting behaviour is needed. When there exists a strong dependency among individual behaviour, the simple sum-up type of aggregation techniques cannot be employed. In the city areas, individuals with various potential meeting demands look for and find the meeting opponents who match their purposes, and repeat the meetings. The meeting progresses are like the chemical reaction of two molecules that meet incidentally and self-create a new material. This kind of meeting process is, as mentioned earlier, a strong non-linear process, in which the meeting center is endogenously determined and the trans-formation of the network structure itself sometimes occurs catastrophically. The development of solid mathematical representation of the stochastic equilibrium for the random matching models is, therefore, the fundamental research for the clarification of the whole aspect of the face-to-face communications in cities. Third, it is also desirable to develop the investigation methodology to grasp the meetings in the society. Unfortunately, almost no actual meeting behaviour on the human network is known. Currently, the most that one can do is to estimate and evaluate one certain aspect of the meetings by using the trip data that is fragmental in most cases. The meeting behaviour can be secured by the investigation of the meeting facility usages and the trip survey for households and companies. Thus, the development of the methodology for combining the traffic behaviour survey and the meeting survey is one of the most important research issues. For example, the methodology to esti-mate the traffic volume in each route in the network, based on the two heterogeneous investigations of 1) the traffic behaviour survey and 2) the screen line survey, was developed by many authors (e.x. Van Zuylen and Willumsen 1980). The issue of connecting the traffic behaviour survey and the meeting survey may become a fun-damental research to understand the interrelationship between the human and transportation networks. While strongly recognizing the importance of the face-to-face communication, there is no tool to approach it. None of the research issues addressed in the above are easy, but they are important enough to be tackled daringly.

5. Information and Traffic Behaviour

5.1 Modelling of Expectations Formation

In the transportation network, individuals form their traffic behaviour on the basis of their own decisions without any mutual pre-negotiation with others. The individual decision is affected by the *results* of the others' decisions. A trip maker, however, cannot know the others' decisions in detail. At most, he or she can only know the traffic situation resulted after using (making a trip on) the network. In the traffic behaviour, a number of individuals decide their behaviour simultaneously, and therefore the communication costs required to have a pre-agreement is prohibitively high. The individual trip maker has to decide his or her behaviour depending on his or her experiences, intuition and other private partial information. These kinds of decisions under the uncertainty may not lead to the preferable situation of the network. Thus, it is important for the public sectors to complement the trip makers' communications by providing the traffic information.

Trip makers have to predict the transportation conditions to be realized in the network under uncertainty. The prediction may be made based on the trip maker's intuition, learning process based on his or her experience, and other private information. Consider a driver who is faced with the choice of route for his or her trip on the network. He or she does not know anything about the traffic conditions for certain, while having private information gathered throughout his or her past trip experiences on the network. He or she predicts the traffic condition such as travel time based on some leading information and the experience. The driver predicts the variables which have significant effects on his or her decision and the prediction is called *expectations*. In order to discuss the role of expectations formation in the traffic behaviour, consider the route choice problem based on a simple random expectation utility model. Model the random expectation utility of a route as follows:

$$EU_t = a + bT_t^s + \varepsilon_t \tag{11}$$

where the parameter a and b represent utility characteristics of the driver. T_t^s is the subjective expectations on the route travel time recognized by the driver. The superscript s shows that the variable is subjective. ε_t is the observation error factor. When choosing a route under uncertain information, the driver is unable to know the exact value of the variable T_t^s *ex ante*. When trying to formulate the driver's behaviour on route choice under uncertainty, a model to describe how the drivers form their predictions on the value (or more specifically, the distribution) of the explanatory variable is required. Therefore, specify this mechanism of prediction by drivers as follows. The driver's subjective expectations on the average travel time of a route at time t, T_t^s, are functions of the driver's experimental information (realized value of his or her past trips) $\left(\tilde{T}_{t-1}, \tilde{T}_{t-2}, \ldots \right)$ and his or her past subjective expectations $\left(T_{t-1}^s, T_{t-2}^s, \ldots \right)$.

$$T_t^s = \Phi\left(\tilde{T}_{t-1}, \tilde{T}_{t-2}, \ldots; T_{t-1}^s, T_{t-2}^s, \ldots \right) \tag{12}$$

This equation shows the mechanism in which the driver's subjective expectations on the average value of the route travel time at time period t are formed based on the experimental information and the past subjective expectations held by the driver, and can be called the *expectations formation mechanism*.

5.2 Expectations Formation Hypothesis

There must be various kinds of behavioural hypotheses on drivers' prediction of the route travel time available. Among others, the representative hypotheses are 1) static expectations hypothesis, 2) extrapolatory expectations hypothesis, 3) adaptive expectations hypothesis, and 4) rational expectations hypothesis. The static expectations hypothesis is based on the idea that the drivers formulate the fixed expectations about the traffic conditions. This hypothesis, while having been employed by many researches in the past, has not explained *how* drivers formulate the static expectations at all.

According to the extrapolatory expectations hypothesis, the prediction at time period t, T_t^s, does not depend only on the travel cost realized in the previous time but also on the degree of its variance. Thus, by representing the realized travel time at time period t - 1 and t - 2 by $\tilde{T}_{t-1}$ and $\tilde{T}_{t-2}$, respectively, extrapolatory expectations can be expressed as follows (Metzler, 1941):

$$T_t^s = \tilde{T}_{t-1} + \eta\left(\tilde{T}_{t-1} - \tilde{T}_{t-2}\right) \tag{13}$$

where η is called the expectation coefficient. The extrapolatory expectations of each time period are assumed to be equal to the sum of the previous level and the difference between the previous level and the level before multiplied by a positive number. If η is positive, the succession of the past tendency is expected, while if negative, the reverse of the past tendency is expected. The behaviour in the extrapolatory expectations is governed by η. While this parameter is to be selected depending on the economic structure underlying the model, the hypothesis does not provide any explanation about this selection.

To overcome the shortcomings of the extrapolatory expectations, the adaptive expectations hypothesis in which the drivers adjust their expectations in response to the error of the expectations at the previous period of time, is proposed (Cagan, 1956): The adaptive expectations hypothesis can be written as follows.

$$T_t^s = T_{t-1}^s + \zeta\left(\tilde{T}_{t-1} - T_{t-1}^s\right) \tag{14}$$

where ζ is called the adaptive coefficient and represents the speed (rate) of the adjustment against the errors in the past. By reformulating the equation (14), we can get $T_t^s = \zeta\tilde{T}_{t-1} + \left(1 - \zeta\right) T_{t-1}^s$. The term T_{t-1}^s is the expected travel time at time period t - 1 formed at the time period t - 2, and can be expressed as $T_{t-1}^s = \zeta\tilde{T}_{t-2} + \left(1 - \zeta\right) T_{t-2}^s$. By adapting the same logic and expanding the equation (14) sequentially, we can get the following formula.

$$T_{t-1}^s = \sum_{k=1}^{\infty}(1 - \zeta)^{k-1}\tilde{T}_{t-k} \tag{15}$$

While extrapolatory expectations depend on the previously realized values up to only two step astern, adaptive expectations depend on the whole time series of the past realization. The adaptive expectations have forms in which the realized values (with various lags) are weighted with the geometrical distribution. There are many forms available for the distributional lags, and the lag with the decrease by geometrical progression is just one special case. The advantage of the geometrical lag is its tractability in the estimation process. The scientific foundation that the model can represent the real behaviour accurately, however, is very week. Also, it is shown that adaptive expectations can not utilize the available information optimally (Sheffrin, 1985).

5.3 Rational Expectations Hypothesis

Muth (1961) proposed the Rational Expectations (RE) Hypothesis in which "as consequences of their long-term learning behaviour, the agents' subjective probability distribution coincides with the objective probability distribution of events". Since then, along with research accumulation, the RE Hypothesis has attracted so far the greatest attention in economics and other research arenas (for instance, Sheffrin (1983), Lucas (1978), Shiller (1978), Radner (1979), Radner (1982) and Lippman and McCall (1982)). Also, the mechanisms of the RE formation through learning have also been researched and developed (see, for example, DeCanio (1979), Bray and Savin (1986), and Fourgeand and Pradel (1987)). Assume that the drivers renew their subjective expectations through the learning behaviour. As a consequence of the learning process, the subjective expectation merges to a certain value. Due to the RE Hypothesis, the drivers' subjective expectations coincide with the actual travel time distribution through their learning behaviours. Once the drivers form the RE, they may lose their incentive to change their subjective expectations. When the driver is inexperienced in the route selection, he or she cannot form the RE. The driver's subjective expectations depends on their accidental history in the past, and therefore, it is impossible to describe the unstable route selection behaviour at a given period of time. If the driver accumulates information and forms the RE, however, his or her subjective expectations can be measured through the probability distribution of the travel time realized in the network. Therefore, the RE Hypothesis provides us with an important tool to measure objectively the subjective expectations that is quite private information and not directly observable (see Kobayashi (1993) for detailed discussion).

The authors propose the route selection model in which the drivers update their expectations successively by Bayesian Learning (Kobayashi 1994). In this chapter, the learning model adjusts the expected value and the variance simultaneously. The learning model of the expected value is given by the following equation:

$$T_{t+1}^s = T_t^s + \frac{1}{v_0 + n^t} \cdot \left(\tilde{T}_t - T_t^s \right) \tag{16}$$

where v_0 is the parameter determined by the initial expectations and n^t, the number of route selections. The expected value at the time period $t + 1$ is adjusted by the expected value at the time t, T_t^s, (the first term in the equation (16)) and the error of the subjective expectations at the time t (the second term). The weight coefficient, $1/(n_0 + n^t)$, is not constant, but approaches to 0 when the value of n^t increases and the adjusted range of the subjective expectations, T_t^s, attenuates. When t in the equation (16) is large enough, then the approximation of $T_t^s \simeq \bar{\tau}_t$ can be used, where $\bar{\tau}_t$ is the sample average of the realized travel times. When the driver chooses his or her route repeatedly an enough number of times, his or her subjective expectations on the average travel time approaches to the realizing objective sample average. Every driver, what ever his or her initial expectations are, eventually forms the RE through his or her learning behaviour. If the driver forms the rational expectations T^*, the following equation holds:

$$T^* = E_\infty[T_t] = \lim_{N \to \infty} N^{-1} \sum_{t=1}^{N} \tilde{T}_t \qquad (17)$$

A numerical example is given below. In this case, the fixed number of drivers repeat their route choices between two alternative routes. The travel time for each route varies along with the volume of the local traffic. The results from the drivers' route selections reflect the travel times of the routes. The results affect the drivers' expectations, and subsequently, affect their route selections. In such a way, both the travel time distribution that is realized in the routes and the drivers' expectations change over the time. Figure 4 shows that for both routes the average travel times realized at each period in time (A, B) and the drivers' subjective expectations (C, D) gradually converge to the stable states (the RE equilibrium). In this simulation, by the end of two hundred iteration, the subjective expectations and the average of the objective travel time coincide and reach at the RE equilibrium.

5.4 Role of Information in Expectations Formation

Recently, some researchs have focused on the induction of the drivers' behaviour by providing information such as the route navigation and the parking information systems (for example, Bonsall (1992), Chang and Mahmassani (1988), Iida, Akiyama and Uchida (1992), and Kobayashi (1993, 1994)). When the traffic information is provided, the traffic information should also be included in the expectations forma-tion mechanism (12) as a variable. The traffic information affects the drivers' expectations and eventually affects the drivers' behaviour. However, when conside-ring the learning behaviour of the drivers in the long term, caution should be paid to recognize the effects of the information. Consider one simple example. Assume that at a certain point in time the governmental agency provides the message to all (or, a part of) drivers to choose a certain route. If all drivers choose the designated route, they may cause congestion in that route. In that case, sooner or later, some drivers will realize that the another route would rather take them to the destination faster. Assume that these kinds of learning repeat and as a consequence the drivers have

Travel time *(min.)*

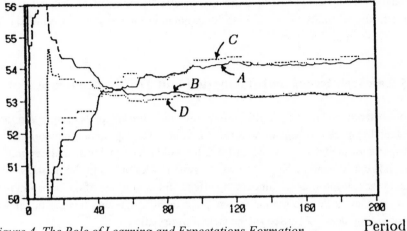

Figure 4. The Role of Learning and Expectations Formation   Period

learned the distribution of the traffic conditions to be realized in the routes. Then, there is a possibility that the public information can no longer induce the drivers' route choices.

In the arena of economics, when the RE equilibrium holds, the neutrality proposition holds, where all information obtained by the agents participating in the market is reflected by the current price in the market, and as a result, the information gained by the agents do not have any effects and therefore any roles as "information" (see Radner (1982), Hellwig (1980) and Grossman (1989)). For the traffic information to have route navigation effects, the rejection of the "neutrality of the traffic information" is a prerequisite. The neutrality of the traffic information can be defined as the situation where each message in the information system does not provide any additional information. Now, let the set of messages that the information system provides be given by η. Also, let the RE which is ultimately realized under the message e be given by $T^*(e)$. When for different messages, e and $e' \in \eta$, if $T^*(e) \simeq T^*(e')$ holds in approximate sense, the traffic information is designated as neutral. If for each message a different RE is not realized, the traffic information does not function as "information" anymore. In this case, the attempt to navigate the drivers' route choice by the traffic information fails in the long run. On the contrary, when the neutrality proposition does not hold, the information can provide a substantial effect on traffic behaviour in the long run.

Similar to the market price being obtained by the demand and the supply, the travel time is obtained from mutual effect of the demand for the routes and the supply conditions. In the case of general market goods, the current price of the goods can be observed in the market *ex ante*, and the market price works as the medium to transfer the information on the future price of the goods. When viewing the case of the traffic as a market, the drivers cannot know the travel time for each route and there does not exist the information to work like a market price. From this critical point of view, Kobayashi (1994) shows by employing the Bayesian learning model that in the traffic market, "the neutrality of information" does not hold and the traffic information can be effective in the long run.

5.5 Research Issues on Information and Traffic Behaviour

The substantial value of the traffic information as "information" is originated from the fact that price mechanism does not work on the transportation network. In the transportation network, we cannot utilize information that can represent the aggregate demand-supply relationship in the whole system, like prices in the markets. This kind of lack of information hinders the effectiveness in usage of the transportation network. The traffic information system plays the role to rectify this inefficiency in the use of the network caused by the lack of information.

The methods to remedy the inefficiency caused by the uncertainty can also be found in systems other than traffic information. For example, the reservation system is an artificial market realized by the rationing of the capacity to accept. By providing artificially a place to exchange information, the uncertainty appearing on the transportation network can be partially internalized. The traffic behaviour by each individual is made definite and the uncertainty decreases. Thus, the effects of the reservation system can also be understood via the role of information. The problems of information and expectations formation relate not only to the route navigation and parking guidance, but also to many other traffic phenomena relevant in much wider fields.

The information reduces the inefficiency caused by the uncertainty and partly ease the traffic congestion. The correction of the uncertainty and the resolution of the externality on the transportation network (i.e. the congestion), are however, essentially two different problems. In fact, it is possible that the information results in, in turn, the decrease in the welfare of the drivers on the network as a whole. The route navigation effect consists of 1) the time-shortening effect enjoyed by the drivers who changed their route, 2) the external effect that the withdrawal of some drivers from the route affects on the drivers remaining in that route being relaxed and 3) the external diseconomy caused from the increased congestion of the route caused by the drivers additionally induced by the information system to choose that route. When the externality of case 3) overbears the economic effects of the cases 1) and 2), then the route navigation may in turn cause a decrease in the total welfare of drivers. The problem of the externality cannot be resolved completely by simply providing the traffic information. Approaches with the combination of congestion fees and peak-load pricing are necessary. The unification of the traffic management policies such as assortment of the information systems with other traffic management measures is eagerly desired.

6. Issues on Refinement of Transportation Infrastructure in Knowledge Society

There is no doubt that the transportation and communication networks are the compounds consisting of the nodes and links. Traditionally, the consolidation of the network infrastructure has concentrated on the refinement of its function on links. In the knowledge society, however, the function of the nodes will increasingly play more important roles. Especially, it is noteworthy that in the human network the meeting functions as the fundamental linkage. The spatially fixed meeting facilities are important network factors constituting the nodes of the physical network. On the contrary, the human network is not spatially fixed in its location. Sometimes we can have a meeting in the transportation means. The TV conference utilizing virtual reality is the technology that links spatially distant locations as if they are in the same node.

The nodes connect different links systematically. In order to fulfill the complicated meeting demand in the human network, it is necessary to construct and reinforce the elaborate and complex network with various nodes. Also, the marketing strategies to provide a new type of traffic services made of the combination of information services and the various conventional traffic services is indispensable. The city is the accumulated place of the nodes constituting various physical networks, and of the nodes connecting the city to other outside worlds. The density and efficiency of the nodes are primary conditions for expansion of the networks in the knowledge society.

The future of the cities in the knowledge society age relies on the urban development strategy of "how we can realize the mutual effects of the human networks and physical networks". In the short run, the human network determines the city potential. The structure of this human network is, however, very vulnerable and unstable. It is possible that one decision making in the human network metamorphoses the fundamental structure of the network in a moment. The examples are the founding, separation and discontinuance of scientific societies and associations. The hub structure of the network may also change drastically. The human network, however, is dependent on the structure of the physical network in the long run. The physical networks evolve slowly. Thus, the physical networks can be regarded as being constant or stable in the short run. On this physical network, the human networks self-organize. When the physical network gradually evolves and approaches and eventually exceeds a certain threshold, however, the human networks in the short term may possibly have catastrophical changes. A node at the comparatively disadvantageous position has a possibility to recover its potentials in the long run, if it holds a highly advanced accessibility in the physical network.

7. Conclusions

In the knowledge society, the importance of face-to-face communication is increasing. While playing an important role in the development of society, economy and culture, this kind of communication inherently holds the conflict that the increase in the time value makes it more difficult to self-organize. The technological innovation of information, telecommunication and transportation systems provide the society with the fundamental means to hold face-to-face communications in the era of increased time value. In this paper, we proposed one promising direction of research on traffic behaviour modelling based on "communication with others", and attempted to abstract about the methodology to explicitly take into account the development of the information/transportation technologies. Of course, this paper, while attempting to get rid of the methodological individualism in the current traffic research paradigm, is still far apart from the position to scrutinize and develop a new comprehensive and systematized methodology. The paper, however, at the least, clarified the importance of traffic behaviour modelling based on the "communication with others", in order to advance the traffic policies in the era of the knowledge society.

References

Beckmann, M. J., 1994, On knowledge networks in science: collaboration among equals, *The Annals of Regional Science,* 28:; 233-242.

Ben-Akiva, M. and Lerman, S. R., 1987, *Discrete Choice Analysis: Theory and Application to Travel Demand,* Cambridge: The MIT Press.

Bonsall, P., 1992, The influence of route guidance advice on route choice in urban networks, *Transportation,* 19: 1-23.

Bray, M. M. and Sabin, N. E., 1986, Rational expectations equilibria, learning and model specification, *Econometrica,* 54: 1129-1160.

Cagan, P., 1956, The Monetary dynamics of hyperinflation, In: Friedman, M. (Ed.), *Studies in the Quantity Theory of Money,* Chicago: University of Chicago Press.

Chang, G. and Mahmassani, H., 1988, Travel time prediction and departure adjustment behaviour dynamics in a congested traffic system, *Transportation Research,* 22B: 217-232.

Daganzo, C., 1979, *Multinomial Probit,* New York: Academic Press.

DeCanio, S. J., 1979, Rational expectations and learning from experience, *Quarterly Journal of Economics,* 370: 47-57.

Domencich, T. A. and McFadden, D., 1975, *Urban Travel Demand: A Behavioral Analysis,* Amsterdam: North-Holland.

Finney, D., 1971, *Probit Analysis,* Cambridge: Cambridge University Press.

Fourgeand, C. C. G. and Pradel, J., 1987, Learning procedures and convergence to rationality, *Econometrica,* 54: 845-868.

Grossman, S., 1989, *The Informational Role of Prices*, Cambridge: The MIT Press.

Hellwig, M. F., 1980, On the aggregation of information in capital markets, *Journal of Economic Theory*, 22: 477-498.

Iida, Y., Akiyama, T. and Uchida, T., 1992, Experimental analysis of dynamic route choice behaviour, *Transportation Research*, 26B: 17-32.

Kobayashi, K., 1993, Incomplete information and logistical network equilibria, In: Andersson, Å. E., Batten, D. F., Kobayashi, K., and Yoshikawa, K., (eds.), *The Cosmo-Creative Society*, Berlin: Springer-Verlag.

Kobayashi, K., Sunao, S. and Yoshikawa, K., 1993, Spatial equilibria with knowledge production with meeting facilities, In: Andersson, Å. E., Batten, D. F., Kobayashi, K., and Yoshikawa, K., *The Cosmo-Creative Society*, Berlin: Springer-Verlag.

Kobayashi, K., 1994, Information, rational expectations and network equilibria, *The Annals of Regional Science*, 28: 369-393.

Le Goff, J., 1980, *Time, & Culture in the Middle Ages*, Chicago: The University of Chicago Press.

Lippman, S. A. and McCall, J. J., 1982, The Economics of Uncertainty: Selected Topics and Probabilistic Methods, In: Arrow, K. J. and Intriligator, M. D. (Eds.), *Handbook of Mathematical Economics*, 1:211-284, Amsterdam: North-Holland.

Lucas, R. E. Jr., 1978, Asset prices in an exchange economy, *Econometrica*, 46: 1429-1445.

McFadden, D., 1974, Conditional logit analysis of qualitative choice behavior, In: Zarembka, P., (ed.), *Frontiers in Econometrics*, New York: Academic Press.

Metzler, L., 1941, The Nature and Stability of inventory cycles, *Review of Economics and Statistics*, 23: 113-129.

Muth, J., 1961, Rational expectations and the theory of price movements, *Econometrica*, 29: 315-335.

Palander, T., 1935, *Beiträge zur Standortstheorie*, Uppsala: Almqvist & Wiksell.

Radner, R., 1979, Rational expectations equilibrium: Generic existence and information revealed by price, *Econometrica*, 47:3:655-678.

Radner, R., 1982, Equilibrium under Uncertainty, In: Arrow, K. J. and Intriligator, M. D. (Eds.), *Handbook of Mathematical Economics*, 2: 923-1006, Amsterdam: North-Holland.

Sheffrin, S. M., 1983, *Rational Expectations*, Cambridge: Cambridge University Press.

Shiller, R. J., 1978, Rational expectations and the dynamic structure of macroeconomic models, *Journal of Monetary Economics*, 4: 1-44.

Schumpeter, J. A., 1908, *Das Wesen und der Hauptinhalt der Theoretichen Nationalökonomie*. Leipzig: Duncker & Humblot.

Van Zuylen, H. J. and Willumsen, L. G., 1980, The most likely trip matrix estimated from traffic counts, *Transportation Research*, 14B: 281-293.

Education, Job Requirements, and Commuting: An Analysis of Network Flows

Björn Hårsman
Department of Infrastructure and Planning
Royal Institute of Technology, Stockholm

and
John M. Quigley
Graduate School of Public Policy and Department of Economics
University of California, Berkeley

Introduction

By now measures of employment "access" and "potential" have been widely diffused in the literature on regional economics and transport planning. Pooler (1995) gives a brief review of accessibility measures, indicating that these concepts date back to the 1930s. According to standard economic intuition, the employment access of a residential area increases with its proximity to concentrations of employment opportunities. The various indices of accessibility which have been proposed merely formalize and quantify this notion.

In this paper, we incorporate the spatial distribution of the demand for educational qualifications and the spatial distribution of the supply of educated workers into this framework. We ask:

1. How *different* are the computed measures of employment access and potential when variations in the spatial pattern of the demand for educated workers are recognized?

2. How *important* is the spatial pattern of the demand for educated workers in explaining variations in employment access and potential?

Following this introduction, we estimate models of employment access separately based upon workers of differing educational qualifications. These results allows us to investigate directly:

3. How *different* is the worktrip and residential location behavior of households of differing educational qualifications?

One reason for addressing these questions is that the knowledge-orientation of society will change the educational profile of the population. There are also reasons to believe that people with high education can choose their working time more freely than those with low education. It is also reported that the possibilities for telecommuting are positively related to the educational level. A more concrete reason is that the results of earlier work (Quigley and Hårsman, 1995) suggested that the travel patterns differ between workers in different industries and with different levels of education.

This empirical analysis is based upon the worktrip behavior of commuters in metropolitan Stockholm, disaggregated into three educational categories. The analysis is undertaken using gravity models of worktrip behavior and employment potential. We also conduct the analysis using more sophisticated models of worktrip behavior, namely the Poisson and the negative binomial relationships. Section II presents the methodology. The statistical results and their interpretation are in Section III.

Spatial Access

The most widely used empirical model of the accessibility of particular residential locations is based upon the gravity concept:

$$T_{ij} = \alpha R_i^\beta W_j^\gamma / d_{ij}^\delta ,$$ (1)

where Greek letters denote parameters.

The data used to estimate equation (1) consist of the matrix of commute flows T_{ij} between origin zones i and destination zones j and the distance or travel times d_{ij} between them. From the elements of the matrix, the number of workers resident in each zone (R_i) can be estimated ($R_i = \sum_j T_{ij}$). Similarly, the number of individuals working in each zone (W_j) can be estimated ($W_j = \sum_i T_{ij}$).

Isard (1960) provides a number of physical and social scientific justifications for the formulation in equation (1). Sen and Smith (1995) provide an exhaustive review. Flows between i and j are positively related to the "masses" of residences and workplaces and inversely related to the "impedance" (travel time) between i and j.

Estimates of the parameters yield a measure of the accessibility (A_i) of each residence zone to the workplaces which are distributed throughout the region (Isard, 1960, p. 510), i.e.,

$$A_i = \sum_j \hat{T}_{ij} / R_i^{\hat{\beta}} ,$$ (2)

where $\hat{T}$ is computed from the parameters of equation (1).

Suppose data are available on worktrip patterns of workers according to k education classes. A straightforward generalization of (1), expressed in logarithmic form, is

$$\log T_{ij} = \alpha' + \beta \log R_i + \sum_k \gamma_k \log W_j + \delta \log d_{ij} \qquad (3)$$

A comparison of (3) with (1) indicates the importance of the spatial disaggregation of workplaces by educational level. Analogously,

$$\log T_{ij} = \alpha' + \sum_k \beta_k \log R_i + \sum_k \gamma_k \log W_j + \delta \log d_{ij} \qquad (4)$$

A comparison of (4) with (3) indicates the importance of disaggregation of educational level by residence.

Finally, a completely disaggregated model can be compared,

$$\log T_{ijk} = \sum_k \alpha'_k + \sum_k \beta_k \log R_i + \sum_k \gamma_k \log W_j + \sum_k \delta_k \log d_{ij} \qquad (5)$$

More sophisticated measures of access recognize that the transport flows to each destination are count variables. The Poisson distribution is often a reasonable description for counts of events which occur randomly.

Assuming the count follows a Poisson distribution, the probability of obtaining a commuting flow T_{ij} is,

$$\mathrm{pr}\,(T_{ij}) = e^{-\lambda ij}\,\lambda_{ij}^{T_{ij}}\big/T_{ij}! \qquad (6)$$

where λ_{ij} is the Poisson parameter. Assuming further that,

$$\exp[\,\lambda_{ij}] = \alpha\,R_i^{\beta}\,W_i^{\gamma}\big/d_{ij}^{\delta}\,, \qquad (7)$$

yields an estimable form of the count model (since $E[T_{ij}] = \lambda_{ij}$). See Smith (1987) for a discussion. Estimates of the parameters similarly yield a measure of the accessibility of each residence zone to workplace in the region,

$$A_i = \sum_j \hat{\lambda}_{ij}\big/R_i^{\hat{\beta}}\,. \qquad (8)$$

As before, if worktrip patterns are available by educational level, this information can be incorporated into (7) in a manner analogous to (3), (4), and (5).

A more general model of the flow count between i and j relaxes the Poisson assumption that the mean and variance are identical. For example, following Greenwood and Yule, Hausman, Hall, and Griliches (1984, p. 922) assume that the parameter λ_{ij} follows a Gamma distribution $G\,(\omega_{ij})$ with parameters ω_{ij}. They show that, under these circumstances, the probability distribution of the count is negative binomial with parameters ω_{ij} and η,

$$\mathrm{pr}\,(T_{ij}) = \frac{G(\omega_{ij} + T_{ij})}{G(\omega_{ij})\,G(T_{ij} + 1)}\left(\frac{\eta}{1+\eta}\right)^{\omega_{ij}}(1+\eta)^{-T_{ij}} \qquad (9)$$

Again, assuming that,

$$\exp[\omega_{ij}] = \alpha R_i^\beta W_j^\gamma / d_{ij}^\delta \quad , \tag{10}$$

yields an estimable form of the count model and the resulting accessibility index for each residence zone.

Again, the availability of worktrip patterns by educational level can be incorporated into (10) in a manner analogous to (3), (4), and (5).

The count models are clearly nested. If η is infinitely large, then equations (9) and (10) specialize to (7) and (8). If η is finite, then the mean and the variance of the count variables are not identical (as assumed by the Poisson representation).[1]

Data, Results, and Interpretation

The model is estimated using data on worktrips for the Stockholm metropolitan area for 1990. The data consist of commuting patterns by educational level among the 26 civil divisions in the metropolitan area. Also available is the average zone-to-zone commute time for the metropolitan area, by civil division. Since modal split is not treated explicitly, we have used commute time by car. Data are available separately for three educational levels corresponding to those with primary schooling, secondary schooling, and post graduates.

Figure 1 indicates the spatial pattern in worksites by educational level in the Stockholm metropolitan area. Figure 2 reports the pattern of residential locations by education level in the region. The patterns are decidedly non random, with the residences of the more highly educated workers concentrated in the northern part of the region.

Table 1 summarizes estimates of the parameters of the gravity model for these three educational groups. For each educational group, we present the parameters of the model based on the non-zero observations (out of 26 x 26 = 676 possible observations). The models are estimated by ordinary least squares. In addition to separate estimates by educational level, the table presents estimates of the model based upon worktrip distributions undifferentiated by educational level.

As the table indicates, each of the models explains a large fraction of worktrip behavior — ranging from 80 to 84 percent of the variance in log worktrips. Not surprisingly, the number of worktrips between jurisdictions varies positively with the number of available residences and workplaces. The number of worktrips is also highly sensitive to the commute time between origins and destinations. The coefficient on commuting time in minutes is large, negative, and highly significant. Importantly, the travel time coefficient declines with increases in educational level. An increased level of education is normally related to a higher income, and higher income to higher time values. Hence, it might be expected that the travel time coefficients

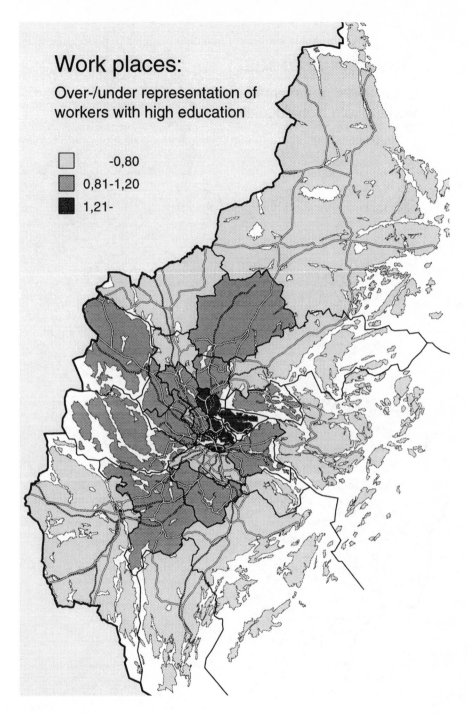

Figure 1

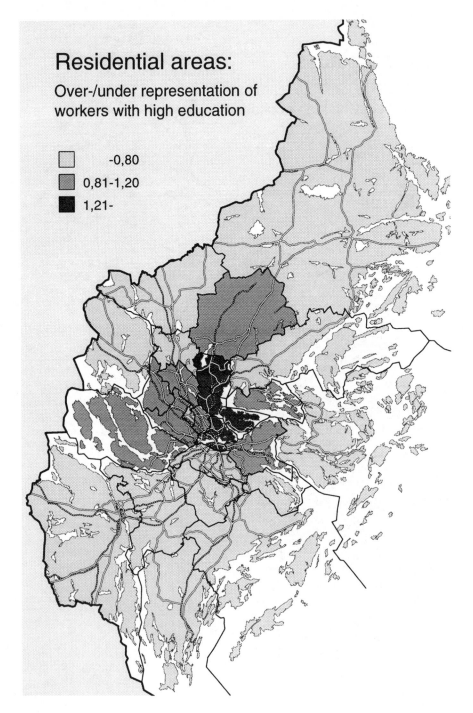

Figure 2

Table 1. Parameters of Gravity Model Estimated Separately by Education Level (t-ratios in parentheses).

$$T_{ij} = \alpha R_i^\beta W_j^\gamma / d_{ij}^\delta$$

	Low	Medium	High	Sum
		Education Level		
α	-0.812	-0.053	0.500	-1.242
	(1.34)	(0.10)	(1.15)	(2.22)
β	0.656	0.571	0.613	0.659
	(13.12)	(13.63)	(17.92)	(15.63)
γ	1.117	1.058	0.938	1.129
	(29.85)	(34.99)	(32.51)	(36.33)
δ	-2.844	-2.533	-2.378	-2.850
	(37.44)	(36.10)	(35.01)	(43.53)
R^2	0.810	0.838	0.841	0.801
Observations	617	580	536	646

increase when the educational level increases. Our results indicate that this effect evidently is more than offset by the strong preference for space and different housing amenities among the best educated.

These general results are confirmed by the more rigorous results reported in Table 2 using the Poisson assumptions. The model coefficients are estimated by maximum likelihood using all 676 elements of the travel time matrix for each educational level. When the Poisson model is used to estimate access, the results are substantially more significant statistically. The t ratios of the coefficients increase by more than ten fold. Again, moreover, the coefficients vary significantly by educational level. The magnitudes of the coefficients are reasonably similar to those estimated by the gravity model. In particular, the effect of travel time in conditioning workplace choice and commute trip behavior declines as educational level increases.

These general results are confirmed by estimates of the negative binomial model. These estimates are not reported.[2]

In Table 3, the three matrices of worktrip behavior are combined in a single estimation. The table presents the coefficients of the gravity model based upon the combined sample of ($626 \times 3 = 2028$) observations. The gravity model is estimated using ordinary least squares on the non-zero observations.

Six models are presented. Model I is identical in form to those presented in Table 1. It includes the number of workers residing in the origin zone, the number of jobs in the destination zone, and the travel time between zones. It is estimated on the 1733 non-zero observations.

Model II disaggregates the workers by origin zone into three educational levels.

Table 2. Parameters of Poisson Model Estimated Separately by Education Level (t-ratios in parentheses).

$$\text{pr}\,(T_{ij}) = e^{-\lambda_{ij}}\,\lambda_{ij}^{T_{ij}}\big/T_{ij}!\,, \quad \exp[\,\lambda_{ij}] = \alpha\,R_i^{\beta}\,W_i^{\gamma}\big/d_{ij}^{\delta}$$

	Education Level			
	Low	Medium	High	Sum
α	1.270	0.861	0.568	0.723
	(31.09)	(16.85)	(8.78)	(23.33)
β	0.571	0.567	0.626	0.553
	(140.94)	(116.31)	(107.29)	(194.18)
γ	1.065	1.031	1.002	1.052
	(354.93)	(300.60)	(229.78)	(519.87)
δ	-2.989	-2.637	-2.492	-2.844
	(425.48)	(272.67)	(179.72)	(541.92)
χ^2	58047	25242	10363	89856
Observations	676	676	676	676

This disaggregation reveals significant differences by educational level. The disaggregation improves the explanatory power of the models by five percentage points.

Models III and IV present disaggregations by destination zone and distance, again revealing significant differences by educational level. The models explain roughly the same fraction of the variance in log worktrip behavior — about 83 percent.

Model V presents a disaggregation by the educational level of workers at origins as well as destinations. Model VI presents a complete disaggregation. Again the models reveal a systematic difference in the importance of travel time by educational level. There is a systematic decline in its influence as education level increases.

Table 4 presents a similar disaggregation using the more complex Poisson representation. Again, the significance levels of the parameters are much higher than for the gravity model. The results are much the same: The disaggregation by educational level at residence places and workplaces "matters" in a statistical sense in the prediction of commuting patterns and traffic flows. There is, moreover, a systematic decline in the importance of travel time in affecting behavior as levels of education increase.[3]

Figure 3 summarizes the partial effect of travel time on trip behavior as a function of educational level. The figure graphs the value of the access measure estimated from the Poisson model, $\hat{\lambda}_{ij}$, using model IV of Table 4. That is, holding the spatial distribution of supplies and demands for education constant for the three groups, it illustrates the decay in worktrip with distance.

Figure 4 indicates the cumulative frequency of worktrips by educational level implied by the same model. It indicates that there are substantial differences in

Table 3. Parameters of Gravity Model Estimated using Combined Samples of Commuting by Education Level. 1788 Observations (t-ratios in parentheses).

	I	II	III	IV	V	VI
α	3.802 (14.07)	0.290 (0.70)	0.336 (1.11)	-0.091 (0.30)	0.245 (0.82)	-0.794 (0.25)
β	0.327 (14.34)		0.590 (23.95)	0.615 (25.36)		
β_L		0.503 (22.56)			0.499 (15.42)	0.612 (16.95)
β_M		0.591 (24.41)			0.565 (17.94)	0.572 (16.74)
β_H		0.677 (25.23)			0.697 (22.48)	0.644 (19.89)
γ	0.864 (45.44)	1.034 (54.63)		1.038 (55.56)		
γ_L			0.936 (52.95)		1.034 (35.48)	1.092 (36.38)
γ_M			1.024 (53.14)		1.056 (39.04)	1.059 (37.85)
γ_H			1.109 (51.61)		1.008 (35.70)	0.958 (32.42)
δ	-2.712 (58.10)	-2.599 (61.60)	-2.610 (61.39)		-2.599 (61.58)	
δ_L				-2.780 (66.72)		-2.891 (46.81)
δ_M				-2.575 (60.78)		-2.531 (41.03)
δ_H				-2.404 (54.47)		-2.317 (35.86)
R^2	0.783	0.826	0.823	0.829	0.826	0.831

commuting behavior by education level — more highly educated workers are more likely to commute across community boundaries (as noted by the intercept) and are more likely to commute longer distances. For example, about 77 percent of workers of the lowest educational level are likely to commute twenty minutes or less, while only about 70 percent of workers of the highest educational level commute twenty minutes or less. At a half hour of commutation, the difference is about three percentage points in the cumulative distributions between the highly educated and the least educated population groups.

Table 4. Parameters of Poisson Model Estimated using Combined Samples of Commuting by Education Level. 2028 Observations (t-ratios in parentheses).

	I	II	III	IV	V	VI
α	2.131 (80.70)	1.181 (42.19)	1.178 (42.04)	1.051 (37.09)	1.177 (42.02)	1.002 (35.08)
β	0.478 (182.23)		0.575 (209.25)	0.582 (211.72)		
β_L		0.549 (204.44)			0.559 (184.13)	0.591 (171.43)
β_M		0.582 (207.65)			0.566 (148.56)	0.567 (143.39)
β_H		0.632 (212.91)			0.628 (137.68)	0.600 (129.03)
γ	1.012 (503.89)	1.043 (518.16)		1.043 (519.35)		
γ_L			1.015 (506.44)		1.030 (378.88)	1.070 (361.16)
γ_M			1.049 (511.95)		1.058 (336.26)	1.029 (303.06)
γ_H			1.096 (507.04)		1.047 (263.90)	0.997 (230.00)
δ	-2.798 (551.29)	-2.818 (539.04)	-2.816 (539.29)		-2.817 (538.91)	
δ_L				-2.881 (542.29)		-2.969 (443.43)
δ_M				-2.772 (520.99)		-2.652 (307.76)
δ_H				-2.645 (486.59)		-2.552 (226.51)
χ^2	106257	95345	95506	94039		93480

It should be emphasized that these comparisons assume that the spatial distribution of worksites and residences are the same for the three educational levels. In fact, these distributions are quite different; thus, the figures by themselves underestimate the importance of educational level on commuting.

The effects of variations in educational level upon commuting and trip-making behavior are quite substantial.

Conclusion

Each one of the three commuting models we have estimated shows that the sensitivity to commute time differs significantly between workers with different levels of education: the higher the education, the lower the influence of commute time.

In addition, the spatial distribution of worksites and residences differs among educational groups. In traditional models of access, it is implicitly assumed that all

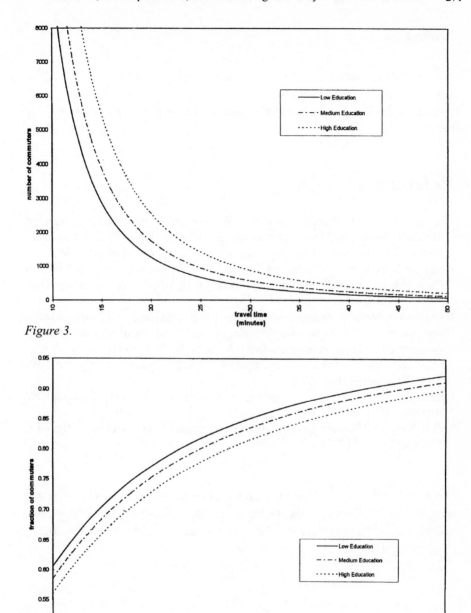

Figure 3.

Figure 4.

workers are equally attracted to all kinds of jobs, and also that all workers have the same chance of getting any job. Our results indicate that it would be more fruitful to differentiate workers according to education in the analysis of traffic flows and employment potential.

Notes

1. It can be shown that the ratio of the variance to the mean is $[1+\eta]/\eta$.
2. These results are available from the authors on request.
3. Again, the results are similar when the parameters are estimated using the negative binomial model.

References

Hårsman, Björn and John M. Quigley, "Worker and Workplace Heterogeneity, Transport Access, and Residential Location," Working Paper 95-233, Center for Real Estate and Urban Economics, University of California, Berkeley, 1993.

Hausman, Jerry, Bronwyn H. Hall, and Zvi Griliches, "Econometric Models for Count Data with Application to the Patents-R&D Relationship," Econometrica, 52(4), 1984: 909-938.

Isard, Walter, Methods of Regional Analysis, Cambridge, MA: MIT Press, 1960.

Pooler, James A., "The Use of Separation in the Measurement of Transportation Accessibility," Transportation Research A, 29(6), 1995: 421-427.

Sen, A. and Tony E. Smith, Gravity Models of Spatial Interaction Behavior, New York: Springer Verlag, 1995.

Smith, Tony E., "Testable Characterizations of Gravity Models," Geographical Analysis, 16, 1984: 74-94.

Smith, Tony E., "Poisson Gravity Models of Spatial Flows," Journal of Regional Science, 27(3): 315-340.

A previous version of this paper was presented at the session in honor of Åke Andersson, World Congress of the Regional Science Association, Tokyo, May 1996. We are grateful for the comments of Lata Chatterjee and for the research assistance of Christian Redfearn.

Informative Contests and the Efficient Selection of Agents

Antoine Billot
Université de Paris 2 & CERAS-ENPC

and
Tony E. Smith
Regional Science Program
University of Pennsylvania

1. Introduction

This paper is based on the tournament model proposed by Araï, Billot & Lanfranchi 1995 [ABL]. Following ABL, a model of candidate selection is constructed in which there exists only limited information about their abilities. In particular, it is assumed that only the reputation earned by candidates in competition with one another is known. Competition is here formalized in terms of *contests* (or tournaments) which may involve individual and/or coalitions of individuals, and in which only *winners* are observed. The specific nature of 'reputation' is left unspecified, and is treated simply as an abstract measure which is increased monotonely for winners and left unaltered for all others. Here, our key assumption is that the winners'gain increases with the reputations of the losers.

In any selection situation (such as the promotion of a candidate) the choice maker, here designated as the *principal*, is assumed to choose that candidate which he/she considers to be the must able candidate, as reflected by current reputation levels. To increase information about relative abilities, the principal is thus motivated to consider contests which are guaranteed to be informative in this sense. More precisely, the principal is assumed to be motivated to construct contests which are *uniformly informative* in the sense that his/her confidence (subjective probability) as to the best candidate is increased regardless of the winner of the contest (where no contest is allowed to end in a draw). In this setting, our main result is to establish the existence of certain 'biased' contests (see, Meyer 1991) which exhibit this property. In each such contest, a candidate with highest current reputation (designated as the *leader*) is allowed to be aided by (i.e. to form a coalition with) one or more agents who are not current candidates, but rather are experienced advisors (with strong reputations) who can enhance the competitive performance for the leader. It is then shown that for any choice of advisors with sufficiently strong reputations, the resulting *leader-biased contest* is always uniformly informative in the above sense. In addition, it is shown that when there is a unique leader with sufficiently strong reputation relative to all other candidates, there exist no biased contest favoring non-leaders which are uniformly informative.

In the next section, we develop the basic notions of contests and reputation profiles. A specific model of *reputation increments* is developed which includes imputations of reputation increments to the individual members of winning coalitions. In section 3, a model of the principal's *perceived ability distribution* is developed in terms of the current reputation levels of candidates. Finally, the notion of uniformly informative contests is developed in section 4, and the leader-biased contests above are shown to exhibit this property.

2. Reputation Profiles and Increments

Given a denumerable set, Ω, of potential *participants* i, let each nonempty finite subset A of Ω be designated as a *coalition* of participants, where each singleton set $\{i\}$ is referred to as an *individual*. The class as all coalitions in Ω is then denoted by $\mathcal{A} = \{A \subset \Omega : 1 \leq \# A < \infty\}$. If $\mathbf{R}_+$ denotes the positive reals then each positive function, $R : \mathcal{A} \to \mathbf{R}_+$, satisfying the monotonicity condition for any $A, B \in \mathcal{A}$,

$$A \subseteq B \Rightarrow R_A \leq R_B , \tag{1}$$

is designated as a possible *reputation profile* on $\mathcal{A}$, and the class of all reputation profiles is denoted by $\mathcal{R}$. Note that $\mathcal{R}$ includes all nonnegative additive set functions, and hence all measures on $\mathcal{R}$. Furthermore, $\mathcal{R}$ also includes all nonnegative super-additive set functions, and hence all game theoretic 'valuation functions' on $\mathcal{A}$. In particular, it is implicitly assumed that adding members to coalitions never reduces the reputation of the group.

Each n-tuple $C = \left(A_1, \ldots, A_n \right)$ of disjoint coalitions is taken to represent a *contest* with *individual participant set*, $A_C = \cup_{i=1}^{n} A_i$. The class of all contests (tournaments) is then denoted by C. If we write $A \in C$ whenever A is a member of C, then we now say that $C \subseteq C'$ iff $A \in C \Rightarrow A \in C'$ for all $A \in \mathcal{A}$. The events of interest are written as $\langle C, W \rangle$ where C is a contest and W is the *winner*.

Coalitions $A_1,\ldots,A_n$ may of course participate in quite different contests, depending on the specification of contest rules, etc... However, since our sole concern is with the possible *outcomes* (winners) of such contests, it suffices to distinguish contests only by their possible outcomes. For simplicity, we assume that any participating coalition, $A \in C$, is a possible winner and in addition, that there is exactly one winner (i.e. that contests are always continued until a unique winner emerges). We shall always use W to denote the winner of a contests C so that the relevant events of interest for our purposes are taken to be pairs $\langle C,W \rangle$ denoting the occurrence of contest $C \in \mathcal{C}$, together with the occurrence of outcome $W \in C$.

In particular we shall be interested in how a given contest-outcome pair $\langle C,W \rangle$ alters the reputations of its individual participants. Since winners are in general coalitions of individuals, it is natural to begin by considering the altered reputations of those coalitions A participating in C. For any given reputation profile $R \in \mathcal{R}$ and event $\langle C,W \rangle$, we denote the change in the reputation of each coalition A resulting from the occurrence of $\langle C,W \rangle$ by $\Delta R_A \langle C,W \rangle$, and designate a family of such changes,

$$\Delta = \left\{ \Delta R_A \langle C,W \rangle : R \in \mathcal{R}, W \in C \in \mathcal{C}, A \in \mathcal{A} \right\},$$

as an admissible *reputation increment* iff Δ satisfies the following five axioms.

Axiom 1 - *For all* $A,B,B' \in \mathcal{A}$ *with* $A \cap (B \cup B') = \emptyset$ *and all* $R \in \mathcal{R}$,

$$R_B \geq R_{B'} \Rightarrow \Delta R_A \langle (A,B) ,A \rangle \geq \Delta R_A \langle (A,B') ,A \rangle. \qquad (2)$$

Axiom 2 - *For all* $C,C' \in \mathcal{C}, A \in C$ *and* $R \in \mathcal{R}$,

$$C \subseteq C' \Rightarrow \Delta R_A \langle C, A \rangle \leq \Delta R_A \langle C', A \rangle . \qquad (3)$$

Axiom 3 - *For each* $\delta > 0$, *there is some* $r_\delta > 0$ *such that for all* $R \in \mathcal{R}$ *and* $A,B \in \mathcal{A}$ *with* $A \cap B = \emptyset$,

$$R_B > r_\delta \Rightarrow \Delta R_A \langle (A,B) ,A \rangle > \delta . \qquad (4)$$

Axiom 4 - *For all* $A \in \mathcal{A}$, $W \in C \in \mathcal{C}$ *and* $R \in \mathcal{R}$, $\Delta R_A \langle C,W \rangle \geq 0$ *and*

$$\Delta R_A \langle C,W \rangle > 0 \Leftrightarrow A \cap W \neq \emptyset. \qquad (5)$$

Axiom 5 - *For all* $A,B \in \mathcal{A}, W \in C \in \mathcal{C}$ *and* $R \in \mathcal{R}$, *and*

$$A \subseteq B \Rightarrow \Delta R_A \langle C,W \rangle \leq \Delta R_B \langle C,W \rangle . \qquad (6)$$

The first two axioms together assert that reputation increments to winners are never decreased by increasing the reputation and/or number of defeated coalitions in the contest. The third axiom is of central importance for our purposes. Taken together with Axiom 2, it asserts that any positive gain in reputation can be achieved by defeating opponents with sufficiently high reputations. These three axioms focus only on reputation increments to winners in contests. The final two axioms cover all other coalitions. In particular, Axiom 4 asserts that the reputation of a coalition is never decreased by the outcome of any contest. In addition, the reputations of all coalitions sharing members with the winning coalitions in a contest always increase. In other words, reputation levels are here regarded as cumulative in nature, and are only increased by winning contests. Axiom 5 is essentially a consistency condition which ensures that the basic monotonicity assumption (eq. 1) is never violated by the outcome of a contest. In particular, if for any prevailing reputation profile, $R \in \mathcal{R}$, we define $R^{C,W} : \mathcal{A} \rightarrow \mathbf{R}_+$ for all $A \in \mathcal{A}$ by:

$$R_A^{C,W} = R_A + \Delta R_A \langle C, W \rangle \tag{7}$$

then, Axiom 5 ensures that this modified profile continues to satisfy the monotonicity condition (eq. 1) and hence always yields a new reputation profile, $R^{C,W} \in \mathcal{R}$.

Finally, it should be noted that from a practical viewpoint, the reputation increments in Axiom 4 are much more subjective in nature than those in Axiom 1 through Axiom 3. They require *imputations* of reputation increments to members of winning coalitions, which involve implicit assumptions about their individual contributions to the winning effort.

3. Perceived Ability Distributions

As mentioned in the introduction, the role of reputation profiles is to serve as information inputs for the principal in some selection situation among a set of possible condidates in Ω. Hence, we now let the class of possible *selection situations, $\mathcal{S}$*, be denoted by $\mathcal{S} = \{S \subseteq \Omega : 2 \leq \#S < \infty\}$ and denote the members of $\mathcal{S}$ as *candidates i* (as distinguished from contestants i belonging to coalitions A participating in contests C). The ability of any candidate i is only partially reflected by his current reputation, R_i. Hence, on the basis of this information, the principal can only form some degree of confidence as to the most able candidate in S. These relative confidence levels can be modeled in terms of *subjective probability* as follows. For any selection situation, S, and current reputation measure, R, let $p_S(i \,|\, R)$ denote the principal's subjective probability (relative confidence) that i is the most able candidate in S. For purposes of analysis, it is convenient to assume that such probabilities are positive (which may equivalently be viewed as restricting selection situations to those candidates who have some chance of being the best). Hence, each family,

$$P = \left\{ p_S \left(i \mid R \right) > 0 \; : R \in \mathcal{R}, \; i \in S \in \mathcal{S} \right\},$$

satisfying the normalization condition that $\sum_i p_S \left(i \mid R \right) = 1$ for all $S \in \mathcal{S}$ and $R \in \mathcal{R}$ is now designated as a possible *perceived ability distribution* for the principal. If for each $\left\{ i, j \right\} \subset \Omega$, we write $p_{ij} \left(\cdot \mid R \right) = p_{\{ij\}} \left(\cdot \mid R \right)$ and let $\left(R_i, R_j \right)$ denote the *restriction* of R to $\left\{ i, j \right\}$, then our behavioral assumptions about such distributions can be stated as follows.

Axiom 6 (Independence) - *For all* $\left\{ i, j \right\} \subseteq S \in \mathcal{S}$ *and* $R \in \mathcal{R}$,

$$\frac{p_S \left(i \mid R \right)}{p_S \left(j \mid R \right)} = \frac{p_{ij} \left(i \mid R_i, R_j \right)}{p_{ij} \left(j \mid R_i, R_j \right)} \tag{8}$$

Axiom 7 (Monotonicity) - *For all* $\left\{ i, j \right\} \in S$ *and* $R \in \mathcal{R}$,

$$p_{ij} \left(i \mid R \right) \geq p_{ij} \left(j \mid R \right) \Leftrightarrow R_i \geq R_j. \tag{9}$$

Axiom 8 (Dominance) - *For all* $\left\{ i, j \right\} \in S$ *and sequences* $\left(R^n \right) \in \mathcal{R}$ *with* $R_k^n = R_k^1$ *for all* n *and* $k \in \Omega - \left\{ i \right\}$,

$$R_i^n \to \infty \Rightarrow p_{ij} \left(i \mid R^n \right) \to 1. \tag{10}$$

Axiom 6 is a version of the 'independence-of-irrelevant-alternatives' axiom in choice probability theory (Luce, 1959) which in the present case asserts that the principal's relative confidence in candidate i versus candidate j depends only on the reputations $\left(R_i, R_j \right)$ of these two candidates (and in particular is independent of the presence or absence of any other candidates). In this context, Axiom 7 asserts simply that the principal always regards the candidate with highest current reputation as the most likely to be best. Similarly, Axiom 8 asserts that as the reputation of any candidate increases relative to all others, the perceived likelihood that he is best approaches complete certainty.

Given these assumptions, we now show that the principal's perceived ability distribution has the following simple representation:

Theorem 1 - *If P satisfies Axioms 6, 7 and 8, then there exists an increasing function,* $v(.) : \mathbf{R}_+ \to \mathbf{R}_+$, *(unique up to a positive scalar multiple) such that:*

(i) $x \to \infty \Rightarrow v(x) \to \infty$, and

(ii) for all $i \in S \in \mathcal{S}$ and $R \in \mathcal{R}$,

$$p_S(i \mid R) = \frac{v(R_i)}{\sum_{j \in S} v(R_j)}. \tag{11}$$

Proof : We begin by establishing the existence of functions $v_i(.) : \mathbf{R}_+ \to \mathbf{R}_+$, $i \in \Omega$, such that for all $i \in S \in \mathcal{S}$ and $R \in \mathcal{R}$,

$$p_S(i \mid R) = \frac{v_i(R_i)}{\sum_{j \in S} v_j(R_j)}. \tag{12}$$

To do so, observe first that if for each $\{i, j\} \in S$ and $x, y \in \mathbf{R}_+$, we let

$$\xi_{ij}(x, y) = \left\{ \frac{p_S(i \mid R)}{p_S(j \mid R)} \; : \; R_i = x, R_j = y, R \in \mathcal{R} \right\}, \tag{13}$$

then it follows at once from Axiom 6 that for all $R, R' \in \mathcal{R}$,

$$(R_i, R_j) = (x, y) = (R'_i, R'_j) \Rightarrow \frac{p_{ij}(i \mid R)}{p_{ij}(j \mid R)} = \frac{p_{ij}(i \mid x, y)}{p_{ij}(j \mid x, y)} = \frac{p_{ij}(i \mid R')}{p_{ij}(j \mid R')} \tag{14}$$

so that (eq. 13) yields a well-defined function $\xi_{ij} : \mathbf{R}_+^2 \to \mathbf{R}_+$, satisfying the following identity for all $\{i, j\} \in S$ and $R \in \mathcal{R}$,

$$\xi_{ij}(R_i, R_j) = \frac{p_{ij}(i \mid R)}{p_{ij}(j \mid R)}. \tag{15}$$

Morover, it also follows from Axiom 6 that for any $\{i, j, k\} \in S$ and $x, y, z \in \mathbf{R}_+$, if we choose $\overline{R} \in \mathcal{R}$ with $\overline{R}_i = x$, $\overline{R}_j = y$ and $\overline{R}_k = z$, then:

$$\xi_{ij}(x, y) = \frac{p_{ij}(i \mid \overline{R})}{p_{ij}(j \mid \overline{R})} = \frac{p_{ijk}(i \mid \overline{R})}{p_{ijk}(j \mid \overline{R})}$$

$$= \frac{p_{ijk}(i \mid \overline{R}) / p_{ijk}(k \mid \overline{R})}{p_{ijk}(j \mid \overline{R}) / p_{ijk}(k \mid \overline{R})}$$

$$= \frac{p_{ik}(i \mid \overline{R}) / p_{ik}(k \mid \overline{R})}{p_{jk}(j \mid \overline{R}) / p_{jk}(k \mid \overline{R})} = \frac{\xi_{ik}(x,z)}{\xi_{jk}(y,z)} \tag{16}$$

which in turn implies that for any $\{i,j,k,g\} \in S$ and $x,y,z,w \in \mathbf{R}_+$,

$$\frac{\xi_{ik}(x,z)}{\xi_{jk}(y,z)} = \frac{\xi_{ig}(x,w) / \xi_{kg}(z,w)}{\xi_{jg}(y,w) / \xi_{kg}(z,w)} = \frac{\xi_{ig}(x,w)}{\xi_{jg}(y,w)}. \tag{17}$$

Note also that for any $\{i,j\} \in S$ and $x,y \in \mathbf{R}_+$:

$$\xi_{ij}(x,y) \times \xi_{ji}(y,x) = \frac{p_{ij}(i \mid x,y)}{p_{ij}(j \mid x,y)} \times \frac{p_{ji}(j \mid y,x)}{p_{ji}(i \mid y,x)} = 1. \tag{18}$$

With these observations, we now show that if for any *fixed* $\{a,b\} \in S$, we define $v_i(.) : \mathbf{R}_+ \to \mathbf{R}_+$, for all $x \in \mathbf{R}_+$ and $i \in \Omega$ by:

$$v_i(x) = \begin{cases} \xi_{ia}(x,1) & \text{if } i \neq a, \\ \xi_{ab}(x,1) \times \xi_{ba}(1,\ 1) & \text{if } i = a, \end{cases} \tag{19}$$

then (eq. 12) holds for these functions. To do so, it suffices to show that:

$$\xi_{ij}(x,y) = \frac{v_i(x)}{v_j(y)} \tag{20}$$

for all $\{i,j\} \in S$ and $x,y \in \mathbf{R}_+$. For then (eq. 12) will follow easily from (eq. 15) and Axiom 6 (together with the normalization condition for probabilities). Hence, suppose first that $\{i,j\} \in \Omega - \{a\}$, and observe that in this case (eq. 20) follows at once by setting $k = a$ and $z = 1$ in (eq. 16) and using (eq. 19). Next, suppose that $j = a$ and $i \notin \{a,b\}$. Then by two applications of (eq. 16) we obtain:

$$\xi_{ia}(x,y) = \frac{\xi_{ib}(x,1)}{\xi_{ab}(y,1)} = \frac{\xi_{ia}(x,1) / \xi_{ba}(1,1)}{\xi_{ab}(y,1)} = \frac{\xi_{ia}(x,1)}{\xi_{ab}(y,1) \times \xi_{ba}(1,1)} = \frac{v_i(x)}{v_a(y)}. \tag{21}$$

The argument for $i = a$ and $j \notin \{a,b\}$ is identical. Finally, suppose that $i = a$ and $j = b$. Then by (eq. 17) it follows that for any $g \in \Omega - \{a,b\}$,

$$\frac{\xi_{ab}(x,y)}{\xi_{gb}(1,y)} = \frac{\xi_{ab}(x,1)}{\xi_{gb}(1,1)} \Rightarrow \xi_{ab}(x,y) = \xi_{ab}(x,1) \times \frac{\xi_{gb}(1,y)}{\xi_{gb}(1,1)}. \tag{22}$$

But by (eq. 17) we also see that:

$$\frac{\xi_{gb}(1,y)}{\xi_{ab}(1,y)} = \frac{\xi_{gb}(1,1)}{\xi_{ab}(1,1)} \Rightarrow \frac{\xi_{gb}(1,y)}{\xi_{gb}(1,1)} = \frac{\xi_{ab}(1,y)}{\xi_{ab}(1,1)}. \tag{23}$$

So that by substituting (eq. 23) into (eq. 22) and using (eq. 18), we obtain:

$$\xi_{ab}(x,y) = \xi_{ab}(x,1) \times \frac{\xi_{ab}(1,y)}{\xi_{ab}(1,1)} = \frac{\xi_{ab}(x,1) \times \xi_{ba}(1,1)}{\xi_{ba}(y,1)} = \frac{v_a(x)}{v_b(y)}.$$

(24)

Hence (eq. 20) holds in all cases, and it follows that (eq. 12) holds for this choice of functions. To complete the proof, observe from Axiom 7 and (eq. 12) that for all $\{i,j\} \in S$ and

$$x \begin{pmatrix} \geq \\ \leq \end{pmatrix} x \Rightarrow p_{ij}(i \mid x,x) \begin{pmatrix} \geq \\ \leq \end{pmatrix} p_{ij}(j \mid x,x)$$

$$\Rightarrow p_{ij}(i \mid x,x) = p_{ij}(j \mid x,x)$$

$$\Rightarrow \frac{v_i(x)}{v_i(x)+v_j(x)} = \frac{v_j(x)}{v_i(x)+v_j(x)} \Rightarrow v_i(x) = v_j(x),$$

(25)

so that $v_i \equiv v_j$ for all i, j. Thus, (eq. 11) follows from (eq. 12) with $v = v_a$ for any fixed $a \in \Omega$. To see that $v(.)$ is increasing, observe again from Axiom 7 that for any $x, y \in \mathbf{R}_+$ and $\{i,j\} \in S$:

$$x > y \Rightarrow p_{ij}(i \mid x,y) > p_{ij}(j \mid x,y)$$

$$\Rightarrow \frac{v(x)}{v(x)+v(y)} > \frac{v(y)}{v(x)+v(y)} \Rightarrow v(x) > v(y).$$ (26)

Similarly, to see that $x \to \infty \Rightarrow v(x) \to \infty$, observe from Axiom 8 that for any sequence (x_n) in $\mathbf{R}_+$ with $x_n \to \infty$, we choose (R^n) in $\mathcal{R}$ with $R_i^n = x_n$ and $R_k^n = 1$ for all n and $k \in \Omega - \{i\}$ then for any $j \neq i$,

$$x_n \to \infty \Rightarrow p_{ij}(i \mid x_n, 1) \to 1$$

$$\Rightarrow \frac{v(x_n)}{v(x_n)+v(1)} \to 1 \Rightarrow v(x_n) \to \infty.$$ (27)

Finally, since (eq. 11) shows that the increasing functions, $v(.)$, is only unique up to a positive scalar multiple, the result is established. □

In view of this representation, it is natural to interpet $v(.)$ as a *value function* for the principal, where $v(R_i)/v(R_j)$ denotes the relative value of reputation levels R_i and R_j in determining his relative confidence as to the best candidate.

4. Uniformly Informative Contests

As mentioned in the introduction, the central purpose of contests is to better inform the principal as to the most able candidate in any selection situation. For example, suppose that in a given selection situation $\{i,j\} \in S$, the current reputations of the candidates i and j are equal $\left(R_i = R_j\right)$, so that by Axiom 7, $p_{ij}(i \mid R) = p_{ij}(j \mid R) = 1/2$. Then, any contest $C = \left(\{i\},\{j\}\right)$ must necessarily improve the principal's confidence as to the best candidate. For, if $W = \{i\}$ then

$$R_i^{C,W} > R_j^{C,W} \Rightarrow p_{ij}(i \mid R^{C,W}) > p_{ij}(j \mid R^{C,W}) \Rightarrow p_{ij}(i \mid R^{C,W}) > 1/2,$$

and similarly if $W = \{j\}$, then $p_{ij}(j \mid R^{C,W}) > 1/2$. More generally, whenever the outcome of a given contest increases the value of $\max_{i \in S} p_S(i \mid R)$, it may be inferred that this outcome provides useful information to the principal. These observations may be formalized as follows. For any selection situation, S, and current reputation profile, R, a contest-outcome pair $\langle C, W \rangle$ is said to be *informative* if:

$$\max_{i \in S} p_S \ (i \mid R^{C,W}) > \max_{i \in S} p_S \ (i \mid R). \tag{28}$$

However, the outcome of a given contest *may not* be informative for the principal. In particular, observe that for any given selection situation, $S \in S$, the most natural contest to consider is a contest among the individuals of S, which we now designate as the *individual contest*, $C_S = \left(\{i\} : i \in S\right)$, for S. But depending on the current reputations $\left(R_i : i \in S\right)$, certain outcomes of C_S may fail to be informative. As a simple illustration, suppose that $S = \{a, b\}$ and $R_a > R_b$ with reputation values, $v\left(R_a\right) = 3$ and $v\left(R_b\right) = 1$. In addition, suppose that $b = W$ for contest $C = C_S = \left(\{a\}, \{b\}\right)$, with $v(R_b^{C,W}) = 4$ (which is consistent with $R_b^{C,W} > R_b$). In this case, $R_a^{C,W} = R_a$ implies $v(R_a^{C,W}) = v(R_a) = 3$, so $p_S(a \mid R^{C,W}) = 3/7$ and $p_S(b \mid R^{C,W}) = 4/7$. But, since $\max_{i \in S} p_S(i \mid R) = p_S(a \mid R) = 3/4 > 4/7 > \ = \max_{i \in S} p_S(i \mid R^{C,W})$ we see that $\langle C, W \rangle$ not only fails to be informative, but actually decreases the principal's confidence as to the best candidate.

In view of this example, it is natural to ask whether there exist contests which are *guaranteed* to be informative for the principal. In particular, if for each $S \in S$ and $R \in R$, we now designate a contest, $C \in C$, *as uniformly informative* for (S, R) iff (eq. 28) holds for all outcomes, $W \in C$, then we seek to construct uniformly informative contests for the principal. In particular, we show that certain contests which favor the candidate with highest reputation exhibit this property. To develop these contests, we first construct for each $S \in \mathcal{S}$ and candidate, $i \in S$, a modified version of the individual contest, C_S, which favors i in the following way. Choose any coalition, $B \in \mathcal{A}$ (possibly a singleton), with $B \cap S = \emptyset$, and allow i to form a coalition with B. If we write $A \cup \{i\} = A \cup i$ and $A - \{i\} = A - i$, then the modification

$$C_S^{B \cup i} = (B \cup i, C_{S-i}) \tag{29}$$

of C_S in which $\{i\}$ is replaced by $B \cup i$ designated an *i-biased contest* for S with *biasing agent* B. In particular, if for any reputation measure, R, we denote the set of *R-leaders* in S by

$$S_R = \left\{ i \in S : R_i = \max_{j \in S} R_j \right\} \tag{30}$$

then, each *i*-biased contest, $C_S^{B \cup i}$, with $i \in S_R$ is designated as a *leader-biased contest* for (S, R). With these concepts we now show that if the reputation, R_B, of the biasing agent B is sufficiently high, then such leader-biased contests are always uniformly informative for (S, R).

Theorem 2 - *For each selection situation, $S \in \mathcal{S}$, and reputation profile, $R \in \mathcal{R}$, there exists a reputation level, $r(S, R)$, sufficiently large to ensure that for any choice of biasing agent B with $R_B > r(S, R)$, each leader-biased contest, $C_S^{B \cup i}$, $i \in S_R$, is uniformly informative for (S, R).*

Proof: Given any leader-biased contest, $C = C_S^{B \cup i}$, suppose first that $W = B \cup i$. Then by Axiom 4, it follows that $\Delta R_i \langle C, W \rangle > 0$ and $\Delta R_j \langle C, W \rangle = 0$, for $j \in S - i$, so that by (eq. 7), $R_i^{C,W} > R_i$ and $R_j^{C,W} = R_j$ for $j \in S - i$. Hence, writing $(R : A) = (R_j : j \in A)$ and:

$$v(R : A) = \begin{cases} \sum_{j \in A} v(R_j) & \text{if } A \in \mathcal{A} \\ 0 & \text{if } A = \emptyset \end{cases} \tag{31}$$

it follows from the increasing monotonicity of $v(.)$ that:

$$\max_{j \in S} p_S(j \mid R) = p_S(i \mid R) = \frac{v(R_i)}{v(R_i) + v(R : S - i)}$$

$$< \frac{v(R_i^{C,W})}{v(R_i^{C,W}) + v(R:S-i)} = p_S(i \mid R^{C,W})$$

$$= \max_{j \in S} p_S(j \mid R^{C,W}). \tag{32}$$

Thus (eq. 28) always holds for $W = B \cup i$, and we may henceforth assume that $W = a \in S - i$. In this case, it again follows from Axiom 4 and (eq. 7) that $R_j^{C,W} = R_j$ for all $j \in S - a$, so that we must have:

$$p_S\big(a \mid R^{C,W}\big) > \max_j \ p_S(j \mid R) \Leftrightarrow \frac{v(R_a^{C,W})}{v(R_a^{C,W}) + v(R:S-a)} > p_S(i \mid R)$$

$$\Leftrightarrow v(R_a^{C,W}) \times \big[1 - p_S(i \mid R)\big] > v(R:S-a) \times p_S(i \mid R)$$

$$\Leftrightarrow v(R_a^{C,W}) > v(R:S-a) \times \frac{p_S(i \mid R)}{1 - p_S(i \mid R)}. \tag{33}$$

It is convenient to express the last inequality entirely in terms of $v(.)$ as follows. If we let $T = S - \big\{a,i\big\}$, then by definition,

$$v(R:S-a) \times \frac{p_S(i \mid R)}{1 - p_S(i \mid R)} = v(R:S-a) \times \frac{v(R_i)/v(R:S)}{1 - \big[v(R_i)/v(R:S)\big]}$$

$$= v(R:S-a) \times \frac{v(R_i)}{v(R:S) - v(R_i)}$$

$$= v(R_i) \times \frac{v(R_i) + v(R:T)}{v(R_a) + v(R:T)} \tag{34}$$

Hence, $p_S\big(a \mid R^{C,W}\big) > \max_j p_S\big(j \mid R\big)$ iff:

$$v(R_a^{C,W}) > v(R_i) \times \frac{v(R_i) + v(R:T)}{v(R_a) + v(R:T)}. \tag{35}$$

To obtain conditions under which (eq. 35) holds, we observe first from condition (i) in Theorem 1 that there is some $\bar{r}_a$ sufficiently large to ensure that:

$$x \ge \bar{r}_a \Rightarrow v(x) \ge v(R_i) \times \frac{v(R_i) + v(R:T)}{v(R_a) + v(R:T)} \equiv \bar{v}_a. \tag{36}$$

Hence, observing that,

$$R_i > R_a \Rightarrow v(R_i) > v(R_a) \Rightarrow v(R_i) + v(R:T) > v(R_a) + v(R:T)$$
$$\Rightarrow \bar{v}_a > v(R_i) > v(R_a) \Rightarrow v(\bar{r}_a) > v(R_a) \Rightarrow \bar{r}_a > R_a, \qquad (37)$$

it follows by setting $\delta = \bar{r}_a - R_a > 0$ in Axiom 3 and setting $r_a(S,R) = r_\delta$, that for any biasing agent B with $R_B > r_a(S,R)$, we must have:

$$R_B > r_a(S,R) \Rightarrow R_{B \cup i} > r_a(S,R) = r_\delta \qquad \text{[by (eq. 1)]}$$

$$\Rightarrow \Delta R_a \langle (B \cup i, a), a \rangle > \delta \qquad \text{[by Axiom 3]}$$

$$\Rightarrow \Delta R_a \langle C, a \rangle > \delta \qquad \text{[by Axiom 2]}$$

$$\Rightarrow \Delta R_a \langle C, W \rangle + R_a > \bar{r}_a \qquad (38)$$

$$\Rightarrow R_a^{C,W} > \bar{r}_a \qquad \text{[by (eq. 7)]}$$

$$\Rightarrow v(R_a^{C,W}) > \bar{v}_a \qquad \text{[by (eq. 36)]}.$$

Thus, (eq. 35) holds and we see that for this choice of $r_a(S,R)$,

$$R_B > r_a(S,R) \Rightarrow \max_{j \in S} p_S(j \mid R^{C,W}) \geq p_S(a \mid R^{C,W}) > \max_{j \in S} p_S(j \mid R).$$
$$(39)$$

Finally, by setting $r(S,R) = \max_{a \in S-i} r_a(S,R)$, we may conclude from (eq. 32) and (eq. 39) that $C_S^{B \cup i}$ will be uniformly informative for (S,R) whenever $R_B > r(S,R)$. □

Remark 1 - *The above definition of $r(S,R)$ can be sharpened by observing that for any $a, b \in S - i$, if $R_a < R_b$ then, $v(R_a) < v(R_b)$ is easily seen from (eq. 36) to imply that $\bar{v}_a > \bar{v}_b$, and hence that it suffices to consider only participants, $a \in S - i$, with minimal reputation R_a in defining $r(S,R)$.*

Remark 2 - *It is also important to emphasize that Theorem 2 makes no assertion about the existence of uniformly informative leader-biased coalitions. In particular if $R_{\Omega-S} < r(S,R)$ then no biasing agents B can exist with R_B sufficiently large. Of course, if R is uniformly bounded away from zero, then the infinite cardinality of Ω guarantees existence. But from a practical viewpoint, biasing agents will generally be restricted to small sets of 'experts'.*

To establish conditions under which only *leader*-biased contests are uniformly informative, we now impose certain regularity conditions on reputation increments with respect to such contests. First, we assume that biasing in favor of a participant (i.e. making it easier to win) never increases his reputation from winning.

Axiom 9 - *For any* $i \in S \in S$ *and biasing agent* $B \in \Omega - S$,

$$\Delta R_i \left\langle C_S^{B \cup i}, \{i\} \right\rangle \leq \Delta R_i \left\langle C_S, \{i\} \right\rangle. \tag{40}$$

In addition, we assume that if the prior reputation of the leader in a given individual contest C_S is sufficiently greater than all other participants, then no other participant can 'overtake' the leader's reputation by winning this single contest.

Axiom 10 - *For each* $S \in S$, *there is some* $\delta_S > 0$ *such that for all* $R \in \mathcal{R}$ *and* $l, i \in S$ *with* $l \neq i$,

$$R_l - \max_{j \in S - l} R_j > \delta_S \Rightarrow \Delta R_i \left\langle C_S, \{i\} \right\rangle < R_l - R_i. \tag{41}$$

A reputation increment Δ is said to be *regular with respect to individual contests* iff Δ satisfies Axioms 9 and 10. In this context, we now show that:

Theorem 3 - *If* Δ *is regular with respect to individual contests, then for any* $S \in S, l \in S$ *and* $R \in \mathcal{R}$ *with* $R_l - R_j > \delta_S$ *for all* $j \in S - l$, *there exists no uniformly informative i-biased contest* $C_S^{B \cup i}$ *with* $i \in S - l$.

Proof: Choose any $i \in S - l$ and $B \in \mathcal{A}$ with $B \cap S = \emptyset$ and suppose that $W = B \cup i$ in contest $C = C_S^{B \cup i}$. Then, by Axioms 9 and 10:

$$R_l - \max_{j \in S - l} R_j > \delta_S \Rightarrow \Delta R_i \left\langle C_S, \{i\} \right\rangle \quad < R_l - R_i$$

$$\Rightarrow \Delta R_i \left\langle C_S^{B \cup i}, \{i\} \right\rangle \quad < R_l - R_i \tag{42}$$

$$\Rightarrow R_i^{C,W} \quad < R_l.$$

This together with $R_i^{C,W} > R_i$ implies that $v(R_l) > v\left(R_i^{C,W}\right) > v(R_i)$. Moreover, since $j \in S - i$ implies that $j \notin W$, and hence that $R_j^{C,W} = R_j$, we must also have $v\left(R_j^{C,W}\right) = v(R_j)$, for each $j \in S - i$. Thus, letting $T = S - \{i, l\}$, we see that:

$$p_S(i \mid R^{C,W}) = \frac{v(R_i^{C,W})}{v(R_i^{C,W}) + v(R_l) + v(R:T)}$$

$$< \frac{v(R_l)}{v(R_l) + v(R_i) + v(R:T)} \qquad (43)$$

$$= p_S(l \mid R).$$

But since $v\left(R_i^{C,W}\right) > v\left(R_i\right)$ also implies that for each $j \in S - i$,

$$p_S(j \mid R^{C,W}) = \frac{v(R_j)}{v(R_i^{C,W}) + v(R:S-i)}$$

$$< \frac{v(R_j)}{v(R_i) + v(R:S-i)} \qquad (44)$$

$$= p_S(j \mid R),$$

we may than conclude from (eq. 43) and (eq. 44) that $\max_{j \in S} p_S\left(j \mid R^{C,W}\right) < \max_{j \in S} p_S\left(j \mid R\right)$. Thus, $C_S^{B \cup i}$ fails to be uniformly informative for $\left(S, R\right)$ under any choice of biasing agent B and $i \in S - l$.□

Remark 3 - *We note finally that Axiom 1 has not be used in either Theorems 2 or 3. However, this axiom is nonetheless central to our behavioral theory of how contest outcomes change the reputations of contestants.*

References

M. Arai, A. Billot & J. Lanfranchi, (1995): "Efficient Selection of Agents under Limited Ability to Rank: Biased Contests and Favoritism", Working-Paper, Ermes, Université Panthéon-Assas, Paris.

R.D. Luce, (1959): *Individual Choice Behavior: A Theoretical Analysis.* Wiley: New-York.

M. Meyer, (1991): "Learning from Coarse Information: Biased Contests and Carrier Profiles", *Review of Economic Studies,* 58, 15-41.

Scientific Collaboration as Spatial Interaction

Martin J. Beckmann
Institute für Angewandte Mathematik und Statistik
Technische Universität, München
and Economics Department
Brown University

and
Olle Persson
Department of Sociology
University of Umeå

Scientific collaboration resulting in joint publication, at one time limited to colleagues in the same institution, has spread to persons at different locations and has in fact become quite common in recent times.

As Andersson and Persson (1993) have shown, "the number of internationally co-authored scientific articles has grown at an average of 14% per year ... in recent decades"

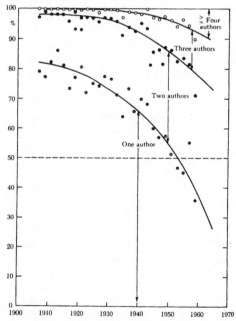

Figure 1. Incidence of multiple authorship as a function of date
(de Solla Price, D. Little Science, Big Science. Colombia UP NY 1963)

From the temporal we turn to the spatial side. In these days of the Internet and of (almost) instant telecommunication, should distance matter to scientists?

A theoretical analysis weighing the advantages of collaboration against its cost (mainly in terms of time) suggests that one should expect an (exponential) distance effect (Beckmann 1994, p.241).

Measurements have been made of collaboration in science among the Nordic countries (Persson 1991) and between the UK, Canada and Australia (Katz 1993).

In this paper we test whether distance matters in scientific collaboration even among the small homogeneous population of Swedish universities: Uppsala, Lund, Göteborg, Umeå, Linköping, all within easy reach of each other. Table 1 shows the data where

i   institution
S_i size = total publications
D_{ij} distance in kilometers
t_{ij} distance in transportation time
x_{ij} number of joint publications between: $i \neq j$ and within: $i = j$ institutions.

Table 1. Total publications and joint publications for six Swedish universities

i	ij	S_i	x_{ij}	D_{ij}	t_{ij}
STO 1		557			
UP 2		1648			
GÖ 3		1379			
LU 4		1990			
UM 5		633			
LI 6		576			
	17		127	0	0
	12		30	200	30
	13		14	3000	60
	14		19	3000	60
	15		18	3000	60
	16		13	1500	60
	22		671	0	0
	23		47	3000	60
	24		76	3000	60
	25		46	3000	60
	26		47	1500	60
	33		804	0	0
	34		74	3000	60
	35		46	5000	120
	36		31	3000	80
	44		794	0	0
	45		53	5000	120
	46		63	3000	60
	55		308	0	0
	56		20	5000	120
	66		243	0	0

We estimated a gravity model with exponential distance effect

$$\ln x_{ij} = \alpha + \beta_1 \ln(S_i S_j) + \beta_2 d_{ij} \tag{1}$$

What should be the appropriate measure of distance? Straight line or air line distance in kilometers, travel cost by the cheapest mode, or minimum travel time?

Experimentation showed travel time to give the best fit, even while using rather rough estimates. This places e.g. Linköping farther away than geographical distance would suggest.

The results were as follows

$$\log x_{ij} = -2.36 + .56 \log S_i S_j - .022 \, d_{ij}$$
$$(-.74)(2.45) \qquad (-4.73) \tag{1a}$$

(t values in parentheses) $R^2 = .628$

It is interesting that β_1 is approximately .5 which means that joint publications are roughly proportional to the geometric average of the publications in the two universities.

Scientific collaboration within the Nordic countries has been studied by O Persson, as noted above, (1991). We now fit a gravity model to his data, introducing a dummy variable F for collaboration across national boundaries, with the following results (In order to include observations of zero collaboration in this log linear equation to avoid bias, we have replaced 0 by 0.4 in equation 2a and 0.1 in equation 2b. t values are shown in parenthesis)

$$\log x_{ij} = -4.16 + .91 \log S_i S_j - 0.0002 \, d_{ij} - 1.08 F$$
$$(-13) \quad (20) \qquad\qquad (-2.7) \qquad (-16) \tag{2a}$$

$$\log x_{ij} = -4.97 + 1.03 \log S_i S_j - 0.00017 \, d_{ij} - 1.16 F$$
$$(-11) \quad (17) \qquad\qquad (-1.68) \qquad (13) \tag{2b}$$

We have also examined 3 786 publications with joint authors in ten cities (each representing one institution) in Taiwan.

Table 2. Matrix of papers produced by authors from pairs of cities. Note that intra city co-authorships are institutional authorships.

City	Chiayi	Chilung	Chungli	Hsinchu	Kaohsiung	Lungtan	Taichung	Tainan	Taipei	Tayuan	Total
Chiayi	19	1	0	2	1	0	4	1	8	0	36
Chilung	1	38	0	1	0	0	0	1	10	0	51
Chungli	0	0	104	10	1	5	2	2	8	1	133
Hsinchu	2	1	12	569	6	11	10	9	49	7	676
Kaohsiung	1	0	2	6	236	2	1	18	42	0	308
Lungtan	0	0	4	9	2	39	1	4	7	1	67
Taichung	4	0	2	10	2	1	185	11	36	0	251
Tainan	1	1	1	8	17	6	12	290	20	1	357
Taipei	10	12	9	44	37	6	39	23	1590	30	1800
Tayuan	0	0	1	7	0	1	0	1	25	77	112

A regression

$$\ln x_{ij} = \alpha + \beta, \ln S_i S_j + \beta_2\, r_{ij} \tag{1}$$

gave the following results

Table 3. Estimates of (1).

	$\hat{\alpha}$	$\hat{\beta}$	$\hat{\beta}_2$	R^2
estimate		0.621	-0.00549	
$\hat{\sigma}$		0.061	0.00229	
t		-7.185	-2.39	
				.842

which reveal once more a significant effect of distance on scientific collaboration even in so small a country as Taiwan.

In "Understanding Patterns of International Scientific Collaboration" Luukkonen, Persson and Sivertsen have studied international collaboration with regard to "scientific size" of countries and to field, but without reference to distance.

In an interesting paper Katz (1993) has estimated the effect of distance on collaboration in the UK, Canada and Australia, where he has scaled the distance variable as a fraction of the largest distance between university pairs in each country.

He concludes that "The frequency of research collaboration between domestic universities in the UK, Canada and Australia, decreases exponentially with the distance separating the research partners". He considers this to be "strong evidence to support the notion that informal "face to face" communication may be an essential ingredient in research collaborations and that factors such as greater geographical distance with the additional travel cost and time involved are impediments to collaborations".

Our findings from Sweden, Taiwan and the Nordic countries are consistent with this.

References

Andersson, Å. and O. Persson, Networking Scientists. *Annals of Regional Science* 27: 11-21 (1993)

Beckmann, M., On knowledge networks in science: collaboration among equals. *Annals of Regional Science* 28: 233-242 (1994).

Katz, J.S. "Geographical Proximity and Scientific Collaboration". *Scientometrics* Vol. 31, No 1 (1994) 31-43.

Luukkonen, T., and O. Persson & G. Sivertsen, Understanding Patterns of International Scientific Collaboration. *Science, Technology & Human Values*, vol. 17 No 1. Sage Publications Inc. (1992)

Persson, O. "Regional Collaboration in Science . Data on Nordic Coauthorships 1988-1990". *CERUM Working Paper* CWP-1991:16.

On Confucius – A Way of Chinese Thinking on Values and Institutions

Wei-Bin Zhang
Institute for Futures Studies, Stockholm

1. Confucianism in China

Confucianism discourages individualism. It is intellectual and rational in character, rejecting the mysticism and incantation common to other religions. The ability of the Japanese to assimilate Western technology and science with astonishing rapidity after the Meiji Restoration was due, at least in part, to their education under Confucianism; Western rationalist thinking was not entirely foreign.

Morishima (1978)

Institutions, customs and other cultural factors play an essential role in affecting economic evolution. On the other hand, these non-economic variables may be strongly affected by economic conditions. There is an intimate interdependence between economic development, institutions and cultural values. In order to examine dynamic processes of the interaction, it is necessary to consider a society as a whole rather than as a system consisting of unconnected parts. This is particularly significant when we are dealing with the evolution of a culture in the long term. For nonlinear science (e.g., Nicolis and Prigogine, 1977, Haken, 1977, Andersson, 1986, Zhang, 1991, 1996), we know that the behavior of nonlinear dynamics may be quite sensitive to small changes in some parameters. It may be quite misleading to fix some aspects of social life, such as morals and institutions, in long-run social and economic analysis. Dynamic interactions between morals, class structures, institutions and economic development have been well observed in human history. The social and economic behavior of people are shaped by their ideas. Their modes of thought affect history.

The institutions and customs of human societies are much related to religions and philosophies. For instance, the establishment of democratic institutions is influenced by thought about freedom and law. It may be argued that ancient religions and philosophies of life still have significant effects on the social and economic evolution of both western and eastern societies. It is thus necessary to examine a society's dominant religions or philosophies in order to understand it.

This study is concerned with Confucianism and its possible implications for the economic and political development of China. Chinese philosophy has emphasized the complementary nature of the intuitive and the rational and has presented them by the archetypal pair, 'yin' and 'yang', which forms a basic characteristic of Chinese thought. Two complementary philosophical traditions - Taoism and Confucianism - were developed in China during the sixth century B.C. These two trends of thought represent the opposite poles of Chinese philosophy. They were seen as poles of one and the same human nature. Confucianism was generally emphasized in the education of children who had to learn the rules and conventions necessary for life in society, whereas Taoism was pursued by older people in order to regain and develop the original spontaneity which had been destroyed by social conventions. Confucianism was mainly adopted by the government; Taoism was mainly developed among the masses.

Confucianism was founded by Confucius. Confucius created a pragmatical rational philosophy which has profoundly framed Chinese civilization. The Confucian tradition consists of a number of texts varying in nature and content which include the Classics, the most ancient commentaries on them, the Analects, and works of the third century B.C. Confucianism goes far beyond the actual personality or teachings of the sage. What is called Confucianism in history contains the thought of other thinkers such as Mencius and Hsün Tzu. The high reputation that Confucius obtained under the Han rulers, and to an even greater extent from the Sung period onwards, was largely due to the theoretical and doctrinal additions made to his thinking by his followers. In this sense, Confucianism may differ from what Confucius taught. For instance, Mencius not only developed some of the ideas of Confucius; he also examined problems not touched on by Confucius.

Confucianism is a philosophy of social organization, of common sense and practical knowledge. Confucianism is not a religion. What most distinguishes Confucius from other founders of religions, such as Buddha, Christ, or Mohammed, is that he inculcated a strict code of ethics. According to Weber, Confucianism, like Puritanism, is rational. But in reality there exists a fundamental difference between the two in that whereas Puritan rationalism has sought to exercise rational control over the world, Confucian rationalism is an attempt to accommodate oneself to the world in a rational manner. Although Confucianism is a rational doctrine that was opposed to any superstitious or even supernatural forms of religion, Confucius' rationality is concerned with the sense of justice - a social attribute. Confucianism had provided

Chinese society with a system of education, and with strict conventions of social etiquette. One of its main purposes was to form an ethical basis for the traditional Chinese family system with its complex structure and its rituals of ancestor worship. It stresses the need for proper behavior of and harmonious relationships among various social actors. It values order, hierarchy, and tradition and assigns the central role of maintaining social control and regulation to the extended family. It gives the pre-eminent status to the family and promotes filial piety, ancestral worship, and collective responsibility.

The Chinese state and the dominant Confucianism managed to preserve social order and institutional continuity for more than two thousand years. During Han times, Confucianism became the official doctrine of the bureaucratic society. Han Kao Tsu - the first Han emperor - offered important sacrifices at the Khung family temple in honor of Confucius in 175 B.C. Later, in A.D. 59, the Emperor Han Ming Ti ordered sacrifices to him in every school in the country, and such rituals transformed Confucius from a model for scholars into the patron saint of the scholar-officials. Confucianism retained the orthodox philosophical position up until the present century. As one of the dominant philosophies in Chinese civilization, Confucianism has complicated implications for almost all other aspects of Chinese life. It has greatly affected the customs, institutions, moral codes, laws, the arts, literature and mentality of the Chinese people in general. Confucianism has had a strong impact on Chinese history. The system has shaped public opinion and directed political movements. Emperors and leaders of thought have been uplifted by or cast down by the system.

This study will not be concerned with all of the important aspects of Confucianism. Such a task would take many years of labor. At this initial stage, we will examine the thought of Confucius, the founder of Confucianism. We will study some aspects of Confucius' thought and provide some insight into the social and economic behavior of Chinese people. Our interpretation of Confucius' thought is entirely based on the Confucian Analects. It should be remarked that the Analects was developed in the form of remarks upon various subjects. The method of argument used in the Analects is quite different from what the modern scientist is used to. It consists of the use of analogy. The use of one thing may throw light on another and the use of one proposition may throw light on another of similar form. In this sense, it is not easy to read Confucius' system as a whole. It is only through careful analysis that one may understand what Confucius meant by saying: "I seek a unity all-pervading." It should be remarked that all the quotations in the remainder of this study are from the Confucian Analects, unless otherwise noted.

2. Confucius

I admire Confucius. He was the first man who did not receive a divine inspiration.

Voltaire (Seldes, 1985)

As with most ancient Chinese thinkers, little is known of the life of Confucius other than what we can glean from the Analects. The Analects forms almost the only reliable source of our knowledge of the thought of Confucius. It collects sayings of the sage, mostly brief and often with little or no context. Many ideas leave room for multiple interpretations.

Confucius (551-479 B.C.) was said to have been descended from a noble family. But he was born in humble circumstances in the state of Lu in modern Shantung. He was poor and fond of learning in his youth. In old age he realized that there was no hope of putting his ideas into practice, so he devoted the rest of his life to teaching. He had a high respect for the ancient sages. He had great respect for the wisdom of the past but did not accept it uncritically. "I am not one who was born in possession of knowledge; I am one who is fond of antiquity, and earnest in seeking it."

Confucius lived in an age which marked the beginning of a process of moral reflection which had been provoked by the crisis of aristocratic society and the decline of ritual. In a society in which human life was cheap, where there was little law and order save what each man could enforce by personal strength, armed followers, or intrigue, Confucius preached peace and respect for the individual. He aimed at an art of life including psychology, ethics, and politics. His ideal men cannot be defined once and for all. They are the subject of approaches which vary individually. He tried to provide a system which identifies personal culture with the public good. He had a large number of students and considered his own function to be the transmission of the ancient cultural heritage to his disciples, though he performed much more than a simple transmission. He interpreted traditional ideas according to his own moral concepts. His own ideas became known through the "Confucian Analects", which were compiled by some of his disciples. Although he considered himself as "A transmitter and not a maker, believing in and loving the ancients" and his thought was affected by the classical teachings, his creativity lies in the creation of the philosophical and social system discussed in this study.

Confucius was dogmatic. He refused to insist on certainties and refused to be inflexible. Confucius did not claim to be either superior in intelligence or in moral qualities. "In letters I am perhaps equal to other men, but the character of the superior man, carrying out in his conduct what he professes, is what I have not yet attained to." "The sage and the man of perfect virtue; - how dare I rank myself with them? It may simply said of me, that I strive to become such without satiety, and teach others without weariness." He had no foregone conclusions, no arbitrary predeterminations and no obstinacy. Confucius was a practical statesman, concerned with the administration of the State. The virtues he sought to inculcate were not those of personal holiness, nor were they designed to secure salvation in a future life, but they were rather those which can lead to a peaceful and prosperous community here on earth. His outlook was essentially conservative, and aimed at preserving the virtues of former ages. He emphasized the significance of stability in social life. He did not take on the subjects of extraordinary things, feats of strength, disorder, and spiritual beings. He was mainly concerned with man in life. "While you do not know life, how can you know about death?". There were four things which he taught: letters, ethics, devotion to soul, and truthfulness. Confucius seems to have enjoyed life. He was fond of learning and music. "In a hamlet of ten families, there may be found one honorable and sincere as I am, but not so fond of learning." When Confucius was in Qi, he heard the music of Shao, and for three months did not know the taste of flesh. "I did not think," he said, "that music could have been made so excellent as this." From the Analects one may be impressed that Confucius was a man whose life was full of joy. The joy may be partially due to the pursuit of the way.

Confucius's thought emphasizes the two complementary sides of human nature - intuitive wisdom and practical knowledge, contemplation and social action. Cooperation, unity and harmony are emphasized in his thought. He was greatly interested in the I Ching which emphasizes that things which oppose each other are complementary. It holds the point of view that the essential spirit is unity and a mutual dependence of opposites. It was said that the leather thongs binding the tables on which his copy of the I Ching was inscribed were worn out three times. He claimed that had he fifty years to spare, he would devote all of them to studying the I Ching. "If some years were added to my life, I would give time to the study of the Book of Changes, and then I might come to be without faults."

3. Learning

> *Let the will be set on the path of duty. Let every attainment in what is good be firmly grasped. Let perfect virtue be accorded with. Let relaxation and enjoyment be found in the polite art.*
>
> *Confucius*

This century the world has been shaped by knowledge to a greater extent than ever before. The technological revolution has transformed the fundamental dimensions of time and space in human life. The great extent of knowledge and the ease of access to it have profoundly changed the mechanisms of wealth creation, distribution, consumption and savings. The main force for change, driven by various technologies, has affected societies throughout the world. Knowledge has become a major instrument in the rivalry between social classes, races and countries. There are dynamic interactions between economic development, creativity, knowledge utilization, and cultural values. Economic development demands educated people and education requires an appropriate combination of family values, school systems, university education, a stable social environment, experiments in working methods and leisure time. Cultural values are significant for creativity and learning. Various cultures place varied "weights" on creativity, learning and (working) morals. One of the main characteristics of Confucius' teachings is the emphasis on learning.

Knowledge is either about man and society or natural phenomena. Confucius' teachings were mainly concerned with man and society. Confucius regarded benevolence, justice, ceremony, knowledge and faith as among the most important virtues. Without knowledge, the other four virtues cannot be perfect. Confucius held that it was benevolence which must be at the heart of humanity. He believed that benevolence had to be tempered with justice and reinforced by knowledge. A simple, spontaneous humanity was not enough. "There is the love of benevolence without the love of learning; - the beclouding here leads to a foolish simplicity. There is the love of knowing without the love of learning; - the beclouding here leads to dissipation of mind. There is the love of being sincere without the love of learning; - the beclouding here leads to an injurious disregard of consequences. There is the love of straight-forwardness without the love of learning; - the beclouding here leads to rudeness. There is the love of boldness without the love of learning; - the beclouding here leads to insubordination. There is the love of firmness without the love of learning; - the beclouding here leads to extravagant conduct."

For Confucius, the pursuit of knowledge is either socially useful or has its own utility. People may learn for different purposes. "In ancient times, men learned with a view to their own improvement. Nowadays, men learn with a view to the approbation of others." Confucius emphasized the pleasure that an individual obtains from learning and taught that learning should be conducted to enrich the heart. "They who know the truth are not equal to those who love it, and they who love it are not equal to those who find pleasure in it."

In the Confucian tradition, the classification of man is not based on race, family background or other social states, but on the knowledge that one obtains. "Those who are born with the possession of knowledge are the highest class of man. Those who learn, and so, readily, get possession of knowledge, are the next. Those who learn after they meet with difficulties are another class next to these. As to those who meet with difficulties and yet do not learn; - they are the lowest of the people." This classification may explain why education has been so highly emphasized throughout China's history. Moreover, as far as learning is concerned, people are equal. As Confucius classified people on the basis of knowledge and each man has capacity to learn, this attitude will not be surprising. "There being instruction, there will be no distinction of classes." Confucius taught that in learning it is the learning incentive that really matters. Although he treated people equally in teaching, this did not mean that he liked to spend his valuable time equally on pupils. He emphasized the efficiency of teaching. He would not waste his time on people who were shallow in understanding. "I do not open up the truth to one who is not eager to get knowledge, nor help out any one who is not anxious to explain himself. When I have presented one corner of a subject to any one, and he cannot from it learn the other three, I do not repeat my lesson."

Confucius considered that man's accessibility to knowledge was not equal. Some might be very talented; while others might be very slow at learning. It seems that there are certain "natural limits" of depth that one may approach in attaining knowledge. The communication of knowledge should be conducted on the basis of capacity of understanding. "To those whose talent is above mediocrity, the highest subjects may be announced. To those who are below mediocrity, the highest subjects may not be announced." "There are some with whom we may study in common, but we shall find them unable to go along with us to principles. Perhaps we may go on with them to principles, but we shall find them unable to get established in those along with us. Or if we may get so established along with them, we shall find them unable to weigh occurring events along with us." Some may obtain knowledge easily, while others may learn with great costs. Knowledge has different effects on personality. "It is by the Odes that the mind is aroused. It is by the Rules of Propriety that the character is established. It is from Music that the finish is received." Although natural talents may vary, the most important factor for knowledge accumulation seems to be one's own efforts. "By nature, men are nearly alike; by practice, they get to be wide apart."

To accumulate knowledge, both thinking and learning are equally necessary. Confucius argued that one should not concentrate solely on learning or thought alone. Perfect knowledge comes from combining learning and thought. "Learning without thought is labor lost; thought without learning is perilous." "I have been the whole day without eating, and the whole night without sleeping: - occupied with thinking. It was of no use. The better plan is to learn." Confucius greatly emphasized applications of learning. Knowledge is not only for the sake of knowledge. Knowledge should be used to solve practical (social) problems. The relation between learning and practicing what one learns is close. "Though a man may be able to recite the three hundred Odes, yet if, when entrusted with a governmental charge, he knows not how to act, or if, when sent to any quarter on a mission, he cannot give his replies unassisted, notwithstanding the extent of his learning, of what practical use is it?" Application of knowledge not only has social utility but also will bring pleasure to the scholar himself. "Is it not pleasant to learn with a constant perseverance and application?".

As Confucius assumed the existence of gaps in learning capacities among people and that the incentive to learn is essential for mutual understanding, one may not be surprised to read that "When we see men of worth, we should think of equalling them; when we see men of a contrary character, we should turn inward and examine ourselves." Confucius distrusted fine words. "Fine words and an insinuating appearance are seldom associated with true virtue." A scholar should not be concerned with clothes and appearance. "A scholar, whose mind is set on truth, and who is ashamed of bad clothes and bad food, is not fit to be discoursed with." He taught that one should learn from others disregarding their social positions. "When I walk along with two others, they may serve me as my teachers. I will select their good qualities and follow them, their bad qualities and avoid them." Rather than suggesting a way to argue one's own point of view, Confucius proposed a "silent" way of learning. The impact of this attitude can still be observed in the behavior of scholars from the Confucian cultures.

4. Moral Codes and Legal Systems

There are no concerns more central to Confucianism than the concern with the ethical gap between norms and actualities or the concern with the capacity of human moral agents to bridge the gap.

B.I. Schwartz (1985)

A civilization is characterized by its moral codes and legal systems. They provide criteria for the punishments and awards for social games. They are essential factors which affect the incentives of human behavior in the long term. Moral codes and legal systems vary over time and space. Their evolution is one of the most significant aspects of human evolution. It may be argued that a basic step in understanding a culture is the investigation of its moral codes and legal systems.

Confucius' thought was much concerned with moral duties, priorities, the purpose and destiny of man and his position in society. Confucius believed that man's nature was fundamentally good. He considered that the natural affection existing between relatives within one family was the cornerstone of social morality. The practice of morality did not lie in people's discharging the dispensations or commands of any transcendent being. Zigong once asked, "Is there one word which may serve as a rule of practice for all one's life?" Confucius said, "Is not RECIPROCITY such a word? What you do not want done to yourself, do not do to others." "Hold faithfulness and sincerity as first principles. Have no friends not equal to yourself. When you have faults, do not fear to abandon them."

For Confucius, it was natural for a man to love his parent or son, and it was only through pushing this affection outward stage by stage that he might love all mankind. It was when the natural human affection found within the family was extended without animosity beyond the confines of the family that human nature reached perfection and the social order was appropriately maintained. Those who acquired this kind of perfect love of humanity were spoken of as men of benevolence, or men of virtue. To become such a man was considered to be the ultimate objective of moral cultivation. Filial piety and the discharging of one's duty as a younger brother were important virtues. One should respect one's parents and take good care of them.

One should act according to the wishes of one's parents. In addition to filial piety to parents, one should obey one's elder brothers and seniors.

In his moral philosophy, man is a centerpiece. The distinction between morality and self-interest is the corner-stone of Confucius' teachings. A man must think for himself. The emphasis on the heart is firmly based on Confucius' morality. What pleases the heart is of higher value than what pleases the senses. It is the quality of the heart that has the highest value. Perfect goodness is more important than perfect beauty. A piece of music is acceptable not merely because of its beauty. Confucius' ideal man embodies the virtue of benevolence and acts in accordance with rites and rightness. "The superior man thinks of virtue; the small man thinks of comfort. The superior man thinks of the sanctions of law; the small man thinks of favors which he may receive." "The superior man is easy to serve and difficult to please. If you try to please him in any way which is not accordant with right, he will not be pleased. But in his employment of men, he uses them according to their capacity. The mean man is difficult to serve, and easy to please. If you try to please him, though it is in a way which is not accordant with right, he may be pleased. But in his employment of men, he wishes them to be equal in everything." The only purpose a man can have and also the only worthwhile thing a man can do is to become good. This has to be pursued for its own sake and with complete in difference to success or failure.

Confucius taught that it is necessary to be benevolent to other individuals as well as to work unstintingly for the welfare of the common people. The reward lies in doing what is good and the punishment lies in what is immoral. But Confucius did not provide one single ideal character. Confucius spoke about the sage, the good man and the complete man. The sage is at such a high level of attainment that it is hardly ever realized. Lower down the classification there is the good man and the complete man. What is a good man is obscure. Confucius also spoke about the gentleman and the small man. The ideal of the gentleman is wider than that of the moral man. More is necessary if we are to have the perfect gentleman. A gentleman pursues morality with single-minded dedication while a small man pursues profit with single-minded dedication. When self-interest comes into conflict with morality, it is self-interest that should give way. It should be noted that self-interest and morality may be in harmony.

For Confucius, whether or not we succeed in being virtuous depends solely on ourselves. "Hold faithfulness and sincerity as first principles, and be moving continually to what is right; - this is the way to exalt one's virtue." Although Confucius made comments on how to act virtuously, he thought that it was neither easy to know virtue nor easy to practice benevolence. "Now the man of perfect virtue, wishing to be established himself, seeks also to establish others; wishing to be enlarged himself, he seeks also to enlarge others. To be able to judge of others by what is right in ourselves; - this may be called the art of virtue." But few people know what is virtuous. "Those who know virtue are few." "Perfect is the virtue which is according to the Constant Mean! Rare for a long time has been its practice among the people." "I have not seen a person who loves virtue, or one who hated what was not virtuous." "I have not seen one who loves virtue as he loves beauty."

Confucius considered that the virtue was more important than law. "In hearing litigations, I am like any other body. What is necessary is to cause the people to have no litigations." He argued that the supreme importance in society was moral rather than law. For him, if the people improved themselves and had a sense of shame, the law would never need to invoked. "If the people be led by the laws, and uniformity sought to be given them by punishments, they will try to avoid the punishment, but have no sense of shame. If they be led by virtue, and uniformity sought to be given by the rules of propriety, they will have the sense of shame, and moreover will become good."

Confucius, like all ancient Chinese thinkers, looked upon politics as a branch of morals. For Confucius, politics was important but only as an extension of morals. The relationship between the ruler and the ruled was considered as a subject of the moral relationship which holds between individuals. "In carrying on your government, why should you use killing at all? Let your evinced desires be for what is good, and the people will be good. The relation between superiors and inferiors is like that between the wind and the grass. The grass must bend, when the wind blows across it." Ji Kang asked how to cause the people to reverence their ruler, to be faithful to him, and to urge themselves to virtue. Confucius answered "Let him preside over them with gravity; - then they will reverence him. Let him be filial and kind to all; - then they will be faithful to him. Let him advance the good and teach the incompetent; then they will eagerly seek to be virtuous."

5. Social Organization and Income Distribution

There was no other way of developing the manifold capacities of man than by placing them in opposition to each other. This antagonism of power is the great instrument of culture, but it is only the instrument; for as long as it persists, we are only on the way towards culture.

J.C.F. Schiller

Social organization that holds men together is an important form of human adaptation to environment. Its discrete structure is deeply dependent on moral codes and legal systems, even though the mechanism of its formation is also related to man's capacity to deal with nature. Social organization conducts the rule of how to ensure that individuals perform their proper functions. It is the basis of the division of labor and the division of consumption. It is an essential condition for individuals to be able to find their proper role.

Confucius' social order was built upon two basic assumptions. The first is that all men are by nature well disposed and have an innate moral sense. Men are therefore educable and can be moved, especially by virtuous example, to do the correct thing. The second assumption is that the ruler's virtuous conduct leads men to accept and follow his authority. "He who exercises government by means of his virtue, may be compared to the north polar star, which keeps its place and all the stars turn towards it."

The Confucian society is a hierarchy with the emperor at the top of the pyramid. Confucian principles of form that impose order on the material flux of the world make the emperor the keystone in the arch of social order. According to Confucius, a virtuous government would strengthen the people by means of morality and serve naturally to bring about order in society by raising the level of virtue among the people. Confucius tried to find a way to secure social justice in a feudal-bureaucratic society. He strongly rejected any idea of constitutional government on the grounds that under the principles of constitutionalism order is imposed upon society by law and those who break the law are penalized, so that people come to think how they can best avoid punishment, and the resulting society has no sense of shame. However, it is essential for a society to be under the sway of the principle of government by virtue of something analogous to the laws found in a constitutional society. This was referred to by Confucius as ceremony - norms established by custom and being less rigid than laws. A society was to be guided by morality and controlled by ceremony. Confucius believed that his principle would mean "people will come to have a sense of moral shame, and to act correctly."

Confucius conceived of individuals and the state as organic wholes. What the state secures is not so much for the individual per se, but for his ability to fulfil his role in the society. It is only when the whole is healthy that it is possible for people to perform their functions and dwell in an efficient and secure state. Confucius taught that every individual has strict duties: a subject must respect his ruler, a wife must respect her husband. "There is government, when the prince is prince, and the minister is minister; when the father is father, and the son is son." "A prince should employ his ministers according to the rules of propriety; ministers should serve their prince with faithfulness." There is a strict division of labor in Confucius' system. One should not interfere in others' official affairs. "He who is not in any particular office has nothing to do with plans for the administration of its duties." One should examine oneself to judge whether one was worthy of one's place. "A man should say, I am not concerned that I have no place; I am concerned how I may fit myself for one. I am not concerned that I am not known; I seek to be worthy to be known."

Confucius advocated universal education and taught that diplomatic and administrative positions should go to those best qualified academically, not socially. The true aim of government was the welfare and happiness of all the people; not brought about by rigid adherence to arbitrary laws but rather by a subtle administration of customs that were generally accepted as good and had the sanction of natural law. Confucius held that whether a ruler deserves to remain a ruler depends on whether he carries out his duty or not. If he does not, he should be removed. He taught that the people's first obligation is to correct their own faults. Confucius considered that the common people were very limited in their intellectual capacity. They cannot understand why they are led along a particular path. "The people may be made to follow a path of action, but they may not be made to understand it."

One of the most important functions of social organization is related to decision mechanisms concerning property and income distributions. Whether wealth and social products should be equally or unequally distributed among the population and what political economic mechanism should be accepted to determine income and property distributions are important in determining the division of labor and the division of consumption.

As shown above, Confucius considered that human talent is distributed unequally. Knowledge accumulation is positively related to human efforts, even though learning efficiency may be individually varied. As individual ability is different, income may be varied. Confucius did not decry concern for profits as a mark of an inferior person. Under just social and economic systems, one's income is related to one's labor. He taught that if the society was governed by virtue, it was a shameful matter to be poor. "When a country is well governed, poverty and a mean condition are things to be ashamed of. When a country is ill governed, riches and honor are things to be ashamed of." The real point is whether the profit is gained by one's ability and with virtue. "Riches and honors are what men desire. If they cannot be obtained in the proper way, they should not be held. Poverty and meanness are what men dislike. If they cannot be avoided in the proper way, they should not be avoided." "Riches and honors acquired by unrighteousness are to me as a floating cloud." "A superior man indeed is Qu Boyu! When good government prevails in his state, he is to be found in office. When bad government prevails, he can roll his principles up, and keep them in his breast." It may be argued that virtue, knowledge, working efficiency, and payment are interrelated in his philosophy. This is perhaps one of the main characteristics of Confucius' thought from social and economic points of view.

In the question of self-interest, the conflict is between profit and rightness. Confucius divided human relationships into two types: one was based on righteousness, the other on personal interest. Different people place different "weights" on righteousness and personal interest. Some men may cooperate toward ends of money, power and vanity. These are mean men who form clubs to pursue selfish interests. Other men cooperate for righteousness. "The mind of the superior man is coservant with righteousness; the mind of the mean man is coservant with gain". The superior men form cooperative groups that can stand the test of time and difficulties.

For Confucius, few people learn for the sake of learning alone. "It is not easy to find a man who has learned for three years without thinking of becoming an official." The incentive to accumulate knowledge is related to social position and income in the long term. Confucius considered some relationship between consumption attitudes and human behavior. "Extravagance leads to insubordination, and parsimony to meanness. It is better to be mean than to be insubordinate." He advised the government to be economic and to use righteous persons. "To rule a country of a thousand chariots, there must be reverent attention to business, and sincerity; economy in expenditure, and love for men; and the employment of the people at the proper seasons." Confucius strongly emphasized long-run advantages in governing the country. "Do not be desirous to have things done quickly; do not look at small advantages. Desire to have things done quickly prevents their being done thoroughly. Looking at small advantages prevents great affairs from being accomplished." Whether a government is good or not is observed from the behavior of the people under its rule. "Good government obtains, when those who are near are made happy, and those who are far are attracted."

6. A Few Further Remarks

Traditional Chinese culture is characterized by the opposite teachings of Taoism and Confucianism. Taoism has been studied from the perspectives of nonlinear science. It may be the time to re-examine Confucianism in the light of nonlinear dynamic political economics with endogenous knowledge. This paper is an initial attempt on this issue.

Confucianism held the orthodox philosophical position in China for many centuries. Although Marxism gained the dominant position in modern China, its historical duration is very short in comparison to Confucianism. On the other hand, it remains to be examined how thoroughly Chinese thinkers understand Marxism and how broadly Marxism has found its applications in reality in China. As far as cultural history as a whole is concerned, the impact of Marxism on China is perhaps insignificant. Indeed, Marxism might have been quite useful for destroying the traditional system and liberating China from colonization. However, it may be more important to examine Confucianism than Marxism in order to study China's industrialization.

The contents and applications of Confucianism are very complicated. Ideas and their social implications are not invariant with the passing of time. Interpretations of a philosophy may vary among different people. The changeability may be either due to the ambiguity of language, the philosophy's internal ambiguity or misunderstanding by the interpreter. As the creation of an idea may be affected by many cultural and historical conditions, it is not surprising to know that a philosophical system may be interpreted in different ways. Marxism may mean different things in different cultures. This is similarly held for Confucianism. Confucianism is interpreted differently in China, Japan and Korea. For instance, Confucian virtues have been emphasized differently in these countries. Benevolence is common in both China and Korea, but there is almost no mention of it in Japan. Loyalty is highly valued in both Japan and Korea, but is not so important on China's list of virtues. Moreover, even a particular concept may be interpreted differently in various cultures. For instance, the meaning of loyalty is not the same in China as in Japan. In China loyalty means being true to one's own conscience. In Japan, although it is also used in the same sense, its normal meaning is essentially a sincerity aimed at total devotion to one's lord, i.e., service to one's lord to the point of sacrificing oneself. In Japan there is no question of the concept of loyalty and faith being considered two sides of the same coin, as is the case in China. Although it is beyond the scope of this paper, it is quite significant to understand why Confucianism had been interpreted differently in these countries. We consider that the climate, geographical conditions, technology, and initial cultural conditions have intimately influenced the interpretations. There is an increasing literature on Confucianism and it is quite beyond my capacity to provide a thorough analysis of the subject. Only a few aspects of the thought of Confucianism's founder have been examined here to provide some insights into the social and economic behavior of the Chinese people. Although it is important to broadly examine the positive and negative effects of Confucius' thought on China's economic development in the future, we have merely provided a few points of view about its implications for China's economic and political development.

China has speeded up its industrialization process since the economic reform was started about one and half decades ago. Over the past fifteen years China has gradually adopted more and more market measures, creating stock markets, freeing prices and privatizing some state enterprises. China has moved towards a market economy while maintaining social stability. One of the most important aspects of the Chinese economic reform is that a great many experiments have been completed without impairing living standards. This process has been characterized as gradual and experimental. Although many serious economic problems remain, the Chinese economy has been improved a great deal since the economic reform was initiated. The economic reforms have resulted in respectable levels of economic growth and improved living conditions.

The fast rate of development may be explained from many perspectives. World peace and international economies, not to mention the large reservoir of scientific knowledge and technology accumulated in the rest of the world, provide a suitable environment for China's economic development. On the other hand, it may be argued that the emphasis on knowledge, hard work and social harmony in Confucius' teachings also play an essential role for the fast economic growth (Zhang, 1977).

One of the main differences between the Confucian and Western cultures is related to individualism. As shown before, Confucius' human value was concentrated on the heart rather than human relationships. The value placed upon the heart is heavily dependent on morality and accumulated knowledge. As far as human life is concerned, it is the heart rather than the human relationships that matters. Confucius created a social system which was intended to produce a harmony of hearts in society. In this sense, it may be argued that Confucius' pragmatical rational system is not incompatible with individualism (if it can be shown that individualism can guarantee a harmony of hearts). As far as individualism is concerned, we think that the Chinese value placed upon individual freedom lies between that of the Japanese and the West. In Japan, the center is the company. In the West, it is the individual. In China, it is the family. The number of the union to be loyal in the Chinese value system lies between the Japanese and Western ones. The significance of this value will become manifest when Chinese people become educated, have more leisure time to become culturally cultivated and are sufficiently relaxed to play cultural games.

Confucius taught that the central task of the state should be primarily concerned with people's living conditions. As living conditions are improved, education should be spread among the people. When Confucius went to Wei, Ran You acted as the driver of his carriage. Confucius observed, "How numerous are the people!" Ran You said, "Since they are thus numerous, what more shall be done for them?" "Enrich them," was the reply. "And when they have been enriched, what more shall be done?" Confucius said, "Teach them." From this comment on the order of economic development and learning, one may gain some hints about the difference between the Confucian cultures and other developing countries. It should be noted that the contents of education in Confucius' thought are different from those in modern times. During China's industrialization improvements in mass education and an increased rate of economic development have been carried out simultaneously. But the cultural value that Confucius placed upon the government giving first priority to people's living conditions is significant for explaining China's political stability.

In comparison to economic reform, the development of democracy has been very slow. Confucius' basic vision of life may be called pragmatical rationality. The essential purpose of the government is for the happiness of the people, rather than a privileged minority. Although Confucianism is not democratic, it may be argued that due to its pragmatic rational character, the friction may not be too great for Confucian cultures to develop democracy. It seems necessary that even China should develop a democratic form of government when it reaches a mature stage of industrialization in which majority of the population is well-off and well educated. By examining Confucianism and theories of democracy, we can see that it is not psychologically difficult for people who have been educated according to pragmatic rational Confucianism to accept democracy in modern industrialized civilization. Indeed, there are dynamics of democratization and marketization in Confucian cultures. Singapore, South Korea and Taiwan all successfully conducted national elections in 1992. This clearly indicates that democracy in Confucian societies is not only possible but also practicable.

References

Andersson, Å.E. (1986) The Four Logistical Revolutions, Papers of Regional Science Association 59, 1-12.

Confucius (1992) The Confucian Analects, in "The Four Book"translated by James Legge and revised and annotated by Zhongde Liu and Zhiye Luo, Hunan: Hunan Publishing House.

Haken, H. (1977) Synergetics: An Introduction. Berlin: Springer-Verlag.

Haken, H. (1983) Advanced Synergetics. Berlin: Springer-Verlag.

Morishima, M. (1978) The Power of Confucian Capitalism. The Observer (London), June.

Nicolis, G. and Prigogine, I. (1977) Self Organization in Nonequilibrium Systems. New York: Wiley.

Schwartz, B.I. (1985) The World of Thought in Ancient China. Mass., Cambridge: The Belknap Press of Harvard University Press.

Seldes, G. (1985, compiled) The Great Thoughts. New York: Ballantine Books.

Zhang, W.B. (1991) Synergetic Economics. Heidelberg: Springer-Verlag.

Zhang, W.B. (1996) Knowledge and Value - Economic Structures with Time and Space, Umeå Economic Studies No. 408. Umeå: Umeå University.

Zhang, W.B (1977) Japan Versus China in the Industrial Race. London: Macmillan (forthcoming).

The Infrastructure Arena – Transportation and Communication Networks

Co-Evolutionary Learning on Networks

David F. Batten
The Temaplan Group
Melbourne

1. Introduction

The inspiration for this paper straddles two streams of scientific thought. First, the notion of emergent macroscopic order through fluctuations, epitomized by the pathbreaking work on self-organization (e.g. Nicolis and Prigogine, 1977) and synergetics (Haken, 1977). Second, the fascinating hypothesis of how towns and cities in Medieval Europe might have responded to the expansion of trade over longer distances in the eleventh century. Championed by the Belgian historian, Henri Pirenne (1925), this mercantile hypothesis has found support in Fernand Braudel's great trilogy (1979) charting the history of Western society.

A simple bridge between these two streams of thought was constructed by Alistair Mees (1975). Using the tools of catastrophe theory, he showed that a slow improvement in communications could make trade easier, triggering a sudden change in regional specialization. But a decade later, a much grander bridge was constructed by Åke E. Andersson. In his Presidential Address to the Regional Science Association in 1985, he broadened the Pirenne-Mees hypothesis to a more general one, namely that a series of revolutionary changes to the world economy (from 1000 A.D. until 2000 A.D.) were triggered by slow but steady changes to its logistical networks.[1] Using the same third-order differential equation system each time, he portrayed the qualitative development of interregional economic relations throughout the world in terms of four logistical revolutions. The slow variable was network infrastructure capacity and the faster variable was production capacity.

In later papers, Andersson has argued that increasing endogeneity of trade and transportation in the long-term development of the economic system implies increasing interdependencies and nonlinearities. These could result in trade and transportation flows which are unpredictable and uncontrollable. The slowly changing collective characteristics of the transportation network and technology, in conjunction with the distribution of mobility, will jointly determine the outcome in terms of transportation flows. He also noted that synergetic decomposition of the problem can sometimes help to find ways out of potentially chaotic situations (see, e.g., Andersson, 1995).

Underpinning the predictability problems associated with trade and traffic flows, as well as the patterns of location and production, are the decision rules invoked by the people who generate these flow patterns. Limiting our discussion to road traffic networks, a series of behavioural questions may be posed. How do travellers perceive and respond to congestion? As the degree of congestion grows, does the traffic's collective behaviour self-organize at or near a critical state beyond each individual's control? Do travellers modify their decision rules in response to being pushed repeatedly towards the "edge of chaos"?[2]

In this paper, some behavioural traits of today's users of logistical networks will be examined. Unexpected outcomes may arise through co-evolutionary learning by a heterogeneous population of travellers. In traffic situations which are complicated or ill-defined, travellers tend to resort to inductive rather than deductive reasoning. Order prevails while each individual can control his travel behaviour and achieve his expected travel time. But uncertainty takes the wheel once the traffic volume approaches a critical density. At the edge of chaos, small perturbations can generate large fluctuations in congestion formation and thus travel times.

Repeated exposure to jamming transitions, which transform the traffic from a freely moving to a jammed state (or the reverse), causes some drivers to search more actively for ways of reducing their travel time. Explorers emerge spontaneously from a seemingly homogeneous traveller population, like mutants in an ecosystem. They are willing to alter their departure times in the absence of a viable alternative route. Being more innovative than other travellers, explorers are responsible for fluctuations which tend to alter the average (deterministic) laws of traffic evolution. The result is a diverse ecology of behavioural hypotheses, which evolves unevenly in response to traffic experiences. As their numbers increase, explorers' more efficient use of the same network leads to an improvement in the network's throughput. Nevertheless, the vast majority of travellers continue to tolerate traffic jams. These sheep are either unwilling or unable to modify their behaviour even if their chosen routes and times accumulate a convincing record of failure.

In brief, this paper explores the dynamics of (congested) traffic as a nested process involving co-evolutionary behaviour at three levels in space and time: the network as a whole, the vehicular flow pattern on each link or route and the driver population's travel behaviour. The evolutionary possibilities at each of these levels are highly interdependent on the characteristics at each level. For example, seemingly small changes in the behavioural ecology of drivers can have profound effects on the final state of the transportation system as a whole.

2. Traffic Behaviour on Urban Transportation Networks

2.1 The Management Problem

Because it is either impossible or undesirable to extend the capacity of many large urban transportation networks, efficient management of the traffic using them has become a high priority. Common management strategies include rapid mass transit systems, car pooling, elaborate parking fees, one way streets, bus and taxi lanes, route guidance systems and congestion pricing. However, large urban transportation systems are actually very difficult to manage efficiently. Due to the complexity of these systems, some traffic control measures may produce counter-intuitive results, outcomes which are quite opposite to their intention.

For example, the addition of a new link between two existing nodes on a congested road network can lead to a *reduced* overall capacity and make all travellers worse off (see Braess, 1968; Cohen and Kelly, 1990).[3] A paradoxical outcome like this is generally explained in terms of the divergence between two benchmark flow patterns - a System Optimum (where one agent routes flows in such a way that the total system cost is minimized) and a User Equilibrium (where many agents select routes which minimize their own travel time). A stable User Equilibrium is reached when no traveler can improve his travel time by unilaterally changing routes. Once this Wardropian condition has been satisfied, no traveller has an incentive to change to another route because there are no other routes which are faster.

On an uncongested network, a User Equilibrium (UE) corresponds to a System Optimum (SO). Wherever a link travel time is not a function of the flow on that (or any other) link, travellers can happily choose their best route without any concern for the behaviour of other travellers. As the flows between origins and destinations increase, however, the UE and SO flow patterns become increasingly dissimilar. Larger flows mean that some links carry an amount of traffic which is nearer to their capacity, while others remain uncongested.

Consider the simple network shown in Figure 1. It includes one Origin-Destination (O-D) pair connected by two routes: a one-lane priority road (route 1) and a two lane freeway (route 2). Hypothetical performance curves for these two routes are shown in the figure, and are given by

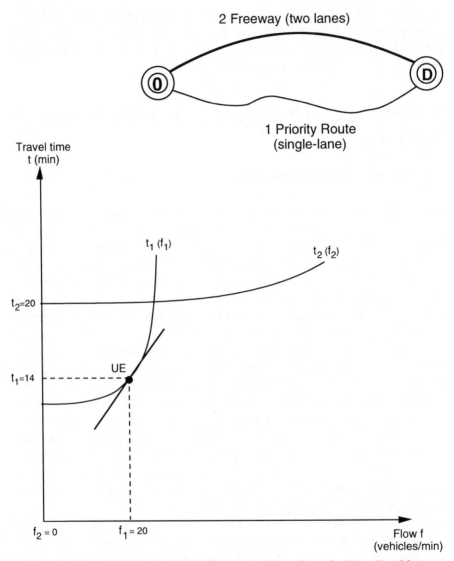

Figure 1: A Two-Link Network Equilibrium Problem where the User Equilibrium is not a System Optimum

$$t_1(f_1) \ = \ 10 \ + \ 0.01f_1^2$$

$$t_2(f_2) \ = \ 20 \ + \ 0.001f_2^2$$

If the total O-D flow is 20 vehicles per minute, the UE solution will be as shown in the figure ($f_1 = 20$, $t_1 = 14$, $f_2 = 0$ and $t_2 = 20$). At this flow level, no traveller will choose to use the freeway. Note, however, that dt_1/df_1 at $f_1=20$ is 0.4. If one vehicle were to shift from route 1 to route 2 every minute (at the UE flow pattern shown in the figure), his travel time would increase by 6 minutes. But the total travel time incurred by the remaining drivers on route 1 would fall by 7.6 minutes. While the UE solution is optimal from the users' viewpoint, it is not a System Optimum. The SO solution to this problem includes some travellers using the freeway as well, since the SO flow pattern is achieved only when

$$3f_1^2 = 1000 + 0.3f_2^2.$$

A failure to recognize this fundamental difference between the normative SO flow pattern and the descriptive UE flow pattern can lead to counterintuitive scenarios like Braess' paradox. In fact, the very existence of this paradox depends on the assumption of independent optimality underpinning a User Equilibrium. Because each individual's choice of route is carried out without consideration of the effect of this action on other network users, there is no reason to expect that the addition of a new link will always decrease total travel time.

Transportation scientists have derived formulae to determine whether Braess' Paradox occurs in a given network (see e.g. Dafermos and Nagurney, 1984). On the other hand, economists have shown that such paradoxical outcomes can be avoided and optimality retrieved by simply modifying the costs incurred by some of the travellers. By levying a suitable flow-dependent congestion fee on each traveller using a particular route on the network, the traffic flow pattern which results from choosing cost-minimizing routes could be made to return to a System Optimum (see e.g. Johansson and Mattsson, 1995).

Can we expect a User Equilibrium to be stable in reality? Furthermore, is Braess' Paradox symptomatic of a broader class of uncertainties associated with network connectivity, link congestability and travellers' behaviour? The current desire to implement automated information systems (like route guidance systems) or management controls (like congestion pricing) highlights a need to resolve some fundamental uncertainties about travellers' behaviour. How do travellers respond in situations where their own behaviour is also dependent on the behaviour of others? Do variations in travel time cause them to alter their decision rules? What might be the behavioural impacts of implementing route guidance systems or congestion pricing?

2.2 Travellers' Expectations and Bounded Rationality

Granting each traveller the ability to recognize his optimal travel mode is a generous but not unreasonable assumption. But if we now add to this the foresight needed to recognize his fastest route and optimal time of departure, all in such a way that his travel time cannot be improved by altering his original decision, it becomes more difficult to believe that each and every traveller possesses the uncanny insight to choose optimally. Furthermore, the UE solution demands that each traveller makes his choice(s) in an identical and correct manner every time, having access to full information (i.e. they know the travel time on every possible route).

Recognizing that such strong assumptions as perfect foresight may not hold in reality, Daganzo and Sheffi (1977) relaxed these restrictions by distinguishing between the travel time that individuals *perceive* and the actual travel time. The perceived travel time may be looked upon as a random variable distributed across the population of drivers. Equilibrium will then be reached when no traveller *believes* that his travel time can be improved by unilaterally changing routes. This condition, known as a Stochastic User Equilibrium (SUE), allows for the fact that travellers possess incomplete information about the state of the traffic system as a whole.[4]

The rational expectations associated with both the UE and SUE solutions are global. As shown by Kobayashi (1993), such equilibria are really fixed points in a rich space of conditional distributions which travellers may form collectively. Growing evidence suggests that the collective behaviour of travellers on a heavily congested network may rarely reach such an equilibrium state (see e.g. Harker, 1988). Each individual traveller's rationality is bounded by the fact that his problem is complicated and he has limited information at his disposal. Even if he turns out to be an excellent optimizer of his own actions, he cannot assume that all the other travellers are able to optimize their behaviour in the face of similar uncertainties. On an extremely dense and sophisticated network, where the choice process is complicated, one is likely to find a handful of travellers who *know* that they have chosen in an optimal manner, some who *believe* that they have chosen optimally, and others who have *no idea* if they have chosen optimally.

Most dynamic network models developed to date have analysed the proximity of solutions to SO, UE or SUE conditions under the influence of choice changes and traffic information systems. They have focused largely on the adverse effects of improved information, such as *oversaturation, overreaction* and *concentration*. With a few possible exceptions (e.g. Mahmassani and Chang, 1986; Ben-Akiva, de Palma and Kaysi, 1991; Ran and Boyce, 1994), relatively little effort has been directed towards developing behaviourally realistic approaches to encompass the dynamic learning mechanisms operating at the individual level.

Because each traveller's beliefs are personal and evolving at different rates, it is also unclear how the traffic system will behave collectively. Travellers cannot rely upon other travellers to behave in a predetermined manner. Instead they are forced into a world of subjective beliefs. Objective, well-defined, shared expectations (like a UE or an SUE) then cease to apply. The truth of the matter is that transportation dynamics on a dense network is the aggregated result of thousands or, in some cases, millions of individual trip-making decisions by an extremely heterogeneous population (producers, shippers, consumers, holidaymakers, etc.). At peak-hour conditions on these networks, travellers often face route and departure time choices which are complicated and ill-defined. Outcomes can be quite unpredictable since subjectivity is the order of the day!

2.3 Heterogeneity and Human Learning on Networks

How might travellers come to terms with such a complicated situation? According to modern psychology, human beings are only moderately good at deductive logic. Therefore we make limited use of it. But we are excellent at recognizing or matching patterns (see e.g. Bower and Hilgard, 1981). In situations which are complicated or ill-defined, we tend to search for patterns which, once recognized, can help us to simplify the problem. We use these patterns to build temporary mental models or hypotheses to work with, carry out *localized* deductions based on our current hypotheses, and then act on them (see Arthur, 1994).

But it doesn't end there. As feedback from the environment comes in, we may strengthen or weaken our beliefs in our current hypotheses, discarding some if they fail to live up to our expectations, sometimes even replacing them with new ones as needed. In other words, wherever we lack full definition of a problem and cannot be sure about the best reasoning to employ, we "paper over" the gaps in our understanding. In the words of Tom Sargent (1993), we act as economic statisticians, using and testing and discarding simple expectational models to fill these gaps. As logic, such behaviour is not deductive but *inductive*. Inductive reasoning is from a part to a whole, from the particular to the general, or from the local to the universal. Far from being the antithesis of "reason" or science, it is precisely the way in which science itself operates and progresses (Arthur, 1994).

An inductive world is a revolutionary change from our traditional world of deductive rationality. It is like an electoral world, where each voter's beliefs are highly individual, largely subjective and mostly private. Moreover, it is a dynamic world All those beliefs or hypotheses that agents form are constantly being tested in a world that forms from their and others' actions and subjective beliefs. This vast collection of beliefs or hypotheses is constantly being formulated, acted upon, changed and discarded; all are interacting and competing and evolving and co evolving; forming an ocean of ever-changing, predictive models (Arthur, 1995).

Can we find any evidence of this type of co-evolutionary learning in the transportation world? Our tentative answer is in the affirmative. Travellers (and researchers studying travellers) have adopted a number of plausible hypotheses (or mental models) to help them cope with a poorly defined situation.

In a well-known paper on the causes of congestion, Downs (1962) recognized two behavioural classes: those with a low propensity to change routes, called *sheep*, and those with a higher propensity to change, called *explorers*. Explorers tend to be imaginative, highly-strung, aggressive drivers who constantly search for alternative options that may save them some time. They are quick to learn and may even hold several hypotheses in mind simultaneously. Sheep are more placid, patient, and prone to choosing the same option (Ben-Akiva, de Palma and Kaysi, 1991). They tend to follow the leader and mostly cling to a particular belief because it has worked well in the past. Sheep are slow learners who must accumulate a record of failure before discarding their favoured option(s).

Recent empirical work by Conquest et al. (1993) confirmed the presence of various kinds of explorers and sheep in real traffic. Using cluster analysis, they found four driver groups among a sample of 4,000 commuters surveyed in 1988. They labelled them *Non-changers, Route changers, Route and Time changers*, and *Pre-trip changers* based on their behavioural responses to traffic information. Non-changers (or Downs' sheep), who made up about one quarter of the sample, were unwilling to modify any part of their commuting behaviours (i.e. departure time, route or mode of transportation) no matter how much traffic information they received. By way of contrast, Route and Time changers were eager to try different strategies in order to reduce their travel time.

Mahmassani and Chang (1986) examined the interdependence of travellers departure time decisions. Depending on the acceptability of the actual arrival time, a readjustment may occur. An important aspect of this readjustment process is the role of prior experience, acquired over a number of days. Their learning process was founded on the notion of an "indifference interval" bounded by threshold values of tolerable earliness and tolerable tardiness.

Kobayashi (1994) also suggested that drivers form their own subjective expectations on traffic conditions from repeated learning. These expectations are conditioned by other (e.g. public) information and updated by observed travel times; in so far as there is sufficient motivation to revise them. Kobayashi classified travellers into two groups based on their propensity for risk: the *risk-averters* and the *risk-neutral* (see Kobayashi, 1983).[5] If the probabilistic functions of travel time are Lipshitzian continuous, he showed that drivers' subjective expectations will converge to a rational expectations equilibrium via this recursive learning procedure.

Given that a rational expectations equilibrium is a fixed point in a richer space of possibilities, and given the heterogeneity of mental models and learning rates among drivers (e.g. sheep and explorers), we might expect the collective behaviour of traffic on a dense network to fluctuate between various equilibrium and nonequilibrium states. In order to expand on the promising start made by earlier researchers, it would be helpful to investigate the conditions under which alternative states might arise in the presence of co-evolutionary learning among travellers. The rest of this paper looks into some of these nonlinear possibilities, with an emphasis on the qualitative aspects of phase transitions at both the physical and the behavioural levels.

3. Co-Evolutionary Learnings on Networks

3.1 A Simple Example

To see that subjective expectation formation matters, and how changing circumstances can breed different expectations, let us return to the simple network introduced in Figure 1. Suppose that, every weekday evening, N_s drivers (vehicles) motor from (a common) downtown to a common suburb using the same single-lane priority route.[6] All have a common departure time (e.g. their knockoff time) and all possess identically rational expectations. They share a common view that their travel time should be no more than t_{max} and, barring any unforeseen circumstances, all N are able to achieve it. Although they are forced to slow down at a narrower section along the way, the route is mostly free-flowing. Because it achieves the desired O-D flow of X, there is general agreement that the system has reached a User Equilibrium.[7] In other words, it has stabilized near the point UE in the figure.

As long as drivers' expectations continue to be borne out by the travel times they experience, there is no reason for any deviating expectations to arise. In this situation, the homogeneous rational expectations equilibrium of the literature is evolutionarily stable. Nobody has an incentive to destroy it. Thus variations in travel time between vehicles remain low.

All is well for several years, the Road Traffic Authority enjoying unprecedented popularity. Gradually, however, a few drivers begin to suspect that their average travel time is increasing. The slowdown is barely noticeable at first because a number of predictable seasonal and monthly fluctuations in flow patterns have clouded the overall trend. But now some of the unexpected delays seem longer than before. With no viable alternative route in sight, one or two more imaginative drivers consider delaying their time of departure for home by 15 or 20 minutes. Owing to family reticence, however, they decide to postpone their decision for a week or two.

Then, on the following Monday, there is another frustrating traffic jam for no apparent reason. A few jams have occurred previously, but the frequency and duration of start-stop-waves on this occasion is disturbing. The seeds of uncertainty are planted in the minds of a few more drivers, each of whom begins to ponder alternative strategies for the future.

What really caused this unexpected jam? The answer is two relatively slow processes. Population and mobility are two variables which are increasing gradually in all societies. The mobility of citizens in many advanced nations has grown by an average of 3-5% per annum throughout this century. When combined with population increases and steadily growing levels of car ownership, a typical urban population of drivers tends to expand at 5-8% per year. This causes higher flow levels between most O-D pairs, pushing travel times up (according to our link performance function in Figure 1). In peak periods, travel times start to exceed t_{max} more frequently.

Whenever the increasingly mobile population of drivers pushes the peak-period flow of traffic to critical levels, it triggers a dynamic jamming transition from free (or laminar) flow to start-stop waves at a critical car density (see e.g. Biham, Middleton and Levine, 1992; Nagel and Schreckenberg, 1992; Nagel and Rasmussen, 1994). Before the transition, the free-flowing travel time is approximately constant and variations from vehicle-to-vehicle are small. Recent simulation work, using cellular automaton models, has shown that the average speed of the traffic drops rapidly once the critical density has been reached (see Figure 2).[8] More importantly, fluctuations in the travel time from vehicle-to-vehicle go up very steeply and reach a peak near the point of critical density (see Figure 3).

The emergent phenomenon is quite striking. When passing from slightly below to slightly above criticality, the traffic changes from a regime where the travel time is predictable with an accuracy of about 3% to a regime where the error climbs to 65% or more (Nagel and Rasmussen, 1994). This highly unpredictable state is the hall-mark of traffic near "the edge of chaos." Travel time predictability can decrease rapidly if the system is pushed towards a regime of maximum flow.

Prior to the onset of jams, each driver expects his travel time to be less than t_{max}. Because order prevails and their shared expectations are fulfilled, the driver population can be regarded as homogeneous. They all base their daily decisions on the same active hypothesis, namely:

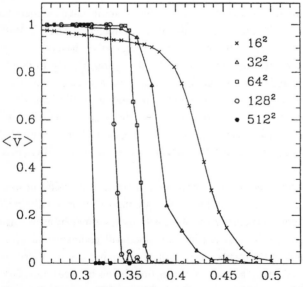

Figure 2: Average Velocity (v) as a Function of Density (p) for five different system sizes (Source: Biham, Middleton and Levine, 1992)

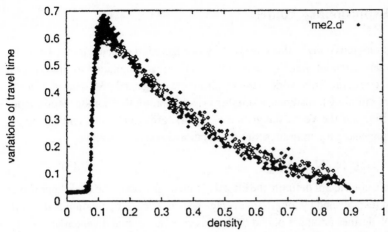

Figure 3: Travel Time Variations as a Function of Density (Source: Nagel and Rasmussen, 1994)

Hypothesis Sl:

"I expect tomorrow's travel time to be much the same as yesterday's and certainly less than t_{max}."

As long as Hypothesis S1 (S for *sheep*) proves to be correct, there is no reason for any of the drivers to reason differently. They can marvel at their foresightedness (although their behaviour is *sheepish* in that none foresee any need to change it). Each time the predictability of travel time is shattered by a jamming transition, however, a small group of drivers loses a little faith in the accuracy of Hypothesis S1. As more drivers realize that their "comfort zone" of travel time has been breached, and that they might need to consider other options in the near future, there is an incentive to expand their own set of working hypotheses. The most popular new hypothesis is:

Hypothesis El:

"I expect that my average travel time could be reduced if I start for home fifteen to twenty minutes later. "

This hypothesis not only focuses attention on possible improvements at the margin but also acknowledges that day-to-day variations in travel time are to be expected. It is a *temporal innovation* because travel time may be improved without altering the choice of mode or route. In terms of Downs' behavioural classification, it is the strategy of an *explorer*. We shall refer to it as Hypothesis E1 (E for *explorer*). As the frequency of traffic jams and the suspicion of longer average travel times increases, a growing subgroup of drivers (the *explorers*) begin to consider Hypothesis El in addition to Hypothesis Sl.[9]

3.2 An Evolutionary Model

The initial repercussions of this emergent hypothesis can be seen with the help of a simple dynamic model. Let N(t) denote the total driver population at time (i.e. on day) t who commute from work to home during the congested peak-period. Prior to the onset of jamming transitions, a simple rate equation for this homogeneous population of drivers is the Verhulst equation of logistic growth (Nicolis and Prigogine, 1977). Suppressing the time subscript for notational convenience, we have

$$\dot{N} = b_s N (N_1 - N) \tag{1}$$

where b_s is a coefficient defining the effective rate of growth in the peak-hour driver population (established by long-term average growth rates in population and mobility), and N_1 defines an upper bound to the driver population and vehicular flow on the network during the evening (peak-period) commute home. This saturation level corresponds roughly to the critical density at which jamming transitions can occur. Equation (1) postulates exponential growth for small values of N, when the density of traffic on the network is low, followed by a slowdown of growth as the critical density is approached.

Suppose that travel time uncertainty and the frequency of jamming transitions have become sufficiently disturbing to a small number of drivers. This group of *explorers* decides to delay their departure time to avoid the peak-period congestion. Let $N_{el}(t)$ denote the number of drivers in this category. After a few more frustrating traffic jams, there are sufficient *explorers* departing later to enable description of their evolution by an equation like

$$\dot{N}_{el} = b_{el} N_{el} (N_2 - N_s - N_{el}) \tag{2}$$

where b_{el} is a learning coefficient controlling the rate of defection by *explorers* and N_2 defines an upper bound on the number of drivers who can use the network for their peak-period commute home, and $N_s = N - N_{el}$. Note that N_1 and N_2 may differ.

We can now describe the behaviour of the peak-hour traffic system using two rate equations. One for the *sheep* who do not defect, namely

$$\dot{N}_s = b_s N_s (N_1 - N_s) \tag{1a}$$

and equation (2) for the *explorers* who do defect. If their expectations are mutually reinforcing, then $N_2 > N_1$, and the group of *explorers* will grow to a finite share of the driver population (restricted only by psychological or external constraints).[10]

What will happen to N_s? Assuming that N_{el}/N cannot grow beyond a fixed ratio r, we reach the simple result that the peak-period driver population approaches a new upper bound where $N_s + N_{el} = N_1/(1 - r)$. The emergence of a stable population of (late departing) *explorers* leads to an improvement in the effective use of the network. Extending the window of commuter time caters for more drivers. Nevertheless, the population of *sheep* will remain close to the edge of chaos.

So the relief provided by the defecting *explorers* is temporary at best. Soon peak period drivers begin to experience travel time uncertainty again and the frequency of jamming transitions increases. Another group of drivers begins to lose faith in Hypothesis S1. Although this group are aware of Hypothesis E1, they either doubt its accuracy or are unable to delay their departure time after work. Instead they can make use of flexible working hours to start and finish work thirty minutes earlier. They favour another hypothesis, which is

Hypothesis E2:
"I expect that my average daily travel time could be reduced significantly if I start for work thirty minutes earlier and leave my workplace thirty minutes earlier than usual."

Let $N_{e2}(t)$ denote the number of drivers in this category. After a few more frustrating traffic jams, there are sufficient new *explorers* leaving 30 minutes earlier in both the morning and the evening to enable a description of their evolution by an equation like

$$\dot{N}_{e2} = b_{e2}N_{e2}(N_3 - N_s - N_{e1} - N_{e2}) \qquad (3)$$

where b_{e2} is a learning coefficient determining the rate of defection by these new *explorers*, N_3 defines a new upper bound on the number of drivers who can make use of the network for their commute home, and $N_s = N - N_{e1} - N_{e2}$.

The emergence of a stable population of (early to work, early back home) *explorers*, N_{e2}, generates a further improvement in the effective use of the network. We may conclude, therefore, that evolution leads to a steadily growing exploitation of time. This exploitation serves to lengthen the window of commuter time, thereby postponing the full repercussions of the original peak-period congestion. Evolution appears as in Figure 4.

A series of temporal innovations by *explorers* (who favour different hypotheses) follows. Each serves to expand the throughput capacity of the network by steering some of these more innovative drivers away from the edge of chaos and thereby lowering their actual and expected travel times. But other drivers remain *sheep* forever. They may alter their active hypothesis in response to repeated jamming transitions, but their travel behaviour does not vary. Some of the hypotheses which evolve in competition with Hypothesis S1 might be:

Hypothesis S2:
"I expect tomorrow's travel time to be much the same as last Monday's but not necessarily less than t_{max}."

Hypothesis S3:
"I expect tomorrow's travel time to be the same as two weeks ago (2-period cycle detector). "

Hypothesis S4: *"Even allowing for the occasional traffic jam, I expect that my average daily travel time should still be less than t_{max} ."*

Each type of driver keeps track of an individualized set of such hypotheses. *Explorers* tend to monitor more hypotheses because they are constantly searching for ways to improve their travel times. But each driver favours one particular hypothesis (his *active* hypothesis), usually the currently most accurate hypothesis within his own set (e.g. Sl, S2, S3, S4, E1, E2).

Just as in Arthur's Bar Problem, the complete set of active hypotheses determines traffic behaviour, and thus the travel times experienced. But the travel time history also determines the set of active hypotheses. Drivers "learn" over time which of

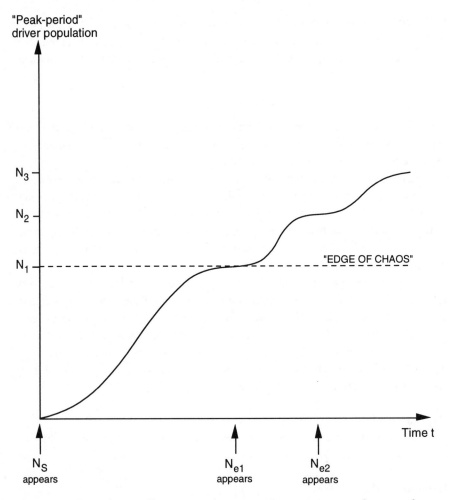

Figure 4: Expanding traffic capacity by temporal innovations on the part of some drivers (explorers)

their hypotheses work best. From time to time, *explorers* discard poorly performing hypotheses and generate new ones in their place. *Sheep* are more conservative and only modify their active hypothesis slightly (e.g. change from Hypothesis S1 to Hypothesis S2) after repeated experience of jamming transitions.

3.3 Temporary Equilibria

The complete set of hunches or active hypotheses forms a kind of driver *ecology* (or knowledge base). A key question of interest is how this ecology coevolves over time. Does it ever converge to some standard equilibrium of beliefs or always remain open-ended, perpetually discovering new hunches and hypotheses?

Because the set of active hypotheses is open-ended, this seems to be a difficult question to answer analytically. One might generate a kind of "alphabet soup" of hypotheses and then proceed by computer experiments. In this way, Arthur showed that a set of hypotheses may self-organize into an equilibrium pattern which is "almost organic in nature.......something like a forest whose contours do not change but whose individual trees do" (Arthur, 1994). However, even the contours of the emergent ecology can be expected to change in our traffic problem, because the driver population is growing in number and the predicted outcome (travel time) may differ significantly for each driver.

As a forerunner to some computer experiments of our own, it may be useful to ponder the validity of earlier results. Wherever cycle-detector hypotheses (like S3) are present, for example, can we expect that cycles will be "arbitraged" away quickly? Not necessarily. The co-evolutionary possibilities are complicated because *sheep* and *explorers* exhibit different elasticities to change. If several *explorers* expect peak period congestion tomorrow because congestion occurred two weeks ago, they are likely to alter their departure time. If several *sheep* face the same prospect, those favouring Hypotheses S3 or S4 will never alter their departure time. Thus cyclical patterns might persist.

However, there are certain conditions under which the driver population might self organize into an equilibrium of beliefs, whereupon such cycles may be expected to disappear. Consider the evolutionary situation shown in Figure 4. The expanding window of commuter time cannot be extended forever. *Explorers* may be innovative, but there are temporal limits to their ingenuity. Eventually traffic chaos will prevail over such a lengthy time period that the pressure for a superior alternative route becomes compelling. Because most drivers are reluctant to spend more than one hour of their daily time budget on commuting, the Government and Road Traffic Authority can only appease such growing discontent with a major alteration to the network itself - such as the construction of a two-lane expressway (see Figure 1). Suddenly, the set of active hypotheses changes radically. *Explorers* may be expected to try out the new expressway if their average travel time has risen above t_{max}. *Sheep* will be slower to change, but gradually they might entertain the following hypothesis:

Hypothesis S5:
"I expect that my average daily travel time could be lowered by using the new expressway. "

At some stage after the new expressway opens, the active hypotheses of *sheep and explorers* may even converge to a common belief such as Hypothesis S5. Given sufficient homogeneity of drivers' beliefs, the standard User Equilibrium of the literature may, once again, be upheld. But, as before, this UE will be temporary at best. Jamming transitions among peak-period traffic on the new expressway will eventually occur, just as they did on the original priority route. This congestion experience spawns a whole new family of *explorers* who actively pursue alternative possibilities (including the old priority route). There is no evidence to suggest that traffic behaviour ever settles down into any stable predictable pattern. Instead the emerging behavioural ecology becomes more complex and contains an even richer population of active hypotheses. This uneven pattern of co-evolutionary learning marches forever onwards, occasionally into but mostly out of equilibrium states.

4. Co-Evolutionary Traffic Management Systems

4.1 Advanced Traveler Information Systems

Under a regime of coevolutionary learning on networks, one is tempted to ask:
"Which drivers should be assisted and what kind of information should be provided if the aim is to reduce traffic congestion?"
This question is important because some information can be counterproductive, causing drivers to change their behaviour in ways that exacerbate congestion. Behavioural phenomena such as oversaturation, overreaction and concentration may negate the benefits of improved information (see e.g. Ben-Akiva, de Palma and Kaysi, 1991). Because these issues are complex in themselves, their full discussion lies beyond the scope of this paper. The impacts discussed here will only be co evolutionary in character.

Consider a situation where a substantial fraction of drivers receive information on traffic conditions. If the driver population is reasonably homogeneous in terms of drivers' willingness to act upon this information, its receipt tends to reduce behavioural variations among drivers because it tends to increase uniformity of perceptions of network conditions around the true values. Consequently, a greater proportion of drivers may select the best alternatives (from their own viewpoint) and will thus tend to concentrate on the same routes during the same periods. The effect of more information in this case could be to generate higher levels of traffic congestion on the preferred routes.

If *sheep* and *explorers* coexist in a driver population, however, the effects of information provision will depend crucially on the nature of the information, when it is provided and who receives it. *Sheep* seldom change their driving decisions based on traffic information. *Explorers* welcome such information which can help them to modify their set of working hypotheses or, in some circumstances, to alter their active hypothesis. Such information will be most effective in particular phases of the co-evolutionary learning process. Referring to Figure 4, for example, the rate of defection of *explorers* (N_{el}) from a congested peak-period to a less congested period could be enhanced by the provision of comparative travel times for different departure times. Providing this comparative information to a limited number of *explorers* would reduce the chances of overreaction or concentration. Travel time information which helps to redistribute commuters on the same route may be more effective than similar information pertaining to alternative routes. Temporal innovations possess inbuilt frictional effects which moderate the rate of defection, whereas route choice changes are less controllable and can easily lead to overreaction or concentration (see Ben-Akiva, de Palma and Kaysi, 1991).

More generally, travel time information pertaining to different departure times and route choices will alter the ecology of active hypotheses which are coevolving over time. But it may not convert the *sheep* into *explorers*. Comparative information on travel times and possible delays could enhance desirable redistributions of traffic if it targets drivers with the correct motivation. Further studies might show whether *explorers* should always be the beneficiaries or whether *sheep* might warrant assistance (or perhaps even stronger inducements) at certain stages.

4.2 Managing Traffic at the Edge of Chaos

In section 3.1, we saw that there is a critical regime around maximal capacity where traffic systems are very sensitive to small perturbations. Whenever the flow of traffic is pushed to such critical densities, it triggers a dynamic jamming transition from free (or laminar) flow to start-stop waves. Fluctuations in the travel time from vehicle to-vehicle go up very steeply and reach a peak near the point of critical density (see Figure 3).

This poses extremely challenging problems for traffic managers. The key difficulty is that traffic as a whole can be driven closer to this critical density (near the edge of chaos) with the aid of on-line adaptive traffic management systems. Yet these systems are supposed to improve traffic efficiency. The seemingly paradoxical result is that higher efficiency causes travel times to become more variable and thus further control measures will have unpredictable consequences.

Is there a way out of this dilemma? One possibility is to keep the density on each road below the density of maximum throughput. If the driver population is homogeneous, this might imply that some drivers have to wait to enter parts of the road network until sufficient capacity is available for them. On the other hand, if *sheep* and *explorers* coexist in a driver population, *sheep* could be made to pay for their intransigence. A peak-period congestion toll could be levied on those drivers who refuse repeatedly to switch to lower density times or routes. On the other hand, innovative *explorers* might be offered switching incentives for earlier or later departures which help to keep densities below criticality.

5. Concluding Remarks

As long as drivers' expectations are uniform and traffic dynamics are deterministic, a rational expectations equilibrium can prevail. This is certainly the case on uncongested networks. Travellers are generally satisfied with the performance of these networks. Standard traffic assignments, such as Deterministic (UE) and Stochastic (SUE) User Equilibria, are evolutionarily stable under these conditions. Nobody has an incentive to destroy them.

The same cannot be said of congested networks. Each time the predictability of travel time is shattered by a jamming transition, a small group of drivers begins to lose a little faith in the validity of their uniformly rational expectations. Breaching their "comfort zone" of travel time on a few more occasions can transform these drivers from placid *sheep*, meekly following the crowd under any circumstances, into aggressive *explorers*, searching desperately for ways to reduce their travel time. Once a few "seeds of discontent" have been sown, *explorers* tend to feed on themselves. Like mutants in an ecosystem, their own growth can become self-reinforcing.

Being more innovative than other drivers, *explorers* are responsible for fluctuations which tend to destroy the deterministic laws of traffic evolution. The traffic dynamics becomes uncertain. A wide variety of decision rules begin to emerge and proliferate among the driver population. Each individual driver's rationality is now bounded by the fact that his problem has become more difficult and he has limited information at his disposal. There is no single hypothesis which can be relied upon any longer. Instead a rich ecology of behavioural hypotheses emerges and evolves over time.

In traffic situations which are complicated or ill-defined, drivers tend to resort to *inductive* rather than deductive reasoning. Order prevails while each individual can control his travel behaviour and achieve his expected travel time. But uncertainty takes the wheel once congestion grows. Near the edge of chaos, small perturbations can generate large fluctuations in congestion formation and thus travel times. The result is a growing population of *explorers* fostering a diverse ecology of behavioural hypotheses, which evolves unevenly in response to traffic experiences. Under these circumstances, one is likely to find a handful of drivers who *know* that they have chosen in an optimal manner, some who *believe* that they have chosen optimally, and others who have no *idea* if they have chosen optimally.

The uneven nature of coevolutionary learning suggests that the collective behaviour of travellers on a congested network may seldom converge to a user equilibrium. Spontaneous emergence of *explorers* from a seemingly homogeneous population of sheep is a nonlinear perturbation which tends to be repeated over and over again. The incentive for repetition is strong. In the true Schumpeterian spirit, each time a new group of *explorers* emerges and evolves, their innovativeness leads to an improvement in the network's throughput. Thus network exploitation becomes more efficient, despite the fact that most of the *sheep* are either unwilling or unable to modify their behaviour even if their chosen routes and times accumulate a convincing record of failure.

In this paper, I have discussed the dynamics of (congested) traffic in terms of its co evolutionary possibilities at three different levels in space and time: the network as a whole, the vehicular flow pattern on each constituent link or route, and the behaviour of users (e.g. mainly *sheep* and *explorers*). Many interdependencies exist between the evolutionary possibilities at each of these levels. For example, seemingly small changes in the behavioural ecology of drivers can have profound effects on the final state of the transportation system as a whole. In these situations, some of the emergent behaviour only becomes apparent with the help of detailed microsimulations.

There is a need to search beneath the traditional aggregate and beyond the usual deductive view of logistical networks in order to unravel their intricate complexities. Different speeds of adjustment poses daunting problems and paradoxes seem to abound. We have seen that the approximation of deterministic, predictable traffic patterns seems less and less reliable the more one approaches a highly efficient (managed) traffic system. The fact that high performance often has the downside of high variability may not just be true in transportation systems. Chances are that it is also true of other socio-economic systems. A higher level simplicity is much easier to think about than some chain of complexities that causes it (see Cohen and Stewart, 1994). The correct representation of a higher-level simplicity must emerge only as an explicit consequence of an accurate representation of its lower-level complexities.

As a step towards a behavioural formalism for the study of logistical networks, we might look more closely at the subjective expectations, multiple hypotheses, and half-hoped anticipations held by travellers. This is a rich and complex world, in which co-evolutionary learning is incessant and sometimes surprising. Beliefs can be mutually reinforcing or mutually competing. Like vintages of technology and products in the marketplace, beliefs are invented, establish a small niche, grow in importance, begin to dominate, mature, fall back, and finally decay. Because they form an ocean of interacting, competing, arising and decaying entities, occasionally they may simplify into a simple, homogeneous equilibrium pattern. But more often than not they produce complex, ever-changing patterns in which non-equilibrium beliefs are unavoidable.

Acknowledgements

The work described in this Chapter was refined while the author was visiting the Politecnico di Torino in April 1996. For providing generous support to facilitate the visit, sincere thanks are due to C.S. Bertuglia and the National Research Council (CNR) of Italy. The author is grateful for the interest shown in the topic by the audiences attending his lectures at the Politecnico di Torino, the Universities of Bologna and Venice, and the 5th World Congress of the Regional Science Association International in Tokyo during May 1996. Thoughtful comments from Åke E. Andersson, David Boyce, Dimitrios Dendrinos, Manfred Fischer, Britton Harris, Kiyoshi Kobayashi, T.R. Lakshmanan, Axel Leijonhufvud, Dino Martellato, Sylvia Occelli, Aura Reggiani and Angela Spence are also acknowledged.

Notes

1. Logistical networks are those systems in space which are used for the movement of commodities, information, money and people.

2. The term "edge of chaos" was coined by Normal Packard, referring to a critical transition point between order and chaos. Many systems tend to adapt towards the edge of chaos because there you gain control (small input/big change) and maximum computational information.

3. The interested reader can find additional examples of paradoxical behaviour discussed in Sheffi (1985), Harker (1988) and Catoni and Pallotino (1991).

4. For an introduction to the analytical aspects of Deterministic (UE) and Stochastic (SUE) User Equilibria, see Sheffi (1985).

5. These two classes correspond closely to Downs' *sheep* and *explorer* categories.

6. The freeway does not exist at this point since all travellers can be carried comfortably on the single-lane route.

7. The point UE could be a deterministic User Equilibrium (UE) or a Stochastic User Equilibrium (SUE), depending on the variance of travel time perception.

8. Fluid-dynamical approaches have also been used to study phase transitions from laminar to turbulent traffic flow.

9. The study by Conquest et al. (1993) showed that commuters have more flexibility to change their time of departure *from* work than to it.

10. In the study reported by Conquest et al. (1993), slightly over half of the respondents were willing to make departure time changes. Women were more willing to adjust than men.

References

Andersson, Å.E. (1985) "Presidential address: the four logistical revolutions," Papers of the Regional Science Association, 59, 1-12.

Andersson, Å.E. (1995) "Economic network synergetics", in Batten, D.F., Casti, J.L. and R. Thord (eds.) Networks in Action, Springer Verlag, Berlin.

Arthur, B. (1994) "Inductive behaviour and bounded rationality", American Economic Review, 84, 406 411.

Arthur, B. (1995) "Complexity in economics and financial markets", Complexity, 1, 20-25.

Batten, D.F. (1982) "On the dynamics of industrial evolution" Regional Science and Urban Economics, 12, M9-462.

Batten, D.F. and L. Westin (1990) "Modelling commodity flows on trade networks: retrospect and prospect," in Chatterji, M. and Kuenne, R. (eds.) New Frontiers in Regional Science, Macmillan, London, 135-156.

Ben-Akiva, M., A. de Palma and I. Kaysi (1991) "Dynamic network models and driver information systems," Transportation Research A, 25, 251-266.

Biham, O., Middleton, A. and D. Levine (1992) "Self-organization and a dynamical transition in traffic flow models", Physical Review A, 46, R6124-R6127.

Bower, G.H. and E.R. Hilgard (1981) Theories of Learning, Prentice Hall, Englewood Cliffs.

Braess, D. (1968) "Ueber ein Paradoxon aus der Verkehrsplanung", Unternehmensforschung, 12, 258 268.

Braudel, F. (1979) The Wheels of Commerce, William Collins and Sons, London.

Casti, J. (1994) Complexification, Abacus, London.

Catoni, S. and S. Pallotino (1991) "Traffic equilibrium paradoxes", Transportation Science, 25, 240-244.

Cohen, J. and F. Kelly (1990) "A paradox of congestion in a queuing network", Journal of Applied Probability, vol.27, pp.730-734.

Cohen, J. and 1. Stewart (1994) The Collapse of Chaos: Discovering Simplicity in a Complex World, Penguin Books, New York.

Dafermos, S. and A. Nagurney (1984) "On some traffic equilibrium theory paradoxes," Transportation Research B, 18, 101-110.

Daganzo, C. and Y. Sheffi (1977) "On stochastic models of traffic assignment," Transportation Science, 11, 253-274.

Downs, A. (1962) "The law of peak-hour expressway congestion," Traffic Quarterly, 16, 393-409.

Haken, H. (1977) Synergetics: An Introduction, Springer-Verlag, Berlin.

Harker, P. (1988) "Multiple equilibrium behaviours on networks", Transportation Science, 22, 39-46.

Johansson, B. and L-G. Mattsson (1995) "Principles of road pricing", in Johansson, B. and L-G. Mattsson (eds.) Road Pricing: Theory, Empirical Assessment and Policy, Kluwer Academic, Boston.

Johnson, J. (1995) "The multidimensional networks of complex systems", in Batten, D.F., Casti, J.L. and R. Thord (eds.) Networks in Action, Springer Verlag, Berlin.

Kobayashi, K. (1993) "Incomplete information and logistical network equilibria, in Andersson, Å.E., Batten, D.F., Kobayashi, K. and K. Yoshikawa (eds.) The Cosmo-Creative Society, Springer, Berlin.

Kobayashi, K. (1994) "Information, rational expectations and network equilibria: an analytical perspective for route guidance systems," The Annals of Regional Science, 28, 369-393.

Mahmassani, H.S. and G-L. Chang (1986) "Experiments with departure time choice dynamics of urban commuters," Transportation Research B, 20, 297-320.

Mees, A. (1975) "The revival of cities in medieval Europe," Regional Science and Urban Economics, 5, 403-425.

Nagel, K. and S. Rasmussen (1994) "Traffic at the edge of chaos", Santa Fe Institute Working Paper No .94-06-032.

Nagel, K. and M. Schreckenberg (1992) "A cellular automaton model for freeway traffic", Journal de Physique I, 2, 2221.

Nicolis, G. and I. Prigogine (1g77) Self-Organization in Nonequilibrium Systems, Wiley, New York.

Pirenne, H. (1925) Medieval Cities: their Origins and the Revival of Trade (English Translation by F.D. Halsey, Princeton University Press, Princeton, 1952).

Prigogine, I. and R. Herman (1971) Kinetic Theory of Vehicular Traffic, Elsevier, New York.

Ran, B. and D.E. Boyce (1994) Dynamic Urban Transportation Network Models, Springer, Berlin.

Sargent, T.J. (1993) Bounded Rationality in Macroeconomics, Oxford University Press, New York.

Sheffi, Y. (1985) Urban Transportation Networks, Prentice Hall, New Jersey.

Vickrey, W.S. (1969) "Congestion theory and transport investment", The American Economic Review, 59, 251-260.

Modeling Adopters´Behaviour in Information Technology Systems: A Multi-Actor Approach

Peter Nijkamp
Department of Economics
Free University of Amsterdam

and
Heli Koski
Department of Economics
University of Oulu

1. Introduction

1.1 Background

The penetration rate of modern information and communication technologies (ICT) is not only determined by the indigenous qualities of these technologies, but also by the adoption behaviour of other actors using the same network technology. This case of network externalities means that the benefits of adopting a new technology are dependent on the adoption rates of others actors in the market (see, e.g., Capello 1994). In addition to cross-sectional interdependencies among actors, the ICT market exhibits sequential externalities related to the timing of adoption of these technologies. The adoption decisions of ICT users are affected both by the adoption behaviour of earlier user generations and by the expected success of technologies in the future.

This paper aims to model the adopters' behaviour in ICT networks from the viewpoint of both cross-sectional and sequential interdependencies. In particular, it will focus on the economic consequences of (in)compatibilities between network technologies. The adoption decision of a new technology is often made between close substitutes, but frequently a network technology is composed of complementary components (see, e.g., Economides and White 1994). For instance, a user may decide whether he buys a mobile phone or chooses a traditional wired telephone system. If a user wants – in addition to a telephone – a telefax machine, he needs a telephone line in order to use the fax. Then, the choice of a telephone may also be affected by the demand for complementary technologies as well as their availability, i.e. the supply side conditions. We will discuss here two types of interdependent network technologies: (i) complementary technologies a user can buy separately and (ii) complementary technologies forming an inseparable composite good. We examine how the supply side decisions to produce compatible or incompatible components or products – as a result of divergent competitive strategies of the suppliers of network technologies[1] – affects the optimal timing of the adoption of new network technologies.

1.2   Aims and Scope

Our paper addresses primarily two strands of literature. First, it aims to investigate the issue of network externalities and network compatibility by highlighting the importance of both cross-sectional and dynamic interdependencies in the economic actors' adoption behaviour in the network technology systems. Second, it also offers a contribution to the broader stream of literature on the diffusion of innovations. These issues have in the past decades received considerable attention from both geographers and economists (see, e.g., Bertuglia et al. 1995, Malecki 1992 and Stoneman 1995). Studies on the geographical diffusion of innovations have for a long time mainly considered two elementary dimensions in the spread of new technologies, viz. time and space (see Hägerstrand 1967). Similarly, the standard economic literature has regarded a diffusion process in two dimensions as a spatial spread of innovation among individuals, firms or population over time. In our paper, we extend the traditional diffusion approach by adding another crucial dimension to the diffusion of innovations; compatibility of technological components. We expect that this third dimension has non-negligible implications for the dispersion of technologies, especially in the case of interdependent network technologies.

Table 1 presents three illustrative examples of the supply of different network technologies. Technology A and B represent both a kind of base technology for which specific technologies C^1 and C^2 provide complementary technologies. Several cases may then be distinguished. First, the supplier(s) of technologies may produce both a base technology and a complementary technology and then integrate them into a single composite good (case (i)). The second possibility is the separate production of base technologies and complementary technologies. In this case, the suppliers of compatible technologies may produce either compatible complementarity technologies with both base technologies (case (ii)) or make a complementary technology compatible only with one (or some) of the base technologies (case (iii)). We will argue that these supply side compatibility and production choices are likely to have prominent implications for the adoption behaviour of economic actors and consequently, for the diffusion speed of network technologies in general.

The prevailing literature provides some theoretical insights into the timing of adoption of network technologies related to the agents' choice among substitute

Table 1. Examples of the supply of network technologies

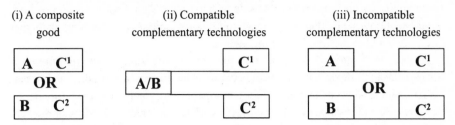

(i) A composite good   (ii) Compatible complementary technologies   (iii) Incompatible complementary technologies

technologies characterized by network externalities (see Choi 1994, Farrell and Saloner 1986, Koski and Nijkamp 1996)[2]. These studies however, do not pay attention to the role of complementarities and compatibility in the adoption of new network technologies. It is also noteworthy that Church and Gandal (1993) consider network externalities related to the complementary technologies, but their static model ignores interdependence between the user generations.

The current literature contains several interesting empirical studies on the adoption behaviour of economic actors in the context of network technologies. Saloner and Shepard (1995) explore the diffusion of automated teller machines (ATM) in the U.S. banking sector in 1970's. Their study focuses on the existence of network effects in the ATM market. Antonelli (1993) provides evidence on the role of externalities and complementarities in the international diffusion of computers and the demand for telecommunications services. He suggests that the high level of interrelatedness and technical complementarity between computers and telecommunications services results in positive externalities. This means that the demand for telecommunications services is positively related to the diffusion of information technologies. Antonelli's data from 30 countries supports his hypothesis. This result indicates that the diffusion of computers is related to the highly complex and interrelated diffusion processes of new network technologies and that the diffusion speed of information technology is likely to exert significant long-run effects beyond the markets for hardware and software.

Gandal (1994) shows that the users attach more value to spreadsheets compatible with the industry standard and to spreadsheets providing links to external databases. His study supports the presence of network externalities in the computer spreadsheet market and stresses the importance of compatibility of technological components to the network users. Another empirical study on the value of compatible network components is provided by Shurmer (1993). His study on the PC software market points out the heterogeneous preferences of network users and consequently, the different relative importance of network externalities of consumers. Gandal et al. (1995) examine the early microcomputer markets. They suggest that the early and late adopters of network technologies may choose different technologies due to consumer heterogeneity and uncertainty about the development of complementary technologies for incompatible substitute technologies. This theoretical proposition is further supported by their empirical exploration of microcomputer markets in the late 1970's and in the first half of the 1980's.

The above discussed studies highlight the significance of complementary components in the adoption and use of network technologies. None of these empirical investigations, however, examines how incompatibility of technological components supplied affect the total demand for these network technologies or the diffusion speed of the new technologies at an aggregate level. Our paper will focus on this issue first by a theoretical analysis and then empirically by exploring cross-national differences in the microcomputer sales in European markets. In summary, our contribution to the existing literature on compatibility and network externalities on the one hand, and to the studies on the diffusion of innovations on the other hand is threefold:

1) Our theoretical model combines both cross-sectional externalities and sequential externalities in the case of complex markets for network technologies offering both substitute and complementary technologies which may be either compatible or incompatible with one another.

2) The paper considers three dimensions – instead of the traditional two ones – of the diffusion of new technologies or innovations: compatibility of technological components, the penetration rate of a new technology, and time.

3) The paper offers empirical evidence on the implications of incompatibility for the diffusion of network technologies.

Section 2 contains a theoretical analysis of the timing of the adoption of complementary network technologies. In section 3, we will first briefly discuss the operating systems market during the past decade and then test the practical propositions – which arise from our theoretical model – on the basis of PC sales data from various European countries. Section 4 concludes the paper with a concise discussion.

---

1. We do not discuss here the potential reasons for and types of the suppliers' divergent compatibility strategies. See, e.g., Economides and White (1994) for a discussion on the firms' horizontal compatibility strategies (i.e. whether to make components compatible with the rivals) and Besen and Farrell (1994) for a discussion on vertical compatibility strategies (i.e. whether to make products compatible with the rivals).
2. We may refer also to Farrell and Saloner (1985) and Katz and Shapiro (1985, 1986) for related studies on technology adoption and standardization in industries characterized by network externalities.

2. Timing of Adoption of Complementary Network Technologies

In this section we will successively discuss separable and inseparable complementary network technologies.

2.1 Complementary Technologies a User Can Buy Separately

Let us assume that the market offers two substitute technologies A and B as well as their complementary technologies C^1 and C^2. We assume that technologies C^1 and C^2 are also substitutes and that the market of technologies C^1 and C^2 exhibits a direct network externality[1] Moreover, we assume that an investment in technology A or B is required before technologies C^1 or C^2 can be utilized (e.g. a user has to buy a computer before he can use a communications programme) and that the market of technology A and B may exhibit only an indirect network externality[2]. We ignore here the effects of an installed user base, information spill-overs and scale economies, since our focus is on the intrinsic nature of network technologies and their effects on the diffusion patterns of new technologies. We distinguish here two cases: in the first case, base technologies A and B are compatible substitutes, whereas in the second case considered, technologies A and B are assumed to be incompatible substitutes.

(i) Base technologies A and B are Compatible Substitutes

We assume that the two types of technologies C^1 and C^2 are incompatible substitutes[3], and that any of them can be linked to either technology A or technology B resulting in a composite good. The possible combinations of complementary technologies are $A+C^1$, $B+C^1$, $A+C^2$ and $B+C^2$ (e.g. A and B are two compatible operating systems both compatible with two communication programmes, C^1 and C^2, which are, however, incompatible with one another). Then, the market of incompatible substitutes C^1 and C^2 exhibits a direct externality implying that an increase in the demand of technology C^1 (C^2) influences the (expected) value of technology C^1 (C^2), and thus its future demand. In addition, the market of complementary technologies $A/B+C^1$ and $A/B+C^2$ exhibits an indirect externality meaning that an increase in the demand of technology A or B is likely to increase the supply of technologies (or network services) compatible with C^1 and C^2 and will thus increase the expected value of these complementary technologies. An increase in the demand of technology A or B does not however, affect the relative shares of the users of technologies C^1 and C^2.

We consider here two time periods, t=1,2 and assume that two users arrive sequentially in the market, user 1 at the beginning of the first period and user 2 at the beginning of the second period. In the first period, user 1 can choose technology A, B, technologies $A/B+C^1$ or $A/B+C^2$ or wait until the second period. If user 1 adopts only technology A/B in the first period (waits until the second period), he will choose between C^1 and C^2 (A/B+C^1 and A/B+C^2) in the second period. Then, the expected values of technologies C^1 and C^2 are likely to increase as well, but the relative shares of the demand of technologies C^1 and C^2 remain unchanged. We assume that the users' preferences may differ across the time periods, but are identical within the period. This results in identical technology choices of users, if user 1 decides to wait until the second period.

We can denote, for simplicity, technologies A and B by one symbol – say A – due to their compatibility. Assume that the stand-alone values of technologies are non-negative and that they are known at the beginning of the periods, but uncertain in the future. We denote the stand-alone value of technologies A, C^1 and C^2 in the first period by α_1, c_1^1 and c_1^2 [24] and in the second period – with the symbols ~ indicating uncertain values of the values conceived – by $\tilde{\alpha}_2$, $\tilde{c}_2^1$ and $\tilde{c}_2^2$, respectively. The external benefit from the compatible technology choice with another user – i.e. the value of network externality related to technologies C^1 and C^2 – is denoted by Δ.

The possible combinations of complementary technologies are A+C^1 and A+C^2. We denote the possible values of the technologies in period 2 by $\Omega = \left\{ \left(\tilde{\alpha}_2, \tilde{c}_2^1, \tilde{c}_2^2 \right); \left(\tilde{\alpha}_2, \tilde{c}_2^1, \tilde{c}_2^2 \right); \in R^3 \right\}$. We define the following subsets in Ω which we will utilize in our calculations below:

$$E_0 = \left\{ \left(\tilde{\alpha}_2, \tilde{c}_2^1, \tilde{c}_2^2 \right); \tilde{\alpha}_2 + \tilde{c}_2^1 + \Delta \geq \tilde{\alpha}_2 + \tilde{c}_2^2 + \Delta \right.$$

$$\left. \Leftrightarrow c_2^1 \geq c_2^2; \left(\tilde{\alpha}_2, \tilde{c}_2^1, \tilde{c}_2^2 \right) \in R^3 \right\}$$

$$E_1 = \left\{ \left(\tilde{\alpha}_2, \tilde{c}_2^1, \tilde{c}_2^2 \right); \tilde{c}_2^1 + \Delta \geq \tilde{c}_2^2; \left(\tilde{\alpha}_2, \tilde{c}_2^1, \tilde{c}_2^2 \right) \in R^3 \right\}$$

$$E_2 = \left\{ \left(\tilde{\alpha}_2, \tilde{c}_2^1, \tilde{c}_2^2 \right); \tilde{c}_2^1 + 2\Delta \geq \tilde{c}_2^2; \left(\tilde{\alpha}_2, \tilde{c}_2^1, \tilde{c}_2^2 \right) \in R^3 \right\},$$

where $E_0 \subseteq E_1 \subseteq E_2$. The complement of subset E_i with respect to space Ω is denoted by E_i^c.

We use backward induction for modelling the actors' behaviour in regard to the adoption of a new network technology. The technology choices of user 2 depend on the choices of user 1 in the first period. If user 1 adopts both technology A and technology C^1 in the first period, user 2 chooses the same combination of technologies if and only if:

$$\tilde{\alpha}_2 + \tilde{c}_2^1 + \Delta \geq \tilde{\alpha}_2 + \tilde{c}_2^2 \Leftrightarrow \tilde{c}_2^1 + \Delta \geq \tilde{c}_2^2. \quad (1)$$

Otherwise, he will choose technologies A and C^2. If user 1, instead, decides to wait until the second period or adopts just the base technology A – which does not exhibit direct network externalities – both users will choose technologies A and C^1 if and only if:

$$\tilde{\alpha}_2 + \tilde{c}_2^1 + \Delta \geq \tilde{\alpha}_2 + \tilde{c}_2^2 + \Delta \Leftrightarrow \tilde{c}_2^1 \geq \tilde{c}_2^2. \quad (2)$$

Now the expected private and social value of adoption of technology A by user 1 in the first period ($V(A)_1$ and $S(A)_1$), the expected private and social value of adopting technology $A+C^1$ in the first period ($V(A+C^1)_1$ and $S(A+C^1)_1$) and the expected private and social value of waiting until the second period ($V(W)_1$ and $S(W)_1$) can be written as follows[5]:

$$V(A)_1 = \alpha_1 + \delta \left[\int_{E_0} (\tilde{\alpha}_2 + \tilde{c}_2^1 + \Delta) dG(\tilde{\alpha}_2, \tilde{c}_2^1, \tilde{c}_2^2) \right.$$
$$\left. + \int_{E_0^c} (\tilde{\alpha}_2 + \tilde{c}_2^2 + \Delta) dG(\tilde{\alpha}_2, \tilde{c}_2^1, \tilde{c}_2^2) \right] \tag{3}$$

$$S(A)_1 = \alpha_1 + \delta \left[\int_{E_0} 2(\tilde{\alpha}_2 + \tilde{c}_2^1 + \Delta) dG(\tilde{\alpha}_2, \tilde{c}_2^1, \tilde{c}_2^2) \right.$$
$$\left. + \int_{E_0^c} 2(\tilde{\alpha}_2 + \tilde{c}_2^2 + \Delta) dG(\tilde{\alpha}_2, \tilde{c}_2^1, \tilde{c}_2^2) \right] \tag{4}$$

$$V(A + C^1)_1 = \alpha_1 + c_1^1 + \delta \left[\int_{E_1} (\tilde{\alpha}_2 + \tilde{c}_2^1 + \Delta) dG(\tilde{\alpha}_2, \tilde{c}_2^1, \tilde{c}_2^2) \right.$$
$$\left. + \int_{E_1^c} (\tilde{\alpha}_2 + \tilde{c}_2^1) dG(\tilde{\alpha}_2, \tilde{c}_2^1, \tilde{c}_2^2) \right] \tag{5}$$

$$S(A + C^1)_1 = \alpha_1 + c_1^1 + \delta \left[\int_{E_2} 2(\tilde{\alpha}_2 + \tilde{c}_2^1 + \Delta) dG(\tilde{\alpha}_2, \tilde{c}_2^1, \tilde{c}_2^2) \right.$$
$$\left. + \int_{E_2^c} (2\tilde{\alpha}_2 + \tilde{c}_2^1 + \tilde{c}_2^2) dG(\tilde{\alpha}_2, \tilde{c}_2^1, \tilde{c}_2^2) \right] \tag{6}$$

$$V(W)_1 = \delta \left[\int_{E_0} (\tilde{\alpha}_2 + \tilde{c}_2^1 + \Delta) dG(\tilde{\alpha}_2, \tilde{c}_2^1, \tilde{c}_2^2) \right.$$
$$\left. + \int_{E_0^c} (\tilde{\alpha}_2 + \tilde{c}_2^2 + \Delta) dG(\tilde{\alpha}_2, \tilde{c}_2^1, \tilde{c}_2^2) \right] \tag{7}$$

$$S(W)_1 = \delta \left[\int_{E_0} 2(\tilde{\alpha}_2 + \tilde{c}_2^1 + \Delta) dG(\tilde{\alpha}_2, \tilde{c}_2^1, \tilde{c}_2^2) \right.$$
$$\left. + \int_{E_0^c} 2(\tilde{\alpha}_2 + \tilde{c}_2^2 + \Delta) dG(\tilde{\alpha}_2, \tilde{c}_2^1, \tilde{c}_2^2) \right] \tag{8}$$

In the above equations, $G(\tilde{\alpha}_1, \tilde{c}_2^1, \tilde{c}_2^2)$ denotes a joint probability distribution of the value of technologies – which will be written shortly as $G(\cdot)$ from now on – and δ is the discount factor. We may also derive the corresponding equations for the adoption decision regarding $A+C^2$, but since this technology combination is symmetric with $A+C^1$, it is sufficient to consider only one of these technology combinations. In period 1, a private agent will choose max $[V(A)_1, V(A+C^1)_1, V(A+C^2)_1, V(W)_1]$, whereas a social planner will choose max $[S(A)_1, S(A+C^1)_1, S(A+C^2)_1, S(W)_1]$.

Next, we compare the private and social incentives to wait until the second period. We can see straigthforward that if $\alpha_1 \geq 0$, then $V(A)_1 \geq V(W)_1$, and thus user 1 prefers the adoption of technology A to waiting. The timing of the adoption of technology A is also optimal from the society point of view, as it can be easily calculated that $[S(W)_1-V(W)_1]-[S(A)_1-V(A)_1]=0$.[6] It can be also easily calculated – since $c_1^1 \geq 0$ by definition – that $S(A+C^1)_1 \geq S(A)_1$ and $V(A+C^1)_1 \geq V(A)_1$. This means that both the social planner and the private agent prefer the adoption of a base technology plus a complementary technology in the first period. We may thus conclude that, in order to compare private and social incentives to wait, it is sufficient to calculate the difference between $[S(W)_1-S(A+C^1)_1]$ and $[V(W)_1-V(A+C^1)_1]$. Note that $\left[E_1 - E_0\right] = \left[E_0^c - E_1^c\right]$ and $\left[E_2 - E_1\right] = \left[E_1^c - E_2^c\right]$.
Then, $[S(W)_1-V(W)_1]-[S(A+C^1)_1-V(A+C^1)_1]$:

$$= \delta \quad \int_{E_1-E_0}(\tilde{c}_2^2 - \tilde{c}_2^1)dG(\cdot) + \int_{E_2-E_1}(\tilde{c}_2^2 - \tilde{c}_2^1 - \Delta)dG(\cdot) + \int_{E_2^c}(\Delta)dG(\cdot)] \geq 0.$$

(9)

(See for more detailed calculations in Annex 1)

The sign of equation (9) is unambiguously positive due to the definitions of $\left[E_1 - E_0\right]$, $\left[E_2 - E_1\right]$ and E_2^c. This implies an inefficiently fast adoption of a complementary technology, but it should be kept in mind that the analysis ignores several factors characteristic to network technologies – like information spill-overs – which are likely to influence on the optimal timing of the adoption of technologies (see Koski and Nijkamp 1996). Our aim here however is not to focus on the question whether the diffusion of new network technologies is likely to be too slow or fast from the society's point of view. Our calculations point out that the divergence between private and social incentives to wait – i.e. the optimal timing of adoption of network technologies – depends merely on the value of and network externalities related to the components which exhibit direct network externalities. The stand-alone values of complementary technologies C^1 and C^2 – not the stand-alone values of base technologies A and B – and network externalities related to these complementary technologies affect the optimal timing of the adoption of the technologies.[7] We could derive analogous results in respect to technology C^2 and taking into account technology B which is compatible with technology A.

(ii) Base Technologies A and B are Incompatible Substitutes

In this second case, we assume that complementary technologies C^1 and C^2 are also incompatible substitutes, but now they are only compatible with specific base technologies such that technology C^1 can only be linked with technology A, whereas technology C^2 can form a composite good only with technology B (e.g. A and B are two incompatible operating systems which are compatible with incompatible communication programmes, C^1 and C^2, respectively). This means that in the first period, user 1 is able to choose technology A, B, technologies $A+C^1$ or $B+C^2$ or to wait until the second period. We can show that even if technologies A and B do not exhibit direct network externalities, the private and social incentives to postpone the adoption decision of complementary technologies depends on the values of relevant factors (e.g. information spill-overs) related to all components A, B, C^1 and C^2 due to the incompatibility of technologies A and B.

Again, the market of technologies C^1 and C^2 exhibits a direct externality and the market of complementary technologies $A+C^1$ and $B+C^2$ exhibits an indirect externality meaning that an increase in the demand of technologies A (B) is likely to increase – due to an increase in the supply and the demand of compatible technologies – the expected value of technology C^1 (C^2). As technology A is incompatible with technology B, the diffusion of complementary technologies C^1 and C^2 is closely related to the diffusion of technologies A and B. Now, it does make a difference whether the early users adopt technology A or technology B and their decisions are also likely to affect the relative shares of the users of technologies C^1 and C^2. We define the following subsets in Ω:

$$E_0 = \left\{ (\tilde{\alpha}_2, \tilde{\beta}_2, \tilde{c}_2^1, \tilde{c}_2^2); \tilde{\alpha}_2 + \tilde{c}_2^1 + \Delta \geq \tilde{\beta}_2 + \tilde{c}_2^2 + \Delta; \ (\tilde{\alpha}_2, \tilde{\beta}_2, \tilde{c}_2^1, \tilde{c}_2^2) \in R^4 \right\}$$

$$E_1 = \left\{ (\tilde{\alpha}_2, \tilde{\beta}_2, \tilde{c}_2^1, \tilde{c}_2^2); \tilde{\alpha}_2 + \tilde{c}_2^1 + \Delta \geq \tilde{\beta}_2 + \tilde{c}_2^2; \ (\tilde{\alpha}_2, \tilde{\beta}_2, \tilde{c}_2^1, \tilde{c}_2^2) \in R^4 \right\}$$

$$E_2 = \left\{ (\tilde{\alpha}_2, \tilde{\beta}_2, \tilde{c}_2^1, \tilde{c}_2^2); \tilde{\alpha}_2 + \tilde{c}_2^1 + 2\Delta \geq \tilde{\beta}_2 + \tilde{c}_2^2; \ (\tilde{\alpha}_2, \tilde{\beta}_2, \tilde{c}_2^1, \tilde{c}_2^2) \in R^4 \right\}$$

Consider first the adoption decision regarding combination $A+C^1$. Now, the respective expected private and social values of adopting technology(/ies) in the first period and the expected private and social values of waiting are:

$$V(A)_1 = \alpha_1 + \delta[\int_{E_1} (\tilde{\alpha}_2 + \tilde{c}_2^1 + \Delta)dG(\cdot) + \int_{E_1^c} (\tilde{\alpha}_2 + \tilde{c}_2^1)dG(\cdot)]$$

$$S(A)_1 = \alpha_1 + \delta[\int_{E_2} 2(\tilde{\alpha}_2 + \tilde{c}_2^1 + \Delta)dG(\cdot) + \int_{E_2^c} (\tilde{\alpha}_2 + \tilde{\beta}_2 + \tilde{c}_2^1 + \tilde{c}_2^2)dG(\cdot)]$$

$$V(A + C^1)_1 = \alpha_1 + c_1^1 + \delta[\int_{E_1} (\tilde{\alpha}_2 + \tilde{c}_2^1 + \Delta)dG(\cdot) + \int_{E_1^c} (\tilde{\alpha}_2 + \tilde{c}_2^1)dG(\cdot)]$$

$$S(A + C^1)_1 =$$

$$\alpha_1 + c_1^1 + \delta[\int_{E_2} 2(\tilde{\alpha}_2 + \tilde{c}_2^1 + \Delta)dG(\cdot) + \int_{E_2^c} (\tilde{\alpha}_2 + \tilde{\beta}_{2\ +} \tilde{c}_2^1 + \tilde{c}_2^2)dG(\cdot)]$$

$$V(W)_1 = \delta[\int_{E_0} (\tilde{\alpha}_2 + \tilde{c}_2^1 + \Delta)dG(\cdot) + \int_{E_0^c} (\tilde{\beta}_2 + \tilde{c}_2^2 + \Delta)dG(\cdot)]$$

$$S(W)_1 = \delta[\int_{E_0} 2(\tilde{\alpha}_2 + \tilde{c}_2^1 + \Delta)dG(\cdot) + \int_{E_0^c} 2(\tilde{\beta}_2 + \tilde{c}_2^2 + \Delta)dG(\cdot)],$$

where $dG(\tilde{\alpha}_2, \tilde{\beta}_2, \tilde{c}_2^1, \tilde{c}_2^2) = dG(\cdot)$. Next, we compare the private and social incentives to wait. As $[S(A+C^1)_1 - S(A)_1] = [V(A+C^1)_1 - V(A)_1] = c_1^1$, it follows straightforward that $S(A+C^1)_1 \geq S(A)_1$ and $V(A+C^1)_1 \geq V(A)_1$. Thus, as above, it is sufficient to calculate merely the difference between $[S(W)_1 - S(A+C^1)_1]$ and $[V(W)_1 - V(A+C^1)_1]$ (see Annex 1 for more detailed calculations):

$$
\delta^{-1} [S(W) - V(W)] - [S(A+C^1) - V(A+C^1)] =
$$
$$
\int (\tilde{\beta}_2 - \tilde{\alpha}_2 + \tilde{c}_2^2 - \tilde{c}_2^1 - \Delta)dG(\cdot) +
$$
$$
\int_{E_0}^{E_2-E_1} (\tilde{\alpha}_2 + \tilde{c}_2^1 - \tilde{\beta}_2 - \tilde{c}_2^2)dG(\cdot) + \int_{E_1} (\tilde{\beta}_2 + \tilde{c}_2^2)dG(\cdot) + \int_{E_2^c} (\Delta)dG(\cdot) \qquad .(10)
$$

Now, the divergence between private and social incentives to wait depends on the values of all components A, B, C^1 and C^2. Also, other relevant factors characteristic to the network technologies – e.g. an installed base of users of any component, scale economies or information spill-overs related to them – affect the optimal timing of the adoption of the complementary technologies irrespective of whether a component exhibits direct or indirect network externalities. Similar conclusions may be found – due to the symmetry of technologies –, if we compare $S(W)-V(W)$ to $S(B+C^2)-V(B+C^2)$.

These results indicate that the compatibility strategies of the suppliers of network technologies may also influence the optimal timing of the adoption of the technologies which do not exhibit direct network externalities. These components – even though they do exhibit indirect network externalities – may be adopted at the optimal time from a social point of view as long as they are compatible with their substitutes. However, if these technologies are incompatible, their adoption may occur too early or too late from the society's point of view.

2.2 Complementary Technologies Forming an Inseparable Composite Good

Consider next the case of complementary technologies which a user cannot buy separately, since the technologies are included as components in a single composite good. Assume that the market offers two incompatible substitutes, AC^1 and BC^2, characterized by direct network externalities. Technology AC^1 (BC^2) is composed of two inseparable components A and C^1 (B and C^2). We assume that technologies A and B are a kind of base technologies, whereas technologies C^1 and C^2 represent accessories increasing the value of a composite good (e.g. car and safety airbag).

The mathematical models for the timing of the adoption of technologies AC^1 and BC^2 are identical to the ones presented in earlier studies (see Choi 1994 and Koski and Nijkamp 1996), apart from the fact that here – instead of two independent incompatible substitutes – we consider two incompatible substitutes which are composed of inseparable components. The complementary technologies which can be bought separately are composed of the two components as well, but the results differ – due to the intrinsic divergencies of the technologies – from the ones presented above. Now, the expected total value of the components and network externalities related to the composite good – rather than the values of externalities related to the single components – determine the optimal adoption time of a new technology. Likewise, the expected information spill-overs regarding the new technologies, the scale economies in production and the installed bases of users determine their contribution to the values of entire products, not just to the values of their components.

In conclusion, we have shown in this section that the intrinsic nature of network technologies – i.e. whether these technologies are substitutes or complements and whether the components of a technology are separable or inseparable – as well as the supply side choices regarding the compatibility of new network technologies or their complements do affect the optimality of the timing of the adoption of a new technology, even if the technologies do not exhibit direct network externalities. In the next section, we will first briefly discuss empirical implications of our theory. Then, we will concisely consider the previous history of the operating systems market (which provides data for our empirical application). And finally, we will present empirical results which test some of the propositions emerging from our theoretical exposition.

---

1. Direct positive (negative) network externality means that existing users' utilities or profits increase (decrease) as additional users join the network. Here, we consider only positive network externalities.

2. Indirect externality arises due to the complementarity of products. As the number of the users of a product increases, it is likely that also the supply and variety of the complementary products increases.

3. Technologies C^1 and C^2 may represent also two groups of technologies or network services such that the technologies within a given group are compatible, but incompatible with the technologies in the other group.

4. The subscript denotes here time, while the superscript denotes the type of a technology.

5. The subscript 1 after the letter combinations denoting private and social values indicates that these values reflect the choices of user 1.

6. It is proven by Choi (1994) that an incentive for society to wait until the second period is higher than an incentive for an individual to wait – in case of technology A – if $[S(W)_1 - V(W)_1 \geq [S(A)_1 - V(A)_1]$.

7. Similarly, taking into account an installed base of users of technologies, scale economies and information spill-overs would result in similar conclusions: only the network size of components C^1 and C^2, the order of magnitude of information spill-overs and scale economies related to technologies C^1 and C^2 matter in determining the optimal timing of the adoption of complementary network technologies.

3. Empirical Evidence from the European PC Market

3.1 Practical Suggestions of the Theory

Our theoretical examination above provides the following practical suggestions for applied work:
1) The (expected) total value of technological components determine the diffusion of network technologies which form a composite good.
2) If complementary technologies are compatible with all technologies supplied on the market, then the adoption of components which do not exhibit network externalities may take place at the optimal time irrespective of uncertainty related to the relative qualities or stand-alone values of their complementary technologies.
3) If complementary technologies are incompatible, the diffusion of all components is affected by the expected diffusion of all other components and their (expected) stand-alone values. Thus, the adoption of technologies which do not exhibit any direct network externalities may take place at the non-optimal time as well.

All of these proposition are however, not directly testable with the databases we have access to. Our data on the microcomputer market provides information on the diffusion of a network technology which does not exhibit (direct) network externalities itself, viz. operating systems. Due to (in)compatibility choices of the producers of operating systems, the market offered mainly two operating system types (in the period of 1985-1994) which were able to run only their own software programs that were incompatible with the programmes run by another operating system. We assume that there exists positive network externalities related to the use of compatible software programs (e.g. via the change of files or information). We do not have direct information on the expectations of economic actors regarding the intrinsic values of different operating systems or their expectations on the value and availability of software programs. However, we have the sales information regarding incompatible operating systems. This data allows a kind of partial empirical analysis of our theory; we are able to test some of the propositions our theoretical examination have pointed out.

We assume that the more heterogeneous the technology choices of actors in an economy are, the more uncertainty the subsequent potential technology adopters face. Since an increase in uncertainty is reflected as a decrease in expected utility, enhanced uncertainty means that the expected value of waiting exceeds the value of adopting the technology in the majority of cases and consequently, more people are likely to postpone their adoption decision. Based on this reasoning, we will test the following hypotheses: (i) the diffusion speed of computers among countries varies with the degree of compatibility of microcomputers sold, and (ii) the higher (lower) the degree of compatibility of microcomputers sold, the higher (lower) the diffusion speed of microcomputers in general. Before we present the results of our empirical examination, we will in the next section shortly discuss the recent history of the microcomputer markets and the data we have used in the empirical estimations.

3.2 The PC Market from the 1980's to 1994

Our empirical exploration considers the microcomputer market from the 1980's to 1994. Several suppliers competed in the operating systems market until 1985. The dominating systems in the computer market at the early 80's were Apple, MS-DOS and CP/M.[1] The early PC users were mostly technically sophisticated hobbyists. In 1984, Apple launched its revolutionary Apple MacIntosh computers. At the same time CP/M continuously lost its market share to MS-DOS and eventually, the CP/M computers were supplanted by 1985. The study of Gandal et al. (1995) indicates that this outcome – the success of MS-DOS compared to CP/M – emerged as a result of the availability of complementary software: the market offered a larger number of MS-DOS compatible software than CP/M compatible software to the computer users. Two operating systems dominated the markets for operating systems from 1985 to 1994: MS-DOS and Apple. New computers were more user friendly and consequently, the diffusion of PCs from expert use to the more general purpose use intensified. It is argued that Apple held technical leadership in personal computer technology for most of the ten years' period 1984-1994[2]. However, it adopted a business strategy which appeared to be unsuccessful and which have been accused to be a main reason for the Apple's relatively small market share in the world PC market: Apple chose to produce machines with closed architecture. This decision meant that the computers with MS-DOS operating system were not able to run the software programs developed for the Apple machines and vice versa. Thus, a consumer buying either an Apple computer or a MS-DOS computer faced the risk of an uncertain supply of compatible software programs in the future. The expected future supply of software was likely to be determined by the relative success and the diffusion of these operating systems. The theory we presented in section 2 indicates that the diffusion of network technologies is likely to be influenced by the incompatibility of technological components – even if they do not exhibit direct network externalities –, when technologies make up a composite good with the complementary components exhibiting direct network externalities. In the spatial context, this means that it is possible – when the markets offer incompatible network technologies – that the speed of diffusion of new technologies in a certain region or country depends on how homogeneous or compatible the technologies chosen by the adopters are. A wider variety of incompatible network technologies may result in a slower diffusion of a network technology in general.

The PC adopters between 1985 and 1994 were "locked-in" to their operating systems' choice in the sense that they were not able to use compatible software programmes and, for instance, to share files with other PC users having different operating system in their computers.[3] We argue that this incompatibility of the operating systems engendered further uncertainty in the PC market and influenced the adoption behaviour of economic agents. In particular, we may expect that the higher the incompatibility of the adopted PCs, the slower the diffusion of computers in general. Better availability of software is often pointed out as an explanation for the success of MS-DOS compared to the Apple. Our aim here however, is not to analyse the reasons for the relative success of divergent network technologies, but to focus on the implications of incompatibility to the diffusion speed of new technologies in the aggregate level. We test our hypothesis on the negative relationship between technological incompatibility and the diffusion speed of new technologies by the PC sales data from various European countries.

The microelectronics industry has experienced drastic technological progress during the last decade. Simultaneously with a continuous decline in computer prices, the markets have constantly introduced better and more efficient microcomputer models. Due to this fast stage of development we are not able to measure directly the diffusion speed of PCs in the relevant countries; we do not have information on the number of new users of computers and the number of disposed machines and consequently, we do not have exact information on the size of an installed base of computers. We will use, instead, the change in the PC sales – or PC sales growth – as a proxy for the change in the penetration rate of microcomputers in an economy.

Our PC sales data is based on the IDC (International Data Corporation) microcomputer sales statistics and covers the period 1985-1994. The countries considered are Finland, Netherlands, Germany, France, United Kingdom, Spain, Sweden and Italy. The database comprises the annual number of Apple computers sold and the annual number of microcomputers sold in total in each country.[4] Since the MS-DOS operating system and its compatible clones have covered – when the market

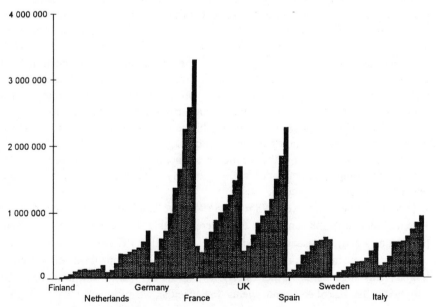

Figure 1. The number of PCs sold (1985-1994).
Note: the data do not include the sales of 8-bit computers.   Source: IDC, 1996.

share of Apple is excluded – almost the rest of the market we suppose that the difference between the total number of computers sold and the number of Apples sold provides a satisfactory proxy for the compatible computer base.

Figure 1 and 2 reflect a general trend in the microcomputer market over the period of 1985-1994: the annual number of computers sold has constantly increased during the period. Figure 1 describes the number of computers sold per year in the sampled countries. In 1985, less than 2 million PCs were sold in total, whereas the corresponding number in 1994 was over 10 million. This means that – on the aggregate – the annual computer sales has increased five-fold in a decade. Figure 1 shows the differences in the volume of computer sales among the countries. The volume of computer sales is however, a rather poor measure for the diffusion of microcomputers in a cross-national comparison. The country sizes vary a lot and the sales volume do not supply any information on the penetration rate of information technology in the economies. In order to be able to compare the relative PC sales of the countries concerned, we divided the number of PCs sold by population[5]. From figure 2 we can see that the divergencies in the computer sales relative to population are less dramatic than the differences in the volume of PCs sold in the sample countries.

We will use here the (logarithmic) difference in the number of PCs sold in relation to the population ("DPC") as a dependent variable for two reasons. First, the series of PC sales per population appeared to be a difference-stationary process, which

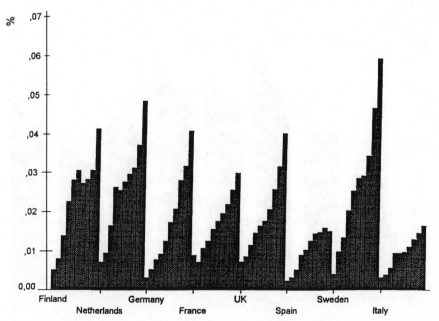

Figure 2. The number of PCs sold per population (1985-1994).
Note: the data do not include the sales of 8-bit computers.   Source: IDC, 1996.

implies that the best method for eliminating the trend is differentiation. We used the Dickey-Fuller test for testing whether the time series is trend-stationary or difference-stationary.[6] In other words, we tested the hypothesis H_0: $\rho=1$ and $\beta=0$ in the following equation:

$$PC_t = \alpha + \rho PC_{t-1} + \beta t + \varepsilon_t,$$

where $PC_t(PC_{t-1})$ describes the PC sales per population at time t (t-1). On the basis of our estimation results, we calculated the following t-test value: t = $(\hat{p}-1)/SE(\hat{p}) \approx -2.02$. This value indicates, when it is compared to the corresponding critical t-value, that hypothesis H_0 cannot be rejected; the series is clearly difference-stationary. Another reason for using differentiated sales data arose from the interpretation of the first difference as a measure of change (or speed). This transformation corresponds to our intention to explore cross-national divergencies in the microcomputer sales growth.

We measure the compatibility of PCs sold by the (log) total number of computers sold per year divided by the number of Apples sold per year. Consequently, a higher value of the variable indicates higher compatibility. Figure 3 suggests an increase in compatibility over time in all countries. The Apple sales data from 1985 to 1987 were available only from Finland and Sweden. We used the first difference of compatibility (COMP1) and the second difference of compatibility (COMP2) as explanatory

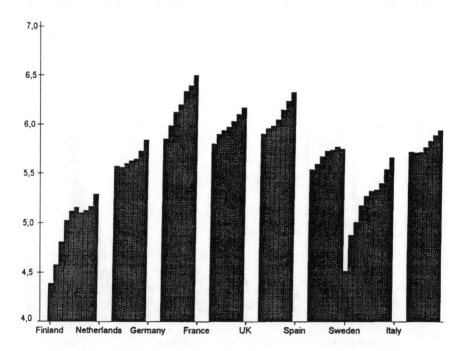

Figure 3. Compatibility of PCs sold (1985-1994).
Note: the data do not include the sales of 8-bit computers.   Source: IDC, 1996.

variables. In addition to the microcomputer sales data, we also used the following potentially relevant variables affecting the diffusion of PCs as explanatory variables:

- PRICE = (log) price index for the office machinery and computers in Finland, 1985-1994 (Statistics Finland). We suppose that the computer prices are negatively related to the demand for microcomputers. We apply the index describing the development of computer prices in Finland for the rest of the countries, as this is in our opinion a rather good proxy for the world wide change in computer prices during the year 1985-1994, so that we may then be able to capture the dynamic relationship of demand for the computers and the PC prices.

It also seems plausible that the demand for microcomputers is positively related to the income level. As the demand for computers is composed of two components – the final demand of the household and the derived demand of the enterprises – we use the following three measures of income to explore this relationship:

- GDP = gross domestic product at market prices in purchasing power parities.[7] Gross domestic product converted in purchasing power standards eliminates the effect of cross-national differences in the price leves and enables a comparison of the volume of goods and services produced and used in different countries. This measure does not separate the demand of the households and firms. Earlier evidence is provided by Antonelli (1993) whose empirical exploration points out a positive relationship between GDP and intensity of sales of telecommunications services.

- EINC = external balance of property and entrepreneurial income in current prices (% of GNP).[8] This measure of net operating surplus comprises total property and entrepreneurial income from production. It is used for exploring the relationship between the demand for computers and entrepreneurial income.

- W = (log) compensation of employees at current prices and purchasing power parities.[9] This measure provides a measure of the households' income. For example, the study of Antonelli (1989) on the regional diffusion of telefaxes and modems indicates that the wage level of the labour force is positively related to the diffusion of information technology.

3.3   Econometric Results

We assumed that heteroskedasticity between the observation groups might be possible in our panel data set of eight countries during the ten years period. However, the Lagrange multiplier test rejected this hypothesis; it favoured the ordinary least squares model to the random effect model in all cases (see results in table 1).[10] It seemed intuitively plausible that three different income indicators may be highly correlated and correspondingly, their estimation in the same model may cause the problem of near collinearity with the unreliable estimates. To explore the relationship between income variables, we calculated the Pearson correlation coefficients. The correlation coefficient between EINC and GDP and EINC and W was over 0.60, and between W and GDP as high as 0.94 indicating near collinearity. Thus, we estimated the income variables separately in the following three models:

$$DPC = \alpha_1 + \beta_1 DCOMP + \beta_2 D2COMP + \beta_3 PRICE + \beta_4 EINC \qquad \text{(Model 1)},$$
$$DPC = \alpha_1 + \beta_1 DCOMP + \beta_2 D2COMP + \beta_3 PRICE + \beta_4 W \qquad \text{(Model 2)},$$
$$DPC = \alpha_1 + \beta_1 DCOMP + \beta_2 D2COMP + \beta_3 PRICE + \beta_4 GDP \qquad \text{(Model 3)}.$$

We will further test the specification of our empirical models by model 4 which excludes all income variables: $DPC = \alpha_1 + \beta_1 DCOMP + \beta_2 D2COMP + \beta_3 PRICE$ (Model 4).

Table 2 presents the estimation results of the models. All models support our hypothesis on the positive relationship between the microcomputer sales growth and compatibility of PCs sold. The first difference of the variable reflecting the degree of compatibility is highly significant, whereas as the second difference of the variable

Table 2. The OLS estimates of the models for the PC sales

Variables	Model 1	Model 2	Model 3	Model 4
constant	20.212	22.429	22.493	27.557
	(2.235)	(2.356)	(2.458)	(3.908)
comp1	4.2610	4.2362	4.2453	4.6922
	(3.165)	(3.169)	(3.207)	(3.894)
comp2	0.72728	0.71811	0.71350	0.79589
	(1.051)	(1.047)	(1.051)	(1.274)
price	-4.4398	-4.8466	-4.8950	-6.0592
	(-2.222)	(-2.341)	(-2.433)	(-3.883)
inc	-0.014726			
	(-0.281)			
w		-0.064032		
		(-0.732)		
gdp			-0.092643	
			(-1.094)	
Nobs	38	38	38	44
R^2	0.66	0.67	0.67	0.72
LM-test	2.18*)	1.84*)	1.66*)	5.09**)
LR-test 41.17	41.69	42.43	57.99	

Note: t-value in the brackets
*) Favours OLS to random effect model by p-value >0.10.
**) Favours OLS to random effect model by p-value >0.01.

appears to be statistically insignificant. The higher significance of the first difference indicates that the adoption behaviour of economic actors in the microcomputer market is more affected by the short-run dynamics or quite recent PC purchasing decisions of other actors than the technology choices made in the past (even two years before). This finding is probably related to the rapid progress in the microelectronics industry, which is reflected as continuously changing information of the microcomputer market the economic actors' use in their adoption decisions.

The price of microcomputers, as expected, is negatively and statistically significantly related to the demand for microcomputers in all models. High coefficients of the estimates of the price variable indicate that the demand for computers is highly price-elastic. However, we should keep in mind that the price variable used describes the change in microcomputer prices in one of the sample countries and may only roughly approximate the real price changes in the other countries. The estimates of all income variables (inc, w, gdp) appear to be statistically insignificant. Even if all models succeed fairly well in explaining the phenomenon both measured by LR-test values and by R^2, it seems that model 4 – which excludes all income variables – provides a better fit than any of the models from 1 to 3. These results indicate that our aggregate income variables are not able to explain cross-national differences in the microcomputer sales among the sample countries.

Our estimation results suggest that the diffusion speed of microcomputers vary with the compatibility of microcomputers sold. Moreover, it seems that incompatibility of technological components is negatively related to the diffusion speed of network technologies in general. When the market offer incompatible microcomputers, the total demand for computers is likely to be hindered. This may happen due to uncertainty related to the technology choices of later user generations and uncertain developments of compatible software programmes. The PC prices play also a remarkable role in the diffusion of microcomputers. The relationships between income variables and the PC sales – at least in the aggregate level – are less apparent. The microlevel data on the economic actors' investment behaviour in information technology systems might provide more accurate information in this respect and may be able to better separate some income groups.

---

1. See Gandal et al. (1995) for a discussion on the early operating systems market.

2. The Economist, December 10th 1994, p. 23.

3. In September 1994, Apple announced that it is willing to license its operating system.

4. The data do not include the sales of 8-bit computers.

5. Source: Eurostat Yearbook 1995.

6. If a series is trend-stationary, the trend may be eliminated by regressing the series – in addition to other explanatory variables – by time. In case of difference-stationary series, the usual least square theory is not valid; differentiation is necessary for eliminating the trend and for obtaining efficient estimates.

4. Conclusions

Our analysis suggests that inefficiencies may emerge in the timing of the adoption of network technologies which do not exhibit (direct) network externalities. The supply of incompatible complementary technologies to the technologies characterized by direct network externalities, may give rise to non-optimal diffusion patterns of network technologies. As long as complementary network technologies are compatible, the diffusion of components which do not exhibit direct network externalities may be optimal from the society's point of view.

Empirical evidence from the European microcomputer market supports our hypothesis that the diffusion speed of computers among countries vary with the degree of compatibility of microcomputers sold. Also, the data suggest that the higher (lower) the degree of compatibility of microcomputers sold, the higher (lower) the diffusion speed of microcomputers in general. This empirical finding indicates that the incompatibility of technological components may have substantial implications for the spread of network technologies in general. Consequently, the issue of compatibility deserves specific attention in the practical technology and network policy decisions.

---

7. Source: Eurostat Yearbook '95.

8. Source: Eurostat Yearbook '95.

9. Source: Eurostat 2 C. National Accounts ESA 1970-1993. Compensation of employees includes all payments in cash and kind by employers in remuneration for the work done by their employees during the relevant period (Eurostat Yearbook '95).

10. A Lagrange multiplier test is based on the OLS residuals. The hypothesis tested is:

$H_0: \sigma_u^2 = 0, \ H_1: \sigma_u^2 \neq 0$ and the LM test statistic is of the form:

$$LM = \frac{nT}{2(T-1)} \left[\frac{\left[\sum_i \left(\sum_t e_{it} \right)^2 \right]}{\sum_i \sum_t e_{it}^2} \right]^2 . \quad \text{(Greene 1993)}$$

Annex 1
(i) Base technologies A and B are compatible substitutes

$$[S(W)_1 - V(W)_1] - [S(A + C^1)_1 - V(A + C^1)_1]$$

$$= \delta[\int_{E_0}(\tilde{\alpha}_2 + \tilde{c}_2^1 + \Delta)dG(\cdot) + \int_{E_0^c}(\tilde{\alpha}_2 + \tilde{c}_2^2 + \Delta)dG(\cdot)$$

$$+ \int_{E_1^c}(\tilde{\alpha}_2 + \tilde{c}_2^2 + \Delta)dG(\cdot)] + c_1^1$$

$$= \delta \cdot \Delta + \delta[\int_{E_0}(\tilde{\alpha}_2 + \tilde{c}_2^1)dG(\cdot) + \int_{E_1 - E_0}(\tilde{\alpha}_2 + \tilde{c}_2^2)dG(\cdot)$$

$$+ \int_{E_1^c}(\tilde{\alpha}_2 + \tilde{c}_2^2)dG(\cdot) - \int_{E_2 - E_1}(\tilde{\alpha}_2 + \tilde{c}_2^1 + \Delta)dG(\cdot)$$

$$- \int_{E_2}(\tilde{\alpha}_2 + \tilde{c}_2^1 + \Delta)dG(\cdot) + \int_{E_2 - E_1}(2\tilde{\alpha}_2 + \tilde{c}_2^1 + \tilde{c}_2^2)dG(\cdot)$$

$$+ \int_{E_1^c}(2\tilde{\alpha}_2 + \tilde{c}_2^1 + \tilde{c}_2^2)dG(\cdot) + \int_{E_1}(\tilde{\alpha}_2 + \tilde{c}_2^1 + \Delta)dG(\cdot)$$

$$- \int_{E_1 - E_0}(\tilde{\alpha}_2 + \tilde{c}_2^1)dG(\cdot) + \int_{E_0^c}(\tilde{\alpha}_2 + \tilde{c}_2^1)dG(\cdot)]$$

$$= \delta[\int_{E_1 - E_0}(\tilde{c}_2^2 - \tilde{c}_2^1)dG(\cdot) + \int_{E_2 - E_1}(\tilde{c}_2^2 - \tilde{c}_2^1 - \Delta)dG(\cdot)$$

$$+ \int_{E_1^c}(\Delta)dG(\cdot)] + c_1^1 \geq 0.$$

where $dG(\cdot) = dG(\tilde{\alpha}, \tilde{c}_2^1, \tilde{c}_2^2)$.

(ii) Base technologies A and B are incompatible substitutes

$$\delta^{-1}[S(W) - V(W)] - [S(A + C^1) - V(A + C^1)]$$

$$= \delta \cdot \Delta + \delta[\int_{E_0}(\tilde{\alpha}_2 + \tilde{c}_2^1)dG(\cdot) + \int_{E_1 - E_0}(\tilde{\beta}_2 + \tilde{c}_2^2)dG(\cdot) + \int_{E_1^c}(\tilde{\beta}_2 + \tilde{c}_2^2)dG(\cdot)$$

$$- \int_{E_2 - E_1}(\tilde{\alpha}_2 + \tilde{c}_2^1 + \Delta)dG(\cdot) - \int_{E_2}(\tilde{\alpha}_2 + \tilde{c}_2^1 + \Delta)dG(\cdot)$$

$$+ \int_{E_2 - E_1}(\tilde{\alpha}_2 + \tilde{c}_2^1 + \tilde{\beta}_2 + \tilde{c}_2^2)dG(\cdot) + \int_{E_1^c}(\tilde{\alpha}_2 + \tilde{c}_2^1 + \tilde{\beta}_2 + \tilde{c}_2^2)dG(\cdot)$$

$$+ \int_{E_1}(\tilde{\alpha}_2 + \tilde{c}_2^1 + \Delta)dG(\cdot) - \int_{E_1 - E_0}(\tilde{\alpha}_2 + \tilde{c}_2^1)dG(\cdot) + \int_{E_0^c}(\tilde{\alpha}_2 + \tilde{c}_2^1)dG(\cdot)]$$

$$\int_{E_2 - E_1}(\tilde{\beta}_2 + \tilde{\alpha}_2 + \tilde{c}_2^2 - \tilde{c}_2^1 - \Delta)dG(\cdot) + \int_{E_0}(\tilde{\alpha}_2 + \tilde{c}_2^1 - \tilde{\beta}_2 - \tilde{c}_2^2)dG(\cdot)$$

$$+ \int_{E_1}(\tilde{\beta}_2 + \tilde{c}_2^2)dG(\cdot) + \int_{E_2^c}(\Delta)dG(\cdot).$$

References

Antonelli, C. (1990). Induced adoption and externalities in the regional diffusion of information technology. *Regional Studies* 24, 31-40.

Antonelli, C. (1993). Externalities and complementarities in telecommunication dynamics. *International Journal of Industrial Organization* 11, 437-447.

Bertuglia, C.S., Fischer, M.M. and Preto, G. (Eds.) (1995). *Technological Change, Economic Development and Space*. Springer-Verlag.

Besen, S.M. and Farrell, J. (1994). Choosing how to compete: Strategies and tactics in standardization. *Journal of Economic Perspectives* 8, 117-131.

Capello, R. (1994). *Spatial Economic Analysis of Telecommunications Network Externalities*. Avebury, Aldershot.

Choi, J.P. (1994). Irreversible choice of uncertain technologies with network externalities. *Rand Journal of Economics* 1994, 25, 382-400.

Church, J. and Gandal, N. (1993). Complementary network externalities and technological adoption. *International Journal of Industrial Organization* 11, 239-260.

Economides, N. and White, L.J. (1994). One-way networks, two-way networks, compatibility, and public policy. Discussion papers EC-. Stern School of Business.

Farrell, J. and Saloner, G. (1985). Standardisation, compatibility and innovation. *Rand Journal of Economics,* 16, 70-83.

Farrell, J. and Saloner, G. (1986). Installed base and compatibility: Innovation, product preannouncements, and predation. *American Economic Review* 76, 940-955.

Gandal, N. (1994). Hedonic price indexes for spreadsheets and an empirical test for network externalities. *Rand Journal of Economics* 25, 160-170.

Gandal, N., Greenstein, S. and Salant, D. (1995). Adoptions and orphans in the early microcomputer market. Working Paper No. 2-95, Tel-Aviv University, Israel.

Greene, W.H. (1993). *Econometric Analysis.* Macmillan Publishing Company.

Hägerstrand, T. (1967). *Innovation diffusion as a spatial process.* University of Chicago Press.

Katz, M. and Shapiro, C. (1985). Network Externalities, competition and compatibility. *American Economic Review* 75, 424-440.

Katz, M. and Shapiro, C. (1986). Technology adoption in the presence of network externalities. *Journal of Political Economy* 94, 822-841.

Koski, H. and Nijkamp, P. (1996). Timing of adoption of new communications technology.

Saloner, G. and Shepard, A. (1995). Adoption of technologies with network effects: an empirical examination of the adoption of automated teller machines. *Rand Journal of Economics* 26, 479-501.

Shurmer, M. (1993). An investigation into sources of network externalities in the packaged PC software market. *Information Economics and Policy* 5, 231-251.

Stoneman, P. (1995), (ed.). *Handbook of the economics of innovation and technological change.* Blackwell.

A Game-Theoretic Analysis of Interdependent Transportation Investment Projects

Folke Snickars
Department of Infrastructure and Planning
Royal Institute of Technology, Stockholm

1. The Transport Investment Problem

The transportation system conveys interdependencies. When analysing the costs and benefits of transport investment projects, it is therefore necessary to address the question of linkages among projects. Such linkages can occur in terms of economies of scale in arising from the combination of projects during the construction phase. Linkages may also arise in supply through interaction among network components, or among the producers of transportation services. Linkages may also emerge in demand through the creation of new opportunities for interaction.

Supply side interdependencies arise from the fact that any link in the transport network is connected to other adjacent links and therefore may be involved in a variety of vehicle routes. Cost savings in investment activity may be obtained by grouping projects to facilitate resource sharing during the investment phase. More important long-term cost savings may be obtained by selecting investment projects which simplify the scheduling and movement of vehicles pertaining to several modes of transportation when the projects have been completed.

Supply side synergy in construction may thus arise from the fact that the total investment cost for a basket of projects may be brought down if some of these projects are completed close to one another in time and space. Supply side synergy also prevail if there are investment projects which on completion will confer joint possibilities for several transport suppliers to ameliorate their transport offers in terms of pricing and scheduling.

The supply side interdependencies of the first kind are basically limited to the time period of investment activity. The second type of interdependency on the supply side is fundamentally tied to the network properties of any transportation system. By choosing the right collection of investment projects, the transport infrastructure providers can create flexibly connected transport networks which allow for freedom of action for the transport suppliers, including individual households, in the organisation of service production. The result will normally be increased efficiency in transportation and, potentially, through market pricing, lower transportation costs for all users. It is by no means self-evident that an investment strategy which aims at removing bottlenecks in the transportation network will be conducive to the creation of supply synergy in the sense understood here.

Demand side interdependencies also arise from the network character of the transportation system. The demand side can be conceived in at least two different ways in this context, signifying either transport producers or ultimate users of transportation services. The transport-producing firms have a double role. They on the one hand demand space and time in the transportation networks for their vehicles. On the other hand they extend movement offers to the purchasers of transportation services. The firms, organisations and individuals fulfilling their transportation needs are the ultimate demanders of transportation services. In some cases, there are no intermediate actors but the final demanders of the transportation services turn their movement requests using their vehicles directly towards the transportation infrastructure. This is the case for private cars and for trucks owned and operated by industrial firms.

The fundamental source of the demand interdependency is the fact that any trip or commodity shipment will utilise a whole path of transport network links, involving possibly several different modal networks and several vehicle types. This means that the traffic on any link in the transport system will be composed from a variety of different trips and shipments. Thus, the demand patterns are influenced in a complex way by transport prices and by temporal resource sacrifices associated with different route and mode choices.

One consequence of this fact is that the benefits of any transport network investment will accrue to a variety of economic actors. The longevity of the transportation networks entail that the stream of benefits provided will be long-term. They will also generically accrue to other regions than the one in which the transport network investment is performed. This simple fact may give rise to inefficiencies in the decisions regarding transport investments, especially in view of the fact that transport investments are normally financed via the national government budget. The strict public financing scheme means that all current inhabitants are contributors to the investment fund while the benefits, and the external costs, accrue to all future users of the network, be they domestic or international, individuals or firms, drivers or passengers.

The network spreads the gains of investments in a fashion which is generally not concomitant with the way that property rights are defined. Planning agencies having control over the operation of different networks, different modes of transportation or geographical regions will neither be able to contain the benefits nor the external effects of the investments within their own jurisdictions. A game situation will arise in which different regional and sector actors see each others as competitors about investment resources both in the public purse, and in the private capital market.

The structure and dynamics of this investment game contain some fundamental ingredients which can be illustrated through formal analysis. What is the structure of the political-economy game which is generic to the negotiated agreements, see e.g. Raiffa (1983), in transport planning?

2. A Simple Transport Investment Game

The initial aim of this paper is to analyse the interdependency among transport investment projects by way of a simple two-region example, involving only one mode of transportation which can be most conveniently thought of as a road network. The reason for the choice of this interpretation is that roads create accessibility along the whole length of a link. We will perform the discussion in terms of work trips although this specification is by no means necessary.

The objective of the analysis is to show how the existence of interdependencies on the demand side might affect the net benefits of alternative investment projects. A further aim is to investigate how the decision-making process preceding the decision to implement the projects might proceed under different interdependency patterns.

The benefits will basically be measured through the use of traditional consumer surplus methods. The cost interdependencies on the supply side will not be focal points of the current analysis. Neither will the synergy in the construction phase be of central concern. The analysis will also be extended to the three-region, or three-actor, case to provide further insight in the mechanisms behind the creation of stalemate situations in investment planning.

The basic situation to be analysed is outlined in Figure 1, which illustrates the transport network conditions in an economically well integrated part of a nation, having similarities with middle Sweden. Then, East would be the Stockholm region, extending both north and south of the CBD area, and West would be the combination of the cities of Eskilstuna and Västerås. The travel between those cities would be treated in the same way as the travel within the Stockholm region. The existence of Lake Mälaren introduces a difficulty in building roads in the north-south direction between the two subregions. This implies that there is an inherent north-south balance question in which actors on the northern side of the lake will favour better road connections in the north and actors on the southern side will favour southerly connections. The result will be a game situation where there will emerge a North coalition opposing a South coalition each attempting to promote their own interests.

These interests can both be to promote investments in their own region and to promote investment in the other region the benefits of which will spill over to their own region.

The game situation drawn up is stylised as compared to the actual decision situation in middle Sweden. The Swedish planning system is designed to have the national government decide about, and pay for, transport system expenditures, be they investments or maintenance costs. The road planning is performed via regional agencies of the Swedish road administration. There is one such agency for the Stockholm region and another one for the rest of the region around Lake Mälaren. These agencies work with the regional interests in identifying projects. During the 1990's there has been an increased interest in infrastructure investments from the municipalities, the regional governments, and the private sector. This has created a situation where the transport planning is more of an open game than before, at east when it comes to the identification of projects. There have also been cases where both national and regional political bodies have committed economic resources to projects, in some cases involving also private capital.

In the current paper we will not be centrally concerned with the financing aspects per se. We will assume that each project, or road maintenance activity, has a cost. The North and South coalitions will have to pay for their on projects. We can imagine that the situation is that the national level has given funds to the regions to spend on transport without co-operating with other regions. We will also introduce a supraregional planning body involving both regions to analyse the total benefits and costs of investment and maintenance activities. The points to be illustrated concern the requisites for the emergence of co-operation in a situation where there are obvious spillover effects.

A total of $Q_{we}(t) + Q_{ew}(t) = Q(t)$ trips per year is taking place between East and West, with either East or West as the home city, as a function of the expected generalised travel times between the two cities. Since travel times are assumed to be the same in both directions it will generally not be necessary to distinguish between trips in either direction. Also, the number of trips out from each region forms a share of the total number of trips made by the inhabitants of the two regions. This share

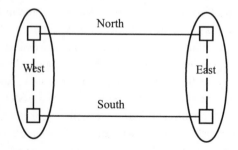

Figure 1. A two-region economic system and a two-link transport network.

will generally depend on the difference between the internal travel time and the travel time between the regions. This effect is taken care of in the analysis by assuming that the trip numbers $Q(t)$ will depend on the travel times. We will not distinguish between generalised travel times and generalised travel costs in the analysis since the value of time will stay the same throughout the analysis.

The number of trips along the northern and southern route is $q_n(t(n),t(s))$ and $q_s(t(n),t(s))$, respectively. It will be convenient to express these patterns as $q_n = p_n Q$ and $q_s = p_s Q$, with p_n and p_s being the chance that a trip will use the northern and southern routes, respectively. The numbers $t(n)$ and $t(s)$ denote the generalised travel times, or generalised travel costs, for the two alternative routes. This is the perceived travel time for travellers in deciding which route to take. It plays the role of a transport price in the analysis since there is no mechanism for changing the price for market reasons. The number of travellers between the eastern and western cities dominate the demand for train transport along the two possible routes. There might also be some trips emanating and ending in the intermediate regions on the two sides of the lake. The number of access points to the transportation system is assumed to be relatively limited.

There is a coalition for transport investments in each one of the two regions, North and South. A supraregional body also exists which may or may not have authority over the two opposing decision-making bodies. Investment in a road link will reduce the generalised travel time in the area of the investment. There will be no travel time effect in the other region. Thus, there is no synergy in supply. Also, there is no difference in the efficiency of the transport planning agencies and road management firms. Two possible project candidates exist, North and South, the cost of the one in the northern region being $c_n(t_0(n)-t_1(n))$, leading to a reduction in the generalised travel cost by $t_0(n)-t_1(n)$, the corresponding numbers in the southern region being $c_s(t_0(s)-t_1(s))$ and $t_0(s)-t_1(s)$, respectively.

The coalitions for transport investments in the Northern and Southern region have independent decision-making powers, and cannot be commanded to perform investments even if they have negotiated agreements with the other regional decision-making body or with the supraregional agency. The supraregional agency has the power to decide by which method benefit calculations should be made. However, it does not have the authority to obtain free benefit or cost information from the two regional agencies. It is in principle possible for the regional decision-making bodies to provide faulty information to each other and to the supraregional agency. The two regional bodies can act on their own, agree to collaborate with each other or interact with the supraregional agency. They can also decide to leave the right to select the most profitable project or projects to the supraregional agency. There are no restrictions in the capital market and funds can float freely between the three actors. There will always be firms willing to operate transport services on the northern and southern links if there is a net economic surplus.

The benefits from transport investments are calculated through the adoption of consumer surplus methods. The benefits of any project are thus measured in terms of generalised-time savings for travellers as well as through reductions in the number of road accidents and diminishing environmental pollution. The regional decision-making bodies measure the benefits to all travellers in their own region. Any project which yields larger benefits than costs may be suggested for implementation. The total net benefits of a project comprise effects in both regions.

Any decision to invest or to simply maintain is to be reported to the supraregional agency at a predetermined point in time. In general, the timing of the investment is not known in advance to other bodies than the one making the decision to go ahead with the project. The decision situation concerns a limited time period of arbitrary length.

There are several possible actions for each one of the agencies in the time-period chosen. Each one of these action possibilities can be associated with an assumed structure of the decision-making situation. In the most simplified case the structure of the transport-investment game may be represented as the following non-cooperative game:

The players are North and South, who decide either to invest in their intraregional transport networks or to maintain the standard, respectively.

Perfect and certain information exists about the costs of the projects and of the demand patterns in the whole system. This information is available to both decision-making bodies. The knowledge includes the future effects on travel patterns of investments in both regions. There is no information bias to the benefit of any of the actors.

North and South simultaneously decide whether to proceed with the transport investment. After these decisions, demand rapidly adjusts to the new conditions. The decision is made at only one point in time.

Investment costs are c_n and c_s. The effects on generalised travel times are $t_0(n)-t_1(n)$ and $t_0(s)-t_1(s)$, respectively.

Trip demand is given by

$q_n = p_n(t(n),t(s))Q(t)$ and $q_s = p_s(t(n),t(s))Q(t)$, with $p_n + p_s = 1$.

$$p_n(t(n),t(s)) = 1/(1 + \exp(\beta[t(n) - t(s)])) \tag{1}$$

$$p_s(t(n),t(s)) = \exp(\beta[t(n) - t(s)])/(1 + \exp(\beta[t(n) - t(s)])) \tag{2}$$

$$Q(t) = q\exp(-\mu t) \tag{3}$$

$$t = p_n t(n) + p_s t(s) \tag{4}$$

The parameter β measures the sensitivity of travellers to travel time changes and μ governs the generation of interregional trips as a function of the expected travel time. It is assumed that the expected travel time is a weighted average of the two link travel times t_n and t_s, the weights being the share of the trips taking the northern and southern tracks, respectively.

Travel time benefits among regions are given by $[N_n(t_0(n)-t_1(n), t_0(s)), N_s(t_0(n)-t_1(n),t_0(s))]$ for the northern region investment, $[S_n(t_0(n), t_0(s)-t_1(s)), S_s(t_0(n), t_0(s)-t_1(s))]$ for the southern region one, and $[G_n(t_0(n)-t_1(n), t_0(s)-t_1(s)), G_s(t_0(n)-t_1(n), t_0(s)-t_1(s))]$ for the case when investments are made in both regions. The first part of the composite benefit refers to the effect in the own region and the second part to the effect in the other region. The benefits are given by the difference in consumer surplus between the reference case and the three composite investment cases. The benefits can be measured via the so-called rule-of-half or via more sophisticated indicators as given by the Hotelling surplus measures, see e.g. Williams and Senior (1977). The payoff function for the two planning agencies North and South may be taken as the total travel time savings for those who pass the northern and southern links, respectively, complemented by an assumption of the economic value of time, SEK per minute, for the travellers. No notice will be taken of other benefits of the investments, as reduced road accidents or diminished environmental pollution. As mentioned above, the investments are made independently of one another, yielding investment costs of c_{an}, C_s and $c_{an} + c_s$, respectively. Formally, the surplus measure for the rule-of-half is given by (5), where underscore denotes vector notation.

$$W(\underline{Q}(\underline{t}), \underline{Q}_0(\underline{t}), \underline{t}, \underline{t}0) = (\underline{Q}(\underline{t}) + \underline{Q}_0(\underline{t}))^T(\underline{t} - \underline{t}_0)/2 \tag{5}$$

The demand schedules are non-linear functions with the basic structure of logit type. The number of trips via the northern track depends both on the northern generalised travel time and on the southern one. A reduction in one of the generalised travel times leads to a shift of the earlier trips towards that track at the same time, more trips are generated in total since the average travel time is reduced. Since that average depends on the new split of trips a strengthening of the initial effect will ultimately occur.

The properties of the demand functions follow from the choice of logit models For instance, the demand for trips on the northern link will vary as a function of the generalised travel times on the northern and southern tracks, respectively. This means that whenever there is a change in the travel time structure, the adjustment behaviour of the travellers as described by the so-called logit demand functions will create a composite volume and distribution effect on trip demand on both links.

3. Equilibrium Analysis of the Transport Investment Game

The structure of the game formulated is so simple that it is enough to use the so-called normal form to analyse its properties. The reason is that none of the players has an information advantage over the other. There are only two decisions, to invest or simply maintain the network that they have. We will assume that investment costs are in addition to maintenance costs so that the reference case implies zero investment costs for both actors.

The details of the benefit calculations will not be kept through the analysis of the two-player game. Instead, they will be condensed into net benefits before investment cost subtractions for the four cases for the two planning agencies. We will use the notation (M_n, M_s) for the non-investment case, assuming as above that they are initially both zero.

The normal form can be represented as shown in Figure 2.

		South	
		Invest	Maintain
	Invest	$(G_n\text{-}c_n, G_s\text{-}c_s)$	$(N_n\text{-}c_n, N_s)$
North			
	Maintain	$(S_n, S_s\text{-}c_s)$	(M_n, M_s)

Figure 2. Choices for the northern and southern coalition and their payoff consequences.

The expected net benefits for North deciding to invest or maintain are $(G_n + N_n)/2 - c_n$ and $(S_n + M_n)/2$, respectively. For South they are $(G_s + S_s)/2\text{-}c_s$ and $(N_s + M_s)/2$. The structure of the game is such that if North invests and takes on the whole investment cost, a benefit will accrue also to South. If for North the quantitative relationship among net benefits is $S_n > (G_n - c_n) > M_n > (N_n - c_n)$, and correspondingly for South, $N_s > (C_s - c_s) > M_s > (S_s - c_s)$, the situation will occur that (M_n, M_s) is the non-cooperative equilibrium of the game. This happens although $(G_n + G_s - c_n - c_s)/2 > (M_n + M_s)/2$ and thus there would be joint gains from cooperation.

The game formulated is a general non-cooperative two-person game. As is well known in the literature there are a limited number of prototype cases of this game, leading to different equilibrium states, some of them dominant and others of Nash type. The generic type of the current game is determined by realistic magnitudes of the various costs and benefits involved. The idea of the numerical example given above is to provide the basis for an assessment of reasonable relationships among numbers. To illustrate this point, the game situation will be illustrated through a numerical example, involving two investments of 1 billion SEK each, giving rise to the benefits stated in Figure 3.

The context described gives rise to a peculiar structure of the game, based on the fact that because of the perfect and symmetric information, both players know that the projects give side-effects in the other region. At the same time they do not have these benefits as a part of their own benefit function. In the completely separated case, none of the projects would be started since they both create lower benefits than costs in their own region. If notice is taken of the effect also in the adjoining region, the northern region project should be performed since it gives rise to a positive benefit minus cost difference.

	North	South
Own region	0.95	0.85
Other region	0.25	0.20
Total benefit	1.20	1.05
Total cost	1.00	1.00

Figure 3. Benefits and costs of two road investment projects (billion SEK present value).

In general, two opposite situations can be conceived as regards the interdependence structure. In one case, the projects create positive side-effects, which will make them profitable in the whole system. In the other case, projects which create positive effects in the own region will worsen conditions in the other region, so that the total surplus might possibly become negative. It may also happen that projects create quite different effects if both are performed than what is computed by adding the effects. There may be positive or negative synergy among projects. There is a tendency for substitution among routes when travel times change. At the same time, such decreases in travel sacrifices will stimulate increased overall travel volumes.

The normal form of the game has the structure given in Figure 4 in the numerical example provided.

		South	
		Invest	Maintain
	Invest	(0.20, 0.05)	(-0.05, 0.25)
North			
	Maintain	(0.20, -0.15)	(0, 0)

Figure 4. Example of transport investment game structure.

The game has a Nash equilibrium in the field (Maintain, Maintain). This is the only solution in which each player does not have an incentive to change the action if the other player does not change his action. On the other hand, if they both agreed to settle for (Invest,Invest), a total surplus of 0.25 billion SEK would result, giving rise to the possibility of both players gaining from collaborating. The problem which still remains is that there is no mechanism for distributing the joint profits among the two agencies.

The example is a special case of the variety of game structures and equilibria that can occur in the transport investment game. There are 64 different 2x2 games. A closer scrutiny into the structure of the payoffs will reveal that only 12 games are different from one another. Among these games four types of games stand out as possibly generic cases for conflicts in relation to transport investment. In the literature they are called pure co-ordination, battle of the sexes, chicken, and prisoner's dilemma, see Rasmusen (1989). The first three games has equilibria both in pure and mixed strategies whereas there is only one Nash equilibrium in the prisoner's dilemma game. It is warranted to ask whether the focal question in the field of interdependent transport investment generically refers to one or several of these classical games.

In pure co-ordination the payoffs are such that using the same strategy is always better for both players, with co-operation generating higher payoffs than defection. This game is very sensitive to the simultanety of the moves. If one player takes the first move, the other player will follow. The game has the two joint strategies as Nash equilibria. However, it is not possible to predict which one of them will be realised, the one with higher values for both or the one with lower values even if both players assess the same one to be the best. In the investment situation this would mean that the net benefit would be highest for each actor group if both projects were fulfilled. The next best for each group would be not to invest at all.

The game called the battle of the sexes has a slightly different payoff structure compared to the pure co-ordination one. Both players are better off choosing the same strategy as the other. However, they disagree on which of the two strategies they prefer. The game again has the two joint strategies as Nash equilibria. The situation reflects a so-called first mover advantage. If one were to decide first, the other player would have a dominant strategy, and vice versa. In the transport investment case it would be best for one group to invest and for the other to only maintain. However, the benefits from sharing the same action would be such that going together would beat the best choice in isolation.

A third conceivable situation which might occur when the two groups of players are very competitive. The highest payoff for each one of them will then be realised if their project is selected and not the other. The game has two Nash equilibria in which the players each have chosen different strategies. The chicken game concept can be clearly illustrated by looking at a situation when no road exists on one side of the lake. Then the other group has a natural monopoly and the persons who live in the southern parts of East and West have to travel longer distances than the ones who live northerly or vice versa. If a road is built on the side where there is none now, the monopoly situation will disappear. The chicken payoff structure also seems to be applicable also to a situation when groups of regional actors attempt to attract indivisible projects which will generate intraregional development gains, for instance, regional universities or airports.

The prisoner's dilemma situation would occur when there are net benefits for both players from investing jointly. There are windfall gains to be attained from defecting when the other actor invests. This will give the spillover benefit to the own region. The Nash equilibrium situation will be the one where net total net benefits are lowest. In the transport investment case this would occur, for instance, when there are cost from having the road investment as accidents or environmental pollution which introduces an incentive to be a free rider making use of the spillover benefits. The prisoner's dilemma game has been extensively studied in the literature, especially when public investments are concerned, also as a dynamic modelling concept, see e.g. Axelrod (1984).

4. A Knapsack Problem Formulation

The problem studied can be seen as a simple knapsack problem in which the value of the basket of one actor depends on the action taken by the other. In this formulation, it becomes obvious that the analysis can be extended to situations in which there are more than two competing projects. To describe this variant of the game model some new notation is needed:

$x(n) = 1$ if North invests, $x(n) = 0$ if North maintains
$x(s) = 1$ if South invests, $x(s) = 0$ if South maintains
$V_n(x(n), x(s)) =$ net value for North under $(x(n),x(s))$
$V_s(x(n), x(s)) =$ net value for South under $(x(n),x(s))$

$$V_n(x(n), x(s)) = [A_n + B_n*x(s)]*x(n) + C_n*x(s) + D_n*(1-x(n)) \qquad (6)$$
$$V_s(x(n), x(s)) = [A_s + B_s*x(n)]*x(s) + C_s*x(n) + D_s*(1-x(s)) \qquad (7)$$

The first part of (6) and (7) refer to the total net benefits of the own investment strategy for North and South. We have introduced a general notation for the components of the effects through the parameters (A, B, C, D). The second benefit component is the spillover effect, i.e. the effect in the neighbouring region of investment decisions in the own region. The third part refers to the benefit from maintaining which is assumed to stay within the own region. It is instructive to see how the above formulation transforms into the normal form of the investment game, see Figure 5.

The formulation illustrates the way that synergy may be created among projects. This synergy may either be complementary or antagonistic. The formulation also

		South	
		Invest	Maintain
North	Invest	$(A_n + B_n + C_n, A_s + B_s + C_s)$	(A_n, C_s)
	Maintain	(C_n, A_s)	(D_n, D_s)

Figure 5. Transformation from the knapsack to the normal form formulation of the transport investment game.

facilitates the extension of the game to more players. There will then be three-way or higher order interaction among projects.

The knapsack formulation may be used to express the existence of a Nash equilibrium in so-called mixed strategies, see also Rasmusen (1989). Direct calculation gives the result that the mixing probabilities will be $[(D_s - A_s)/(D_s + B_s), (D_n - A_n)/(D_n + B_n)]$, respectively. There is a mixed-strategy equilibrium if these expressions yield probabilities between zero and one under the further condition that there is more than one pure-strategy equilibrium. The maximum expected payoff for (North, South) will be given by the expression $[D_n + (D_n - A_n)*(C_n - B_n)/(D_n + B_n), D_s + (D_s - A_s)*(C_s - B_s)/(D_s + B_s)]$. If it is assumed that the net benefit from maintenance is zero the mixing probabilities will be given by the ratio between the net benefit in the own region divided by the synergy from the combined projects in each region. Obviously, one of these expressions must be negative for an internal point to be generated. It should also be noted that the vale of the maximum payoff is independent of the benefit components in the other region.

Assuming that the game is symmetric, the parameter set $(A, B, C, D) = (0, 2, 1, 2)$ will give rise to the stylized game pure co-ordindation. The mixed strategy equilibrium has (50, 50) as the mixing probabilities (expressed in percent) for (North, South), respectively. When $(A, B, C, D) = (0, -4, 2, 1)$ the situation will be typical of the chicken game with mixing probabilities (20, 20). Setting $(A, B, C, D) = (-3, 2, 2, 0)$ will yield the parameter setup for the prisoner's dilemma game which has only one Nash equilibrium. Finally, the structure of the battle of the sexes game is such that symmetry between players cannot be present. If North has the parameter setup (0, 2, -1, 1) and South has parameters (-1, 2, 0, 2) the battle of the sexes setup will arise. The resulting mixing probabilities will be (67, 33) for (North, South), respectively.

Another extension which can be formulated using the value functions (6) and (7) is to specify the role of the supraregional planning agency. Its value function might be, for instance, the following:

$$V(x(n), x(s)) = V_n(x(n), x(s)) + V_s(x(n), x(s)) \qquad (8)$$

$$V(x(n), x(s)) = [A_n + C_s + D_n]*x(n) + [A_s + C_n + D_s]*x(s)$$
$$+ [B_n + B_s]*x(n)*x(s) \qquad (9)$$

The formulation (9) illustrates the quadratic nature of the synergy. If the supraregional planning agency has a budget for investment and maintenance the knapsack problem for the two-region problem can be formulated as an integer quadratic program.

The main result is that the interdependency arising from spillovers in demand may give rise to strategic behaviour which is detrimental to the efficiency in decision-making, see also Kremer et al (1991) and Edwards (1992). The analysis points at a role for the supraregional level as an actor incurring government spending to prov-ide incentives for the independent regional planning agencies to move away from their inefficient strategic behaviour.

5. Concluding Remarks

The analysis outlined here can be extended in a variety of directions, both theoretical and empirical.

A thorough analysis of interdependent projects in the framework of the demand functions formulated can be performed. This analysis presupposes that consumer surplus calculations can be done for any separate or joint investment programme. The benefit calculations can be extended to the supply side, both by explicitly considering producer surplus values, and by introducing interdependencies on the supply side. The game can be compared to the consumer surplus calculations performed by the supraregional agency, taking both regional benefit streams into notice at the same time.

The game structure can be made different by assuming that the temporal development is governed by other rules. Such rules would imply that decisions are taken at different moments in time by the two agencies, see also Weibull (1996). They could also mean that the game is repeated several times, e.g. as a part of a planning system in which yearly program updates are performed. Also, the assumption of perfect information concerning the demand development can be removed. The game will then be complicated by Nature, in the form of i.e. travel demand for the two links, producing events which are dependent of the timing of the actions of the players.

The investment problem can be formulated as a knapsack problem for the supraregional agency in which the projects to be placed in the knapsack are dependent on one another. This will generate non-linear benefit functions in the decision variables determining whether one or both projects will be performed together. The resulting problem will be of quadratic assignment type, possibly involving several local optima.

References

Axelrod, R, 1984, The Evolution of Co-operation. Basic Books, New York

Edwards, J, 1992, Indivisibility and Preference for Collective Provision. Regional Science and Urban Economics, vol 22, pp 559-577

Kremer, H, Marchand, M and Pestieau, P, 1991, Investment in Local Public Services. Nash Equilibrium and Social Optimum. Working Paper, CORE, University of Louvain

Raiffa, H, 1983, The Art and Science of Negotiation. Belknap Press, Cambridge, Massachusetts

Rasmusen, E, 1989, Games and Information - An introduction to Game Theory. Blackwell, Oxford

Weibull, J W, 1996, Evolutionary Game Theory. Princeton University Press, New Haven

Williams, H and Senior, M, 1977, Accessibility, spatial interaction and the spatial benefit analysis of land use - transportation plans. In Karlqvist, A et al (eds), Spatial Interaction Theory and Planning Models. North Holland, Amsterdam

Roadway Incident Analysis with a Dynamic User-Optimal Route Choice Model

D. E. Boyce
Department of Civil and Materials Engineering
University of Illinois at Chicago

D.-H. Lee
Institute of Transportation Studies
University of California at Irvine

and
B. N. Janson
Department of Civil Engineering
University of Colorado at Denver

1. Introduction

Intelligent Transportation Systems (ITS), also known as Intelligent Vehicle Highway Systems (IVHS), are applying advanced technologies (such as navigation, automobile, computer science, telecommunication, electronic engineering, automatic information collection and processing) in an effort to bring revolutionary improvements in traffic safety, network capacity utilization, vehicle emission reductions, travel time and fuel consumption savings, etc. Within the framework of ITS, Advanced Traffic Management Systems (ATMS) and Advanced Traveler Information Systems (ATIS) both aim to manage and predict traffic congestion and provide historical and real-time network-wide traffic information to support drivers' route choice decisions. To enable ATMS/ATIS to achieve the above described goals, traffic flow prediction models are needed for system operation and evaluation.

This chapter describes a large-scale dynamic route choice model and solution algorithm for predicting traffic flows off-line for ATMS and ATIS; the solution algorithm, DYMOD, was developed by Janson (1991a, 1991b) and Janson and Robles (1995). A variational inequality (VI) formulation provides an improved theoretical basis for the solution algorithm (Lee, 1996). To solve the VI model, a multi-interval, time-varying-demand route choice problem with fixed *node time intervals* (denoted as the diagonalized or *inner* problem) is formulated. This problem is equivalent to a sequence of static route choice problems. Then, the shortest route travel times and *node time intervals* constraints are updated in an *outer* problem with temporally-correct routes and time-continuous traffic flow propagation.

To generate time-dependent traffic characteristics for a large network, realistic traffic engineering-based link delay functions such as Akcelik (1988) functions are applied for better estimation of link delays at various types of links and intersections. Unexpected capacity reducing events causing nonrecurrent traffic congestion are analyzed with the model. Route choice behavior based on anticipatory and non-anticipatory network conditions are considered in performing the incident analysis, extending the capability of this model to contribute to the evaluation of ATMS and ATIS.

Section 2 describes the mathematical formulation and solution algorithm of the model. The link travel time functions are described in Section 3. The ADVANCE Network representation, travel demand and input data preparation are presented in Section 4. Computational results and output analysis are reported in Section 5. Conclusions and a summary are provided in the last section.

2. Model Formulation and Solution Algorithm

The link-time-based variational inequality model of the dynamic user-optimal (DUO) route choice problem solves for the following travel-time-based ideal DUO state (Ran and Boyce, 1996):

> *Travel-Time-Based Ideal DUO State:* For each O-D pair at each interval of time, if the actual travel times experienced by travelers departing at the same time are equal and minimal, the dynamic traffic flow over the network is in a travel-time-based ideal dynamic user-optimal state.

The link-time-based variational inequality model can be stated based on the following equilibrium conditions:

$$\Omega_{ra}^{d*} = \pi_{ri}^{d*} + \tau_a\left[d + \overline{\pi}_{ri}^{d*}\right] - \pi_{rj}^{d*} \geq 0 \qquad \forall \ a = (i,j), r, d; \ (1)$$

$$x_a^{rs*}\left[d + \overline{\pi}_{ri}^{d*}\right]\Omega_{ra}^{d*} = 0 \qquad \forall \ a = (i,j), r, s, d; \qquad (2)$$

$$x_a^{rs*}\left[d + \overline{\pi}_{ri}^{d*}\right] \geq 0 \qquad \forall \ a = (i,j), r, s, d. \qquad (3)$$

where

Ω_{ra}^{d} = difference between the minimal travel time from zone r to node i $\left(\pi_{ri}^{d}\right)$ plus the travel time on link a $\left(\tau_a\left[d + \overline{\pi}_{ri}^{d}\right]; \ a = (i,j)\right)$, and the minimal travel time from zone r to node j $\left(\pi_{rj}^{d}\right)$, for vehicles departing from zone r in time interval d

π_{ri}^{d} =   minimal actual travel time from zone r to node i for flow departing in interval d

$\overline{\pi}_{ri}^{d}$ =   number of time intervals traversed in π_{ri}^{d}

$x_{a}[t]$ =   total flow of vehicles on link a in time interval t

$x_{a}^{rs}[d]$ =   flow of vehicles from zone r to zone s on link a that departed in time interval d

$\tau_{a}\left(x_{a}[t]\right)$ =   actual travel time on link a with flow x_{a} in time interval t

t =   a time interval

N =   set of all nodes

Z =   set of all zones (i.e., trip end nodes)

K =   set of all links (directed arcs)

T =   set of all time intervals in the full analysis period (e.g., 18 ten-minute intervals for a three-hour analysis period)

Therefore, the VI formulation is as follows:

$$\text{VI:} \quad \sum_{d}\sum_{rs}\sum_{a}\Omega_{ra}^{d*}\left\{x_{a}^{rs}\left[d + \overline{\pi}_{ri}^{d*}\right] - x_{a}^{rs*}\left[d + \overline{\pi}_{ri}^{d*}\right]\right\} \geq 0 \qquad (4)$$

where * denotes the equilibrium solution, and the above variables satisfy definitional, flow conservation, nonnegativity and flow propagation constraints (Lee, 1996).

The algorithm for solving the proposed variational inequality model of the DUO route choice model is shown in Figure 1. The *outer* step solves for the *node time intervals* $\left\{\alpha_{ri}^{d}[t]\right\}$, a [0,1] indicator of whether the flow departing zone r in time interval d has crossed node i in time interval t, and shortest route travel times $\left\{\pi_{ri}^{d}\right\}$ with fixed link flows $\left\{x_{a}[t]\right\}$. Then, the algorithm solves the route choice problem using the Frank-Wolfe method with fixed *node time intervals* $\left\{\alpha_{ri}^{d}[t]\right\}$ and shortest route travel times $\left\{\pi_{ri}^{d}\right\}$. The adjustments of link capacities proceed between the *inner* and *outer* steps to account for capacity changes caused by spillback queuing effects, signal timing changes, incidents and other events. The algorithm terminates when the number of changes of *node time intervals* between two consecutive *outer* iterations is less than an acceptable threshold.

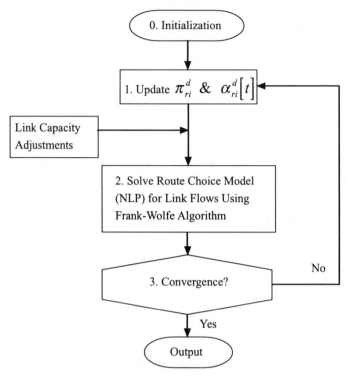

Figure 1: Flowchart of the Solution Algorithm

Steps of the algorithm are described as follows:

Step 0: Initialization. Input all network data, temporal trip departure rates and initial link flows. Initial link flows are optional and can be set to zero. However, the static user-optimal flows reduced to the chosen time interval duration may be good initial values. Calculate initial $\left\{\alpha_{ri}^{d}[t]\right\}$ and $\left\{\pi_{ri}^{d}\right\}$ with initial link flows. Set outer iteration counter m = 1.

Step 1: Update Node Time Intervals and Shortest Route Travel Times. For fixed link flows $\left\{x_{a}[t]\right\}$, update the node time intervals $\left\{\alpha_{ri}^{d}[t]\right\}$ and shortest route travel times $\left\{\pi_{ri}^{d}\right\}$ to maintain temporally-correct, time-continuous routes and flow propagations.

Step 2: Solve Route Choice Problem. Solve the DUO program using the Frank-Wolfe algorithm with the optimal values of node time intervals $\left\{\alpha_{ri}^{d}[t]\right\}$ from Step 1, which simplifies to a sequence of static route choice problems.

Step 3: Convergence Test for Updating Iterations. Sum the total number of node time interval differences (NDIFFS) between iterations m − 1 and m. Compare each $\left\{\alpha_{ri}^{d}[t]\right\}^{m}$ to $\left\{\alpha_{ri}^{d}[t]\right\}^{m-1}$. If NDIFFS ≤ allowable percentage of all node time intervals, then stop. Otherwise, return to Step 1.

To determine $\left\{\pi_{ri}^d\right\}$ and $\left\{\alpha_{ri}^d[t]\right\}$, the following procedure is used in Step 1:

1. Find shortest route travel times, $\left\{\pi_{ri}^d\right\}$.

2. Reset values of $\alpha_{ri}^d[t]$ as follows:

$\quad$ if $\left(\overline{\pi}_{ri}^d \leq t\Delta t\right)$ and $\left(\overline{\pi}_{ri}^d \geq \left(t - 1\right)\Delta t\right)$

$\quad$ then $\alpha_{ri}^d[t] = 1$; otherwise $\alpha_{ri}^d[t] = 0$

$\quad$ perform for all $r \in Z, i \in N; d = 1, \cdots T$ and $t = d, d+1, \ldots T$

$\quad$ note: $\displaystyle\sum_{t=d}^{T} \alpha_{ri}^d[t] = 1$

3. Set $\pi_{rr}^d = d\Delta t, \ \forall \ r \in Z, d \in T$

4. Enforce the first-in-first-out (FIFO) conditions:

$$\pi_{ri}^d = \max\left(\beta_{ri}^d, \pi_{ri}^{d-1} + h\Delta t\right) \ \forall \ r,i,d \ \text{and} \ \pi_{ri}^0 = \pi_{ri}^1 - \Delta t \quad (5)$$

$$\theta_{ri}^d[t] = \left[\left(\pi_{ri}^d - \left(t - 1\right)\Delta t\right)/\Delta t\right]\alpha_{ri}^d[t] \quad \forall r,i,d,t \quad (6)$$

$$\mu_{ra}^d[t] = \left[\theta_{ri}^d[t]\tau_a\left(x_a[t]\right) + \left(1 - \theta_{ri}^d[t]\right)\tau_a\left(x_a[s]\right)\right] \alpha_{ri}^d[t] \ \forall r,a,d,t, \ s = t - 1$$

$$\quad (7)$$

$$\left\{\beta_{rj}^d - \max\left[\pi_{ri}^d, (t-1)\Delta t + \Delta\tau_a^t[s]\right]\right\}\alpha_{ri}^d[t] \leq \mu_{ra}^d[t]\alpha_{ri}^d[t] \quad (8)$$

$$\forall \ r,a,d,t, \ s = t - 1; \ \text{where} \ \Delta\tau_a^t[s] = \tau_a\left(x_a[s]\right) - \mu_{ra}^d[t]$$

where

$\beta_{ri}^d \qquad =$ time at which the last flow departing zone r in time interval d crosses node i via its shortest route, less FIFO delay time at node i

$\theta_{ri}^d[t] \qquad =$ fraction of a time interval t that the last flow departing zone r in time interval d crosses node i

$\mu_{ra}^d[t] \qquad =$ average travel time on link a of the last flow departing zone r in time interval d

$h \qquad\qquad =$ minimum fraction of time interval separating flows departing in successive time intervals

$\quad$ Equations (5)-(8) impose FIFO trip ordering between all O-D pairs according to their travel times in successive time intervals. Vehicles are assumed to make only one-for-one exchanges of traffic positions along any link, which is acceptable and expected in aggregate traffic flow models. For detailed descriptions of the FIFO conditions used, see Lee (1996).

3. Link Travel Time Functions

Mathematical functions are used within the route choice model to estimate link travel times for given flow rates. The choice of the delay functions involves several criteria: (1) the desired mathematical properties of the function to satisfy the condition for a unique solution of the model; (2) the cost and limited availability of road data; (3) the computational effort required by the model; and (4) the desired accuracy of the travel time estimates generated by the model.

Delay functions selected for this study can be classified by road type and intersection type. First, delay functions for signalized intersections are presented, next for unsignalized intersections, and then for freeway-related facilities.

The specific delay function for links at signalized intersections applied in the study has the following form (Akcelik, 1988):

$$d = \frac{0.5C(1-u)^2}{1-ux} + 900T\gamma \left[x-1+\sqrt{(x-1)^2 + \frac{8(x-0.5)}{cT}} \right] \tag{9}$$

where

d	=	average delay per vehicle (second/vehicle)
C	=	signal cycle length (second)
$u = g/C$	=	green split
g	=	green time (second)
$x = v/c$	=	flow-to-capacity ratio
T	=	duration of the flow (hour)
g	=	1, for $x > 0.5$; 0, otherwise

The first term, called the *uniform delay*, was originally developed by Webster (1958). It reflects the average delay experienced by drivers in *undersaturation conditions*, that is when the arrival flow does not exceed capacity. In oversaturation conditions, $x=1$ is used in the uniform delay term. The second term of Equation (9) is called the *overflow delay*. It reflects the delay experienced by the vehicles when the flow rate is close to or exceeds capacity. Temporary overflow at an intersection may also occur when the average arrival rate is lower than the capacity, due to a random character of the arrival pattern. The earliest delay functions (e.g., Webster, 1958) were based on the steady-state model and were defined only for undersaturation conditions. Figure 2 shows an example of an Akcelik function.

Delay functions for unsignalized intersections, whether major/minor priority intersections or all-way-stop controlled intersections, are discussed next. For major/minor priority intersections, formulae developed by Kimber and Hollis (1979) are adopted by a related study of an asymmetric route choice model. While the delay function developed by Kimber and Hollis is being tested by that study in the

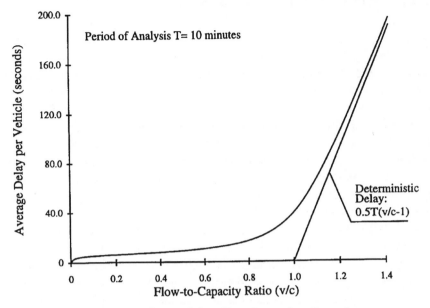

Figure 2: Steady-State Delay Model vs. Time Dependent Formulae

large-scale network, the BPR function is temporarily used for estimating delays at major/minor priority intersections.

As for the delay function for all-way-stop intersections, the following exponential delay model is used (Meneguzzer et al., 1990):

$$d = \exp [3.802(v/c)] \tag{10}$$

where d is the average approach delay, v is the total approach flow, and c is the approach capacity. Note that the form of this exponential function, which is relatively flat at low flow-to-capacity ratio but becomes very steep as the degree of saturation increases, reflects the operational characteristics of all-way-stop intersections well. Kyte and Marek (1989) found that approach delay is approximately constant and in a range of five to ten seconds per vehicle for approach flows up to 300 to 400 vehicles/hour, but increases exponentially beyond this threshold. An increase in conflicting and opposing flows has the effect of reducing this threshold. In addition, it should be noted that Equation (10) is suitable for use in a network equilibrium model, since it is defined for any flow-to-capacity ratio.

Several types of freeway-related facilities occur within the test area: basic freeway segments, ramps and ramp-freeway junctions, weaving sections and toll plazas. A sophisticated scheme of delay functions has been developed for individual freeway-related facilities (Berka et al., 1994). Those functions are expected to be incorporated into the solution procedure in the future. The BPR function is adopted for freeway-related facilities.

4. The ADVANCE Network

ADVANCE (Advanced Driver and Vehicle Advisory Navigation Concept), a field test of ATIS was recently concluded by the Illinois Department of Transportation (IDOT) and the Federal Highway Administration (FHWA), in collaboration with the University of Illinois at Chicago, Northwestern University, and the IVHS Strategic Business Unit of Motorola, Inc.

The ADVANCE Test Area is depicted in Figure 3. It is located in the northwestern suburbs of the Chicago area and covers about 800 square kilometers. Dense residential communities, office centers, regional shopping centers, subregional government centers, and the O'Hare International Airport are located in the ADVANCE Test Area. The network topology of the test area is almost a regular grid with a few diagonal major arterials directed towards the Chicago CBD. The freeway system includes I-90, I-94, I-190, I-290, I-294, IL-20 and IL-53. Except for the remote northwest corner, the freeways serve nearly all parts of the Test Area. The southwest quadrant is characterized by modern, multi-lane arterials designed for high volumes. In Figure 3, collectors, arterials and freeways are drawn with lines of different widths, freeways being the widest line. The heavy black line indicates the boundary of the ADVANCE Test Area.

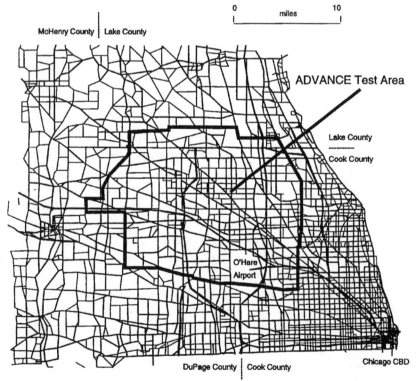

Figure 3: Test Area and Network in the Northwestern Suburbs of Chicago

The ADVANCE Test Area is divided into 447 zones, originally specified by the Chicago Area Transportation Study (CATS), to assign time-dependent travel demand. Daily trip tables based on CATS estimates for 1990 were factored to represent travel demand for five time-of-day periods (night, 12 am to 6 am; morning peak, 6 am to 9 am; mid-day, 9 am to 4 pm; afternoon peak, 4 pm to 6 pm; and evening, 6 pm to 12 am). Each time-of-day period is divided into 10-minute intervals used for solving the proposed dynamic route choice model. The 10-minute departure rates for each origin zone are derived from the time-of-day half-hour departure rates obtained from CATS. Trips originating and/or terminating outside the test area are represented by zones on the boundary. An auxiliary analysis of these flows is based on route flows generated by a static route choice model for a larger region encompassing the ADVANCE Test Area (Zhang et al., 1994).

In a conventional route choice model, the network is coded so each intersection is represented as a single node and each approach is represented as a single link. With conventional network representation, 7,850 approach links and 2,552 nodes are included in the ADVANCE Test Area (Table 1).

Table 1: Size of the Test Network

Number of Links	Number of Nodes	Number of Zones
7,850	2,552	447

Table 2: Frequency of Links by Facility Type

Type of Facility	No. Of Links
Arterial/Collector	4,061
Tollway/Freeway	197
Freeway Ramp	202
Toll Plaza	14
Freeway Weaving Section	11
Centroid Connector Links	2,491
Approach Links	874
Total	7,850

Table 3: Intersection Frequency by Number of Legs and Control Type

	Signalized	Priority	All-way-stop	Total
Three-leg	257	174	51	482
Four-leg	558	52	60	670
Five-leg and more	7	0	0	7
Total	822	226	111	1,159

Table 2 lists the frequency of links by the facility type. Table 3 presents the breakdown of arterial/collector intersections by the number of the legs and control type. To account for better link flow/delay relationships and potential queue spill-back effects, each turning movement is coded as a separate link called an intersection link. For example, a typical four-leg intersection with two-way approaches without any turning restrictions (U-turn excluded), four approach nodes, four exit nodes and twelve intersection links are required in this expanded network representation (see Figure 4).

The expanded intersection representation procedure is applied only to nodes representing an intersection of arterials or collectors. Nodes that do not need to be expanded are freeway nodes, no-delay intersections, and other nodes not representing intersections. Because of the expanded network representation, the network size increases about three times in comparison with conventional network representation. To that end, 22,918 links and 9,700 nodes are actually modeled for solving the proposed dynamic user-optimal route choice model. Note that the detailed delay functions by turning movements are applied only within the actual ADVANCE Test Area.

In order to utilize the traffic engineering-based link delay functions, the following data need to be available. Besides the typical input data such as link capacity and free flow travel time, types of turning movements (left, through, right), link facility types (e.g., centroid connector, freeway, tollway, arterial), intersection categories, types of traffic control at intersections (e.g., signalized, priority, all-way-stop), cycle length, saturation flow, and signal timing split are required with this regard.

A static asymmetric user-optimal route choice model for ADVANCE Network was used to generate the above information (Berka et al., 1994). That static asymmetric user-optimal route choice model is the basis for generating link travel times for the static link travel time profiles of the ADVANCE Project.

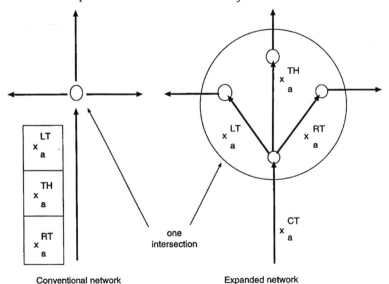

Figure 4: Expanded Intersection Representation

5. Computational Results and Analysis

The model is implemented on the CONVEX-C3880 at the National Center for Supercomputing Applications (NCSA), University of Illinois at Urbana-Champaign. The Convex-C3880 is a vector shared memory machine consisting of 8 processors (240 MFLOPS per processor peak) and 4 Gbytes of memory and 60 Gbytes disk space. For the ADVANCE Network with morning peak (18 ten-minute intervals) travel demand, 512 Mbytes of memory and nearly 60 CPU hours are needed to reach a very fine convergence (see definition in the next subsection) from a zero flow initial solution. A modified version of DYMOD can be executed with much less memory (only 55 Mbytes are needed for the same problem size as described above) and thus can be implemented on a workstation; however, it requires higher CPU times and more disk space.

Five global network performance measures are adopted to monitor the solution process of the Convergent Dynamic Algorithm and to assess the traffic condition over the ADVANCE Network. Definitions of these global network performance measures are listed as below:

1. Average travel time

$$\bar{c} = \frac{1}{R}\sum_a\sum_t c_a[t]x_a[t]$$

where

$\bar{c}$	=	average travel time (minutes)
R	=	total flow during the analysis period (trips/period)
$c_a[t]$	=	travel time on link a at time interval t (minutes)
$x_a[t]$	=	flow on link a at time interval t

2. Average travel distance

$$\bar{\ell} = \frac{1}{R}\sum_a\sum_t \ell_a x_a[t]$$

where

$\bar{\ell}$	=	average travel distance (miles)
ℓ_a	=	length of the link a (miles)

3. Network space mean speed

$$\bar{S} = \bar{\ell}/(\bar{c}/60)$$

where

$\bar{S}$	=	network space mean speed (mph)

4. Average flow-to-capacity ratio

$$\bar{x} = \frac{1}{K}\frac{1}{T}\sum_a \sum_t \frac{x_a[t]}{C_a[t]} x_a[t]$$

where

$\bar{x}$	=	average flow-to-capacity ratio
K	=	number of links in the network
T	=	number of time intervals
$C_a[t]$	=	capacity of link a at time interval t

5. Convergence Index

The convergence index monitors the change in node time intervals between two consecutive outer iterations. The solution algorithm terminates if the change in node time intervals (NDIFFS) is less than nodes x zones x intervals x X. This index indicates that only X changes of the total node time intervals are allowed for the last flow that departed from each zone over the total analysis period. With perfect convergence, the changes of node time intervals equal to zero.

Table 4 shows the selected characteristics of the final solution for morning peak period of the ADVANCE Network and separate road classes (collector, arterial and freeway related facility). Figures 5, 6, 7 and 8 show the variations of the network performance measures among the *outer* iterations.

For the convergence measure, *nodes* = 9,700, *zones* = 447, *intervals* = 18 and X = 0.001 were used in this solution. Note a rather small value was chosen for X indicating a fine convergence of the algorithm is desired. After 10 *outer* iterations of Step 1, NDIFFS equals to 20,537 showing the algorithm has converged at the level of $X \approx$ 0.00026. The rate of change of *node time intervals* is displayed in Figure 9 which indicates that this model was solved quite smoothly by the solution algorithm for the ADVANCE Network. The flow-to-capacity ratio was 0.68 using this dynamic model versus 0.78 using the asymmetric static model (Berka et al., 1994).

Table 4: Solution Characteristics for Morning Peak Period by Road Class

Link Class	Travel Distance (miles)	Travel Time (minutes)	Mean Speed (mph)
Collector	1.18 (1.34)	4.20 (5.97)	16.86 (13.47)
Arterial	6.03 (6.83)	17.42 (21.65)	20.77 (18.92)
Freeway	2.82 (3.01)	3.62 (3.95)	46.74 (45.73)
All Classes	10.02 (11.18)	25.24 (31.61)	23.81 (21.22)

(.) results from Berka et al. (1994)

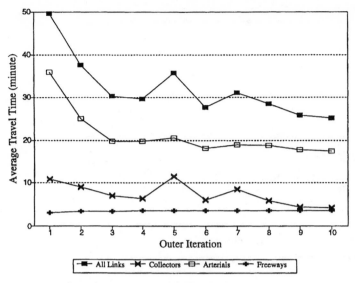

Figure 5: Predicted Average Travel Time

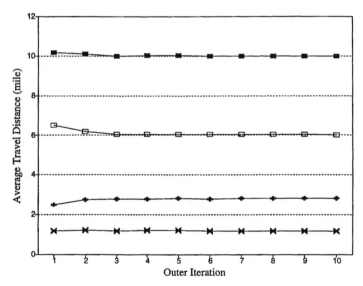

Figure 6: Predicted Average Travel Distance

Unfortunately, link flows and link travel time data are not available for the ADVANCE Network, either in general, or more specifically for the O-D matrix used in this solution. These data, as well as route flow data, are urgently needed to advance the state of the art of network modeling for ITS.

The problematic dynamic user-optimal assumption that complicates the modeling of many route choice options is that route choice decisions must be based on travel

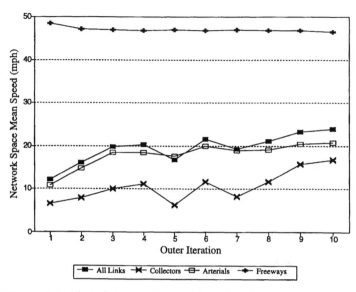

Figure 7: Predicted Average Space Mean Speed

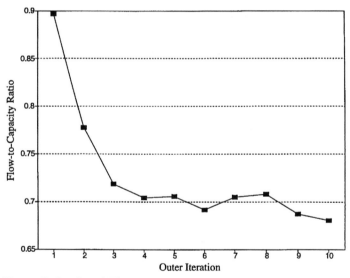

Figure 8: Predicted Flow-to-Capacity Ratio

costs which are temporally-consistent with future link flows at time of link use. The assumption is appropriate for recurrent trips and traffic conditions and is also acceptable for scheduled events (e.g., ballgames, concerts, detours, road constructions) and for predicted weather conditions. However, this behavioral assumption is inconsistent with *unexpected* events (e.g., stalled vehicles, dropped objects, accidents) at future times because drivers have very limited capability to be informed about the times and locations of such events before they encounter unusual queuing delays

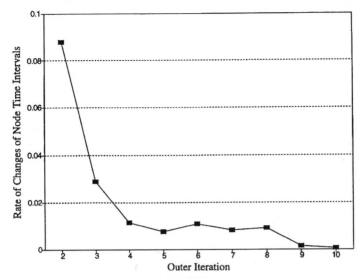

Figure 9: Rate of Change of Node Time Intervals

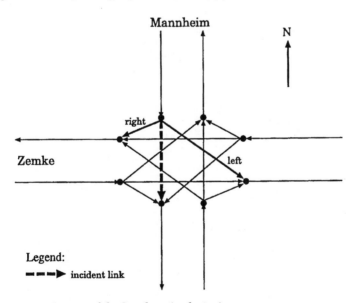

Figure 10: Layout of the Incident Analysis Area

caused by those incidents. Enroute diversions are thus expected to happen when incidents are encountered by drivers.

The solution algorithm DYMOD was modified to model incident-related enroute diversion. First, base-level link flows of all trips assuming usual (incident-free) conditions are generated. Then, the portion of trips assumed to choose a diversion strategy is removed from those base-level flows and reassigned onto the network according to then-current travel times. Through this approach, drivers who choose

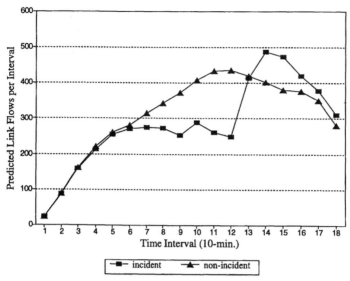

Figure 11: Predicted Flows of the Incident Link

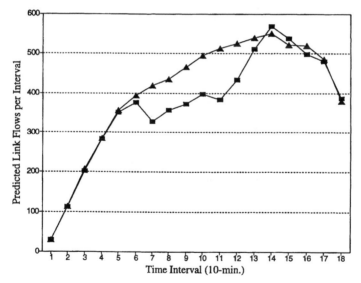

Figure 12: Predicted Flows of the Upstream Link

non-diverting strategy are actually based on fully anticipatory travel times for route choice. This approach might violate temporal continuity of trips to some extent in those time intervals in which the non-diverting base-level flows use the links are affected by the enroute diversion flows. However, it becomes less severe as the non-diverting portion of drivers increases. Further, this approach also provides an estimate when enroute diversions are guided by in-vehicle information.

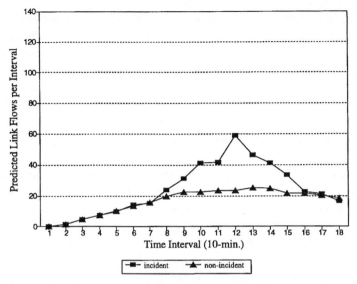

Figure 13: Predicted Flows of the Right-Turn Movement

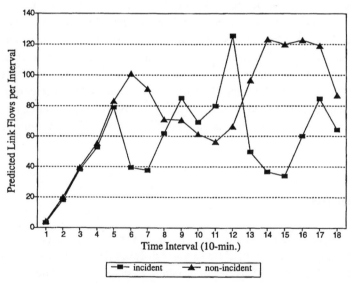

Figure 14: Predicted Flows of the Left-Turn Movement

A key advantage of DYMOD is that link-specific capacity adjustments can be input exogenously or generated endogenously over time intervals when they occur. Exogenous link capacity adjustments specified to the program when detected or expected to occur might be caused by accidents, weather conditions, time-of-day restrictions for special events, road construction or signal timing changes (if signalization attributes are given outside from the program). Link capacity changes endogenously because spillback queues reduce capacities of upstream links, ramp metering or signal timing changes (if actuated signal timing mechanism is presented and integrated). Capacity adjustments are made between the *inner* and *outer* iterations.

To demonstrate the effects of enroute diversions that resulted from incidents, we placed an incident on a through-movement link of a major intersection which caused a 50% capacity reduction for one hour (through 7 AM to 8 AM) and 50% of drivers were assumed to choose the diversion strategy. Note that the scale of capacity reduction can also be specified over intervals to account for the gradually capacity loss caused by the incident.

Figure 10 shows the layout of the incident intersection. It is a four-leg signalized intersection without any turning restrictions (U-turn excluded). Figure 11 exhibits the flow profile of the incident link along time intervals. Obviously, the flows on the incident link are lower than the non-incident flows during the incident periods. Due to the incident removal, the incident flows become higher than the non-incident flows again following the removal of the incident. The upstream link has a similar result as shown in Figure 12. However, significant flow diversions appear in the right-turn and left-turn links during the incident periods (see Figure 13 and 14). The diversions on the right-turn link are quite obvious; the incident flows are higher than the non-incident flows during the incident period and a few subsequent intervals. The flow profile of the left-turn link is somewhat complex but still reasonable. The incident flows begin to exceed the non-incident flows at the second incident interval because of the normal left-turn delay. The incident flows drop soon and return to the usual used links following the clearance of the incident because of the capacity restoration on the incident link. Based on this modeling capability, this model can be utilized as a tool to assess an area-wide incident management strategy and resulting flow pattern.

6. Conclusions

This chapter describes a VI formulation of a dynamic user-optimal route choice model for predicting real-time traffic flows for Advanced Traffic Management Systems (ATMS) and Advanced Traveler Information Systems (ATIS). The model formulation and solution algorithm are described. Model refinements, extensions and computational results from the ADVANCE Network are also presented. Realistic traffic engineering-based link delay functions are utilized for better estimation of traffic dynamics.

The proposed model is capable of modeling the effects of enroute diversions caused by highway incidents such as stalled vehicles, dropped objects and traffic accidents, etc. Using Janson's DYMOD algorithm, various time-dependent traffic characteristics are obtained and analyzed. We believe this is the largest dynamic route choice model implementation which has been solved thus far.

To date, very few traffic flow prediction models are suitable for ATMS and ATIS applications. Although not yet fully validated, DYMOD is able to predict time-dependent traffic characteristics for a large-scale traffic network which are reasonable and internally consistent. Eventually, dynamic route choice models should be integrated into a traffic control and management center to support the decisions on the adjustments of arterial signal timing, ramp metering, incident management and future route guidance strategies, etc.

Acknowledgments

The authors are pleased to acknowledge the following sources of support in the preparation of this chapter: the Illinois Department of Transportation and the Federal Highway Administration through the ADVANCE Project; the Institute of Transportation of the Ministry of Communications, Taiwan, Republic of China, in support of the ADVANCE Project; the National Institute of Statistical Sciences; the National Center for Supercomputing Applications, University of Illinois at Urbana-Champaign, for the use of computational facilities. The last two organizations are supported by the National Science Foundation, Washington, D.C.

References

Akcelik, R. (1988) Capacity of a shared lane, *Australian Road Research Board Proceedings*, 14(2), 228-241.

Berka, S., Boyce, D.E., Raj, J., Ran, B., and Zhang, Y. (1994) *A Large-Scale Route Choice Model with Realistic Link Delay Functions for Generating Highway Travel Times*, Report to Illinois Department of Transportation, Urban Transportation Center, University of Illinois, Chicago.

Janson, B.N. (1991a) Dynamic traffic assignment for urban networks, *Transportation Research*, 25B, 143-161.

Janson, B.N. (1991b) A convergent algorithm for dynamic traffic assignment, *Transportation Research Record*, 1328, 69-80.

Janson, B.N. (1995) Network design effects of dynamic traffic assignment, *Journal of Transportation Engineering/ASCE*, 121, 1-13.

Janson, B.N. and Robles, J. (1995) A quasi-continuous dynamic traffic assignment model, *Transportation Research Record*, 1493, 199-206.

Kimber, R.M., and Hollis, E.M. (1979) Traffic Queues and Delays at Road Junctions, *Transport and Road Research Laboratory Report*, 909, Crowthorne, Berkshire.

Kyte, M. and Marek, J. (1989) Estimating capacity and delay at a single-lane approach all-way-stop controlled intersection, *Transportation Research Record*, 1225, 73-82.

Lee, D.-H. (1996) Formulation and Solution of a Dynamic User-Optimal Route Choice Model on a Large-Scale Traffic Network, Ph.D. thesis in civil engineering, University of Illinois, Chicago.

Meneguzzer, C., Boyce, D.E., Rouphail, N., and Sen, A. (1990) *Implementation and Evaluation of an Asymmetric Equilibrium Route Choice Model Incorporating Intersection-Related Travel Times*, Report to Illinois Department of Transportation, Urban Transportation Center, University of Illinois, Chicago.

Ran, B. and Boyce, D. E. (1996) *Modeling Dynamic Transportation Networks*, Springer-Verlag, Heidelberg.

Webster, F.V. (1958) Traffic Signal Settings, *Road Research Technical Paper*, 39, Road Research Laboratory, Her Majesty's Stationery Office, London.

Zhang, Y., Hicks, J. and Boyce, D.E. (1994) Trip Data Fusion in ADVANCE, *ADVANCE Working Paper*, 43, Urban Transportation Center, University of Illinois, Chicago.

Transportation in Knowledge Society

T. R. Lakshmanan
Bureau of Transportation Statistics
U.S. Department of Transportation

> "What hath God wrought!"
>
> Samuel Morse, launching the telegraphic era, 1844

Introduction and Overview

The history of transportation is one ever increasing speed, comfort, convenience, safety, reliability, and continuously dropping costs of travel. Most people attribute this historical progression of transportation to technological progress in the vehicles and infrastructures. It is true that from the invention of the wheel, the harness, the sailing ship etc. in early times, the pace of technology advances in vehicles and propulsion systems quickened in recent times, as the steam engine, the electric street car and railroad, the internal combustion engine, the jet engine, containerization, and the "mega-ship" have arrived to improve the quality, and lower sharply the costs of travel. Parallel advances in transport physical infrastructures – tunnels, suspension bridges, railroads over all kinds of terrain, the U.S. Interstate System, and modern airports and marine terminals – have also contributed. Third, the development of appropriate *non-material infrastructure* – knowledge, technical standards, procedures, laws, and policies that guide the governance of transportation— has provided social coordination of the vehicles and the physical facilities and made possible a remarkable level of cheap, efficient, and safe transport – even as all types of travel are expanding dramatically.

What has not been, however, generally noted is the important and recently increasingly role of a fourth component of the transportation system – namely the variety of information and knowledge technologies embodied in transportation capital and infrastructure. These technologies vastly improve the speed, reliability, and safety of transportation, while sharply dropping its costs. They provide information and knowledge vital for transport operations, enhancing their responsiveness and efficiency, and enabling and indeed driving innovations in transportation.

Such *knowledge providing and enabling* technologies have historical antecedents in the sextant and the chronometer which permitted precise global navigation in the 18th century, the telegraph which promoted transcontinental rail operations in the last century, and the radio and radar critical to navigation in this century. Currently there is an efflorescence of new information technologies which are transforming in a major way the transportation industries and the scope of services they offer.

In the last several decades, the term Information Technologies or IT is used to refer to a *broad array of specific devices, functionalities, and supporting tools associated with the sensing, generation, processing, communicating, and presenting information.* The contemporary IT – represented by dramatic technological developments in the computer hardware and software, telecommunications, navigation and positioning surveillance, sensing, tagging, data exchange and data fusion, etc – offer key capabilities, many embodied in hardware, which can be applied for a broad variety of transportation applications. As applied to transportation, IT form nodal elements of physically distributed transport networks, offering many new broad capabilities – in the form of efficient, faster, safer air, rail, truck or port operations, computerized transport documentation, "quick response" intermodal freight transport, and the development of intelligent passenger and freight "transport governance systems" – all of which increase personal mobility and lower costs of freight transport firms enabling them to offer transport solutions customized to knowledge-intensive enterprises serving national and global markets.

This paper discusses:

- the nature of these knowledge technologies,
- the factors underlying their rapidly expanding application in the transportation system, and
- the variety of benefits of the rapid dissemination of transport IT accruing to travelers, transport enterprises and to the customers of these enterprises in the emerging Knowledge society.

The latter benefits appear in two forms: *first,* as consequences of transport information technologies in terms of a) the *availability and supply* of new transportation services with sharply more attractive cost and service attributes, and b) the redefinition of mobility and access needs that shape the *demand* for transportation services based on the substitution of telecommunication for physical movement; *second,* as improvements in the quality and variety of transport services offered and the creation of new forms of customer, cost, and competitive relationships – resulting from the operational and strategic uses of IT by a transport firm – occurring in the firm itself (e.g., the role of computerized reservation systems in an airline's quality of customer service, flexibility, reliability, response times, yield management, market positioning, etc.), as well as in the production enterprises the transport firms serve (global sourcing arrangements, faster response innovation systems, networking of firms, etc.).

The paper begins with a brief survey of the major elements and the key attributes of information technologies in transportation. An important attribute of IT is its primary complementary role as an enabler, rather than a replacement for existing hardware, a factor which reduces major obstacles to new technology and indeed promotes rapid and continuously evolving applications of IT in transportation. Another favorable factor is the ongoing broad restructuring and strategic changes in the production and service industries, which in turn provides a hospitable environment for pervasive applications of IT in transportation. The paper then explores the implications of IT on the *supply* of a variety of 'intelligent' passenger and freight transportation services, in particular the potential for expanding *the horizons* for mobility and access. Then the paper discusses the evidence and speculations on the *complementarities* and *substitutions* between information technologies and transportation and the likely influences on the *demand* for transportation mobility and access. Finally, the paper surveys the available information on the ways in which the improvements created by IT in the cost/service characteristics of transport enterprises are affecting the quality, variety and performance of services that can now be offered by non-transportation industries, and the new ways the latter industries can function.

Information Technologies and Transportation Services

Information Technologies in Transport History

There has long been a degree of interactivity and synergy between information and transportation technologies. As noted earlier, non-transportation technologies providing information on location and other variables of interest to transport operations, control and management have greatly aided over the years the efficiency of provision of transportation services. At the same time, the ease and efficiency of conveyance of information depended for a long time on the capabilities of the transportation system.

In the history of this interaction between transportation and information technologies, the telegraph provides a major discontinuity. In the pre-telegraphic era, information was provided in a physical form, as exemplified by mail, newspaper, book, the human messenger, etc. Information could travel as fast as the available means of transportation would permit at that point in time. Thus in 1799, the news of President George Washington's death took about two weeks to reach southern New England from the new nation's capital and over three weeks to get to Southwest Ohio (Fig. 1). Three decades later, in 1830, the "express relay" of President Jackson's "state of the Union" message took only a day and a half to reach southern

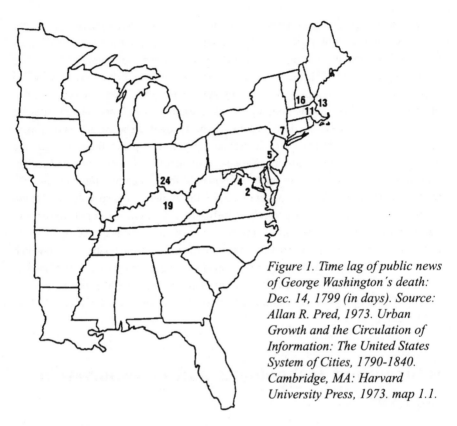

Figure 1. Time lag of public news of George Washington's death: Dec. 14, 1799 (in days). Source: Allan R. Pred, 1973. Urban Growth and the Circulation of Information: The United States System of Cities, 1790-1840. Cambridge, MA: Harvard University Press, 1973. map 1.1.

New England and over two days to arrive in southwest Ohio (Fig. 2).The advent of the telegraph, fifteen years later, was a radical event. The telegraph, for the first time in history, made possible *not only almost instantaneous communications between distant locations possible but also separated the message from the messenger.*

The impacts of the telegraph, arriving contemporaneously with the railroads, were far-reaching. Not only were earlier forms of long distance transport adversely affected, but there were revolutionary impacts on the size, scale, and manner of provision of transportation and information services, and indeed on the structure of the overall 19th century economy (Chandler, 1973). In 1861 mail service via the Pony Express between Missouri and California was terminated after one year's operation when the transcontinental telegraph became operational.

The railroad offered the capability to move an unprecedented volume of goods at an unprecedented speed on a precise schedule (measured not in weeks but in days and hours). The telegraph system, built on the railroad right-of-way, became a critical instrument in assuring fast, safe, and efficient movement of trains carrying a wide variety of goods shipped from hundreds of origins to as many destinations – particularly critical in the continental rail system in the U.S., where more single-track roads were built than in the U.K. The telegraph also made it possible for

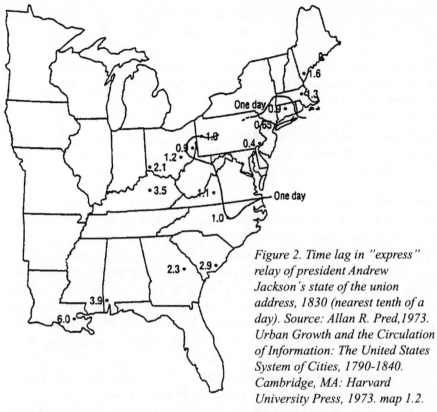

Figure 2. Time lag in "express" relay of president Andrew Jackson's state of the union address, 1830 (nearest tenth of a day). Source: Allan R. Pred, 1973. Urban Growth and the Circulation of Information: The United States System of Cities, 1790-1840. Cambridge, MA: Harvard University Press, 1973. map 1.2.

enterprise owners to use the new power for instantaneous information delivery to *manage the enterprise from a distance.* American railroads became during the 1850s pioneers in modern management. Larger enterprises with new managerial hierarchies (to run multiple departments) developed and led to the large modern corporation, first in the railroad industry and later in other industries in the economy – an organizational development that Chandler (1973) attributes to the emergence of U.S. as a major industrial power in late 19th century.

The New IT: Key Elements

"Information technology" is typically used as a collective term with which to refer to a very broad array of specific devices, functionality's and supporting tools associated with the sensing, generation, processing, communicating, and presenting information. IT has now progressed to the point of dealing almost entirely with information in digital form – captured in strings of zeros and ones, thereby permitting levels of precision and freedom from error unattainable with the analog technologies of earlier years. Exploitation of the power of digital communication and information processing is an essential characteristic of many IT applications.

Most transportation uses of IT depend upon the often-transparent and inseparable integration of multiple specific technologies and capabilities. In many cases, cost and performance improvements in any one such technical area make a whole new range of applications or services available. The potential for IT to affect mobility and access in modern society is rooted in a small set of building blocks:

- Telecommunications
- Computer Hardware and Software
- Navigation and Positioning Systems
- Surveillance, Sensing and Tagging Technologies
- Data Exchange and Fusion Capabilities.

Telecommunications

One of the most visible aspects of the information revolution is the ease and relatively low cost of telecommunications. The advent of satellite-based and fiber optic communications channels, cellular telephone systems, and network linkages epitomized by the internet have made a reality of "anywhere-to-anywhere" communications, often at a remarkably low cost. The ability to make real-time operational information available throughout a transportation system, no matter how geographically dispersed, and to combine with it data from other sources, carries powerful benefits for operators, managers and users. A very important subset of this function is electronic data interchange (EDI), which can occur over a very short-range: between trucks and wayside inspection stations, highway vehicles and toll collection stations, or cargo containers and port logistic systems.

Telecommunications services are also playing another, potentially even more fundamental role: replacement of physical transportation. Telecommuting, tele-shopping, video conferencing, and other "tele-substitutions" are already, although in relative infancy, beginning to be a significant factor in shaping mobility and access needs and how they are met.

Computer Hardware and Software

The many-orders-of-magnitude reduction in the cost of computing power in the last few decades now permits the inclusion or embedding of highly sophisticated sensors and computer-based control systems in "smart" consumer items such as automobiles. "Smart cards" – that can include the functions of credit and ATM card, identification card, transit pass, and even "digital cash," are now becoming common.

This processing power can be used for pattern recognition, generation of speech and images, data storage and retrieval, exercise of highly sophisticated algorithms for traffic control and management functions, and complex simulations of transport operational systems. Concepts of artificial intelligence, expert systems and fuzzy logic represent software approaches that can be of great effectiveness in particular situations. Enormous data bases can be already accessed directly using compact disc storage technology.

Navigation and Positioning Systems

Navigation has always been central to long-distance transportation, particularly for marine and aviation modes. From the invention of the sextant and chronometer to radio-based systems of the present (e.g., RADAR, radio beacons, LORAN, and OMEGA), improved means of determining position have represented major transportation advances. Other techniques, used singly or in combination, increase the tools from which the system designer can choose: automated beacons, gyroscope-based inertial navigation, and even sophisticated dead reckoning. The area is now being transformed by operational availability of the Global Positioning System (GPS), based on 24 earth-orbiting satellites emitting precisely timed signals. GPS now makes possible convenient real-time three-dimensional position information throughout the world with great accuracy and at very low cost.

As the practical precision of these technologies has improved – dropping from hundreds of meters or more down a meter or less – and prices fall, new transportation and other applications become potentially feasible, such as route navigation for highway vehicles and landing aircraft under zero-visibility conditions. Many applications involve position determination equipment on the vehicle, with that information then communicated (often via satellite links) to a central computer for fleet or traffic management functions. Precise knowledge of location can support functions as disparate as placing and retrieving containers in a port storage area, identifying the whereabouts of a stranded motorist for response vehicles, and assuring safe separation of trains on shared tracks.

Just as the 18th century navigator needed charts and astronomical almanacs in order to derive position using a sextant and chronometer, the GPS and other modern systems fulfill their potential only in concert with digital maps of high precision, based on a standardized geodetic reference system. More generally, geographic information systems (GIS) are finding many applications in transportation, where they support operational, maintenance, planning, system analysis and other functions.

Surveillance, Sensing and Tagging

Many IT applications in transportation depend upon a vehicle's being able to characterize elements of its environment, or on the ability of some element of the fixed infrastructure to identify and exchange information with vehicles, cargo or individuals. Closed circuit television – monitored by computers – is increasingly playing a role in monitoring freeway traffic to detect incidents and congestion, railroads tag freight cars for wayside identification, and aircraft continually exchange identification numbers, altitude and other information with the air traffic control system. High performance computer chips and sophisticated pattern recognition software can increasingly carry out complex sensing and surveillance tasks. Tagging of freight-carrying vehicles and of the cargo itself is revolutionizing logistic processes, as tags become capable of storing situation-specific data (such as container contents and destination) and of being interrogated over long distances – via satellite in some cases. A truck with a GPS-enabled tag and cellular or satellite communications capability can be tracked anywhere in the country, reporting status of the vehicle and cargo wherever it goes. Here is the emergence of "on-board intelligence" in the freight system or the intelligent freight transportation system.

Another aspect of surveillance is the use of vehicle-mounted sensors, such as radar, to detect potential hazards, coupled to computers and displays that provide warnings and suggested responses. Collision-avoidance systems of this type have been used in aircraft for many years, and Federal and corporate research programs are currently very active in the area of highway vehicle crash prevention.

Data Exchange and Fusion Capabilities
Above and beyond the specific transportation functions that various information technologies support, the broad impact of IT on transportation is to produce a much more "knowledge-rich" system, in which all participants – users and providers alike – can make decisions and perform tasks on the basis of greatly enhanced understanding of the likely consequences. The combination of often-disparate information from diverse sources is central to many IT applications, including real-time operations, inventory and process management, system characterization, and provision of user services. Beyond technical interoperability and establishment of institutional and organizational relationships for effective data exchange, the new discipline of "data fusion" has emerged – the design of computer systems that can facilitate convenient, flexible and adaptable blending of information from a wide range of independent sources.

Factors Underlying the Rapid Diffusion of Transport IT

Two sets of factors underlie the on-going rapid progress of the new information technologies in the transportation sector:
- From the supply side, certain inherent traits of these technologies, and their enabling and complementary role vis-a-vis transportation technology facilitate rapid applicability of IT;
- From the demand side, an enormous variety of market-driven opportunities for IT investment are developing in the hospitable economic environment provided by the transforming American economy, which is "dematerializing", globalizing, and restructuring in response to technological and institutional (e.g. deregulation, privatization, etc.) changes.

Enabling and Complementary Role of Transport IT
The application of IT differs in many ways from the traditional means by which new propulsion, vehicle and physical infrastructure technologies have been incorporated into transportation operations. This is in due part to the inherent nature of IT, and in part to its role in transportation. Special characteristics that shape the evolutionary process include the following:

- *Rapid Technical Evolution.* Driven by very broad markets across all economic sectors, information technology is developing very rapidly – dropping sharply in cost while simultaneously increasing in performance. Not only are the basic information technologies advancing, but understanding of how best to use them is also still a major source of improvements by itself.
- *Enabling Complementary Role and Continuous Evolution.* The rapid technical evolution of IT, and its primary role as an *enabler, rather than a replacement* for existing hardware, implies that systems not only can evolve continuously, but are virtually required to. Information technology improves the performance – extending the reach, power, and quality of service – of existing transportation hardware. In contrast to the spurts of innovation typically associated with periodic injections of new transport technology (requiring replacing older still functional hardware) – such as characterized the diesel locomotive or jet aircraft – barriers to introduction of new technology are much reduced. As the costs of its innovation thus come down, transport IT becomes more attractive.
- *Process Reengineering.* IT often involves changes in how a system operates. Thus, realization of significant benefits from the introduction of the hardware and software of information technology typically call for new "orgware" – redesign of operational processes throughout the enterprise, accompanied by substantial redefinition of work functions. As for all innovations, the pace of change is determined primarily by ability to incorporate the new technology. But IT is so pervasive that the required level of organizational change can be significant – as widely experienced in the more general infusion of IT into business activities of all types during the last two decades
- *Ubiquitous Applications.* The information technologies applied in the transportation sector are generally those also used across the economy and society for an enormous range of purposes. This characteristic has provided the broad markets that sustain the observed high rate of improvement in cost and performance characteristics, and also facilitates large-scale integration and optimization across the transportation enterprise and its suppliers and users.

Further, the commonality of information technologies in virtually all sectors enables and fosters an ever tighter linkage among transportation and other business processes. The initial focus of the ubiquitous bar code – the Uniform Product Code – was improved speed and accuracy in accomplishing retail sales transactions. Success in this effort led to linkage of point-of-sale computers (no longer simple "cash registers") to inventory data bases and supported a shift to greater reliance on "just-in-time" logistic practices and marketing strategies. (That linkage can be implemented at a corporate level, with the local purchase price based on query to a central computer hundreds or thousands of miles distant.) With this information infrastructure in place, supermarkets are now beginning to offer remote computer-based ordering, via Internet or PC-modem connections, coupled to home delivery – a step which, if successful, has significant transportation implications.

The Hospitable Economic Environment for Transportation IT

In the context of rapid technological change and changes in economic institutions, a number of interrelated supply and demand-related factors are ushering in major structural changes in the American economy. These ongoing economy-wide changes, as detailed below, are transforming and in turn being transformed by transportation activities. Such economy-wide changes include: the increasing role of information and knowledge in the economy, the long-term lowering of intensity of material use and energy use (hence the appellation "dematerializing economy"), the globalization of markets, resources, and leisure destinations, the emergence of more market-oriented relationships among economic actors, and the beginnings of seamless integration of (intermodal) transportation into the overall production processes. Such broad changes offer a fertile ground for transport sector investment in information technologies, which would facilitate adaptation to a restructuring economy.

An Increasingly "Dematerializing" Economy

As real incomes increase in an affluent, industrialized country such as the US, con-sumer demand increasingly shifts towards information, knowledge, and services. The emergence of this knowledge-intensive Information Economy has been noted for some time (Machlup,1962, Porat and Rubin 1977) – though with little comment on the transportation implications. Increasingly higher value and (lighter) goods as well as services, embodying a high level of knowledge, become high value generators and become progressively more important in such "post-industrial" societies. In these societies some rapidly growing high value-adding industries move only a small amount of tonnage while accounting for a disproportionately large share of total value of freight carried.

Moreover, in recent times, the advent of new production technologies has led to more efficient production processes and energy-saving processes. This has the effect of further lowering the material and energy intensity of the overall US economy. As a consequence, as noted in Figure 3, the long-term trend in this century in the US is clearly a drop in the intensity of use of industrial materials – as measured by kgms/ \$GDP in constant dollars – such as steel. cement, ammonia, aluminum, paper, and chlorine. Further, the energy use per unit of GDP has also been dropping over time. While the material and energy intensity may be falling, in a large and diverse industrial economy such as the US large volumes of energy materials, other natural resources and primary products still move.

Globalization of Markets, Resources, Leisure Destinations

With the recent expansion of international trade (growing three times as fast as GDP in the US in the last quarter century), and the development of flexible production systems, economic activity is increasingly geared to serving global markets, assemble production inputs from supply points around the globe, and provide for the growing demand for leisure destinations all over the globe. This development was further

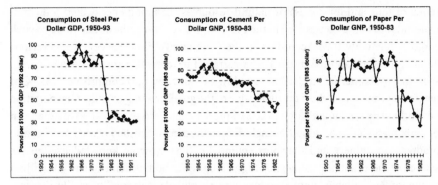

Figure 3. Trends in U. S. consumption of selected materials per dollar GDP, 1950-1993.

helped by the collapse of the Soviet Union and the tempering of a previously hostile, polarized world. Markets for raw materials, components, capital, labor and final products are increasingly global and flexible. An inherent requirement of this environment is ever-more responsive and efficient global transportation of both goods and people. From the transportation perspective globalization has been made possible, initially by deployment of steamships, more recently accelerated by jet aircraft, containerized shipping, and satellite and fiber communications.

Increased Market-oriented Strategies: Privatization, Deregulation, Restructuring and Partnerships

Globalization and the intense competitiveness produced by a multiplicity of sources and markets is generating new relationships among all players in the economic system, public and private. These relationships take a variety of forms such as partnerships and networks. The resulting organizational networks are enabled by associated communication and transportation links.

At the same time, the boundaries between the private and public realms in transportation in U.S. and elsewhere are undergoing a long-term shift – not only are transport services being deregulated, but infrastructure services are also being and privatized.

Incorporation of Transportation into Production Processes and Products

Competitive pressures on price and service, accompanied by just-in-time manufacturing, outsourcing and on-demand delivery, and supply management have focused attention on seamless integration of transportation functions into the overall production process. This integration and incorporation of transport into production offers increasing opportunities for the use of IT in transportation.

The transportation implications of the above "dematerializing economy" trends are already evident in the 1993 Commodity Flow Survey conducted by the Bureau of Transportation Statistics (BTS). In the US economy, while a large quantity (in terms of tons and ton-miles) of natural resources including energy, and industrial materials are being transported in the growing US economy, the dollar value of such cargo is relatively low. According to the Commodity Flow Survey, if the various commodities transported in the US are arrayed in terms of value per ton of cargo, such materials (defined as $1000 or below of value per ton) comprise of 95.7% of the tonnage but only 46% of the total value of the goods moving over the transportation networks. The remaining cargo (4.3% of the tons and 54% of the value) are made up of eight high value commodities including instruments, finished leather and apparel products, machinery and transport equipment. The growing knowledge-intensive sectors of the economy using such high value commodities demand and are willing to pay for fast, reliable, quick response service, global reach, quality, etc. IT in transport is eminently suited for providing such characteristics and hence the growing investments in IT.

The Transformation of Transportation by IT

Transformation of Services: A Framework

To understand better the broad impacts of IT on the *emerging quality and quantity of transportation* services it may be useful to clarify the *systemic* nature of transportation services. Societal needs for mobility—and hence transportation—have traditionally been met by a combination of *four* basic system elements:
1. *vehicles of all kinds,*
2. *the physical or material infrastructure* (roads, rail, airport, etc) used by the vehicles,
3. *information capital and network infrastructure* and,
4. a *non-material infrastructure*: or an operational framework – people acting within a framework of processes and knowledge (policies, rules, regulations, institutions, etc) – that design, build, operate and maintain the vehicles and infrastructures, and that plan and shape the governance and future evolution of the entire enterprise (Fig. 4).

To a large degree, the history of transportation has been driven by transportation technology – particularly by the technology of vehicles and propulsion systems – the steam engine applied to marine and rail transport during the industrial revolution and global expansion, the electric street car a major force in creating early suburbs in the late 19th century, the internal combustion engine in the 20th century, and recently, the jet engine, "mega-ship," and containerized freight .

Physical infrastructure has kept pace with the vehicle and propulsion advances, as is most evident in achievements such as tunnels and suspension bridges, railroad lines crossing deserts and mountain ranges, airports that are virtually small cities, highly efficient marine ports, and the network of modern highways that comprise the US Interstate System.

Several enabling information-related technologies not inherently associated with transportation (e.g. the telegraph, radar, and the new IT) as well as the stock of information have also been central in supporting the movement of goods and people.

The system context or the non-material infrastructure within which the above three elements are used is less obvious, but represents a complex web of knowledge, institutions, skilled personnel, policies, rules, regulations, etc., used in the Governance of transport. Over the years, this web has evolved, supporting global aviation and marine transport, and surface transportation that reaches across continents, and the development of technical standards and procedures that yield an unprecedented level of efficiency and safety in the face of dramatically increased volume of passenger and freight traffic.

The quantity, variety, and quality of the transportation services are jointly determined by these four transportation system elements and the interactions among them. Strategic decisionmakers (both in the public and private sectors) concerned with improvements in transportation ought to consider all the elements including information infrastructure.

As the information technologies experience rapid incorporation into all facets of the creation, management and use of transportation services, the mobility of goods and people in the US is being affected ever more strongly. In the closing decades of the 20th century, information technology has become a primary element driving the

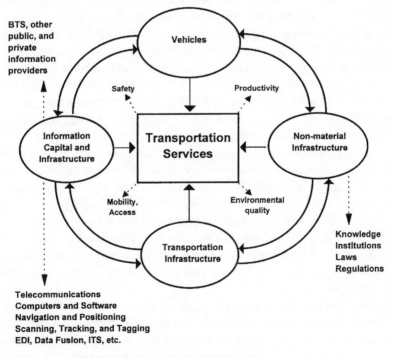

Figure 4. Components of the tranportation system.

evolution of transportation supply and demand. As a motivater and enabler of innovation in transportation, IT may likely ultimately have an impact on system performance and use comparable to that of the propulsion system advances from steam power to the jet engine. IT is already a major piece of the system by which societal needs for mobility and access are satisfied. Application of these technologies not only provides an information "overlay" to improve operation, control and management of the physical elements of transportation, but also is transforming virtually every organizational process involved in the creation of transportation services.

Information technologies are being adopted throughout the economy in ways that are fundamentally affecting the nature of society's needs and desires for mobility and access. Today, people across the world constantly choose among fax, e-mail and overnight express for movement of documents, or debate whether an inter-city trip might be avoided with a telephonic conference call – such changes affecting commercial operations and the daily life of individuals, at every step from research through manufacturing of goods and operation of services.

But beyond those more visible effects, *fundamental changes are occurring in how transportation works and is used.* Applications of IT lowers the cost and raises the quality of service of operations such as the acquisition and processing of operational, financial and related data directly supporting higher system capacity, greater labor and capital productivity, improved efficiency, more effective resource allocation, and better integration of transportation services.

Detailed understanding of system operations also enables design and implementation of innovative operational concepts and practices. The overnight package delivery service. for example, could hardly exist on its present scale without myriad computers, digital tags on each item, and extensive communications links. It is difficult to conceive of operating the Nation's commercial aviation system – with well over one million airliner seats to be filled daily at market-driven prices for several thousand origin-destination pairs – without the technology that underlies and links modern computer reservations systems.

More recently, information technologies are beginning to transform the interface between the service provider and the customer or user. The implementation of Intelligent Transportation Systems (ITS) includes as a basic component the provision of real-time status information for local highways and transit systems, potentially with guidance as to optimal choices for any particular trip. Any individual with access to the World Wide Web can find the status of an overnight package in seconds. With a highly controlled and visible transportation system, manufacturers can safely integrate their suppliers into "just-in-time" production systems and tailor their outputs to short-term customer needs.

Less direct, but potentially of greater impact, *are the ways in which information technologies alter the need to travel, or the type of transportation consumed.* A film can *be* seen by traveling to a theater, accepting what is available from television broadcasters, traveling to a video film rental store, or staying home but ordering a specific choice for direct cable delivery. A trip for a business meeting may be replaced by a joint telephone call or video conference, and personal purchases may be made by traveling to a store, or by ordering (over telephone or Internet) and relying on a delivery service. Aided by modem-equipped computers, fax machines, and on-line services, a growing segment of the population is finding it feasible – and often preferable – to work from home, either as a telecommuting employee or as an independent business or contractor.

The full impacts of this IT evolution on transportation will emerge only slowly in coming decades, and often *will not be separable from the effect of other societal changes*. These impacts are described in two parts in what follows. First, the next section surveys the new transportation services made possible by IT in the form of the new supply of transportation services created by IT, and the changing demand for transportation services (e.g., Telesubstitution – substitution of communication for travel), induced by IT. Second, an inkling of the broader societal changes being generated by the new IT-related transportation services can be provided by the emerging major structural and other changes in the national and global production and consumption systems, which define the demand for transportation services. Such economy-wide changes pertain and of personal travel *to structural economic shifts*, global reach of product markets, and production inputs.

The Supply of IT-Induced New Transportation Services

IT is changing the availability and supply of transportation services in many ways, with concomitant changes in the ways that needs for mobility and access are satisfied. These derive from the IT-induced effects – reduced cost, improved capacity, shorter trip times, expanded availability, better service and convenience, new functionality's, and by minimization of adverse environmental and other social impacts.

Cross-Modal and Intermodal Transportation

Urban/Suburban Passenger Transportation
One of the largest and most focused areas in which information technology is being applied to transportation is road and public transit operation in urban and suburban areas. A broad partnership between public agencies at all levels and the private technology and transportation industries is stimulating major investments in "intelligent transportation systems" (ITS) ranging from R&D, through operational tests and demonstrations, to "real-world" deployment. Centralized and coordinated *management of highway traffic and transit operations* – bus and light, heavy, and commuter rail – is a central ITS element, as are advanced systems for providing

convenient and comprehensive information to travelers concerning current and projected status of all components of the local transportation system. Electronic collection of tolls can greatly reduce peak-hour delays at highway toll booths in urban areas. Examples of such ITS deployment in the US are: the 18-mile Santa Monica Smart Corridor project in Los Angeles (a 20% reduction in intersection delay,11-15% decrease in motor travel time, and substantial fuel savings by 1998), and the Kansas City transit system, which for a $2.3 million investment in AVL (automatic vehicle location) technology avoided $1.5 million investment in new rolling stock, saved $400,000 per year in operations and improved on-time performance.

Only limited success has been achieved over the years in bridging the gap between taxicabs (high level of service but relatively high cost) and fixed-route buses (inflexible and often infrequent service at low cost), with only limited success. Merely providing reliable and accurate information at bus stops (and by phone or Internet) could make service significantly more attractive. Demand-responsive para-transit services (e.g., dial-a-ride, shared-ride taxi, and vanpooling) have been used across the US since the 1970's, but the cost-service characteristics have generally restricted them to specialized functions, such as transportation for the elderly and handicapped. However, technology advances in position-location, navigation, automated dispatching (centralized or on-board), and local communications may yield more efficient and less labor-intensive realizations that greatly expand the viability of this concept.

Freight Transportation and Logistics
IT is also rapidly being applied to the broader management of freight logistics. Combinations of computer and satellite communications technology now make it technically possible to track the location of a shipment or vehicle virtually anywhere in the world, and for on-board and central computers to exchange any information relevant to the operation. The widespread adoption of "just-in-time" manufacturing and delivery to wholesalers and retailers virtually necessitates the tracking individual shipments and continual adjustments throughout the enterprise to respond to changing circumstances. The net effect is an increasingly tight integration of the entire chain from raw materials and components to manufacturing and delivery to the final customer. Only by providing the ability to track and control shipments precisely (sometimes referred to as "in-transit asset visibility") can the transportation system achieve the flexibility to respond to the continual changes and quality-of-service demands that characterize the modern global economy.

Data describing origin, destination, contents, etc. – whether encoded on the tag or contained in a linked database – can substantially improve the speed and reliability of modal transfers while minimizing errors. Even for the large quantity of goods moving without explicit tagging, extensive networks for electronic data interchange permit paperless exchange of invoices, customs documents, bills of lading and other materials can significantly reduce delays and costs for all involved parties.

Air Transportation

The computer reservation system (CRS), which permits air travelers to arrange flights and related services on virtually all airlines, covering the major and minor airports of the world, within minutes – keeping track of price, availability, seat assignments, and meal preference is more than a convenience; it has become a necessary element in operating the aviation system at its current and anticipated levels. Beyond the basic seat reservation function, the CRS is key to airline yield management techniques, in which pricing is continually adjusted to match airline offerings to market demand for air travel.

The other IT aviation application is the air traffic control system – an enormously complicated mix of hardware, software, procedures and people that lies at the heart of both safety and air system performance. The advances in this area have been sufficient to support more than a tripling of domestic passenger-miles flown between 1970 and 1990, with steady improvement in safety and efficiency. By aggregating information on all flights in the US airspace, now coupled with weather data, it has become possible to move from air traffic control to air traffic management, with flights held at originating airports to avoid expensive and irritating in-flight holds ("stacking") in the vicinity of congested terminal areas. One highly sophisticated IT-based advance – now being examined for highway applications – is onboard collision avoidance technology, providing pilots with alerts of other aircraft potentially on a collision course.

IT pervades modern aviation. Plans and systems are currently being developed for transition to a conceptually new approach – "free flight" – in which aircraft location is determined on-board by augmented GPS technology, rather than by ground radar facilities, with greatly increased flexibility given to crews in choosing routes. Further, the "glass cockpit" (reflecting the adoption of computer-driven displays of information in place of traditional electro-mechanical gauges) is one of the most visible manifestations of the many automated functions that contribute to safety and reduced pilot workload, and have permitted an economic gain through reduction to two- rather than three-person crews in large airliners.

The design process for the latest entry into the commercial fleet, the Boeing 777, was greatly facilitated by extensive use of computer-aided-design (CAD) technology, engineering-design flight-deck simulators, and formation of widely-separated but digitally-linked subsystem engineering teams.

Rail Transportation

The current renaissance in US railroading has been driven by a powerful combination of regulatory reform, innovative operating concepts, and application of technical advances. Railroads must deal with an extreme dispersal of moving and fixed assets – involving not only the trains a given company operates and its track network, but also the free exchange of rolling stock with other railroads. In the 1970's they implemented optical tags and scanners to identify freight cars as the entered terminals, and more recently have moved to a microwave transponder system. Classification yards, in which several thousand freight cars are received in inbound trains and re-assembled into outbound traffic, have long been highly automated in their operation. Traffic management (train dispatching) is now often accomplished from centralized control centers responsible for thousands of miles of rail system, with graphic display of track occupancy and infrastructure status.

Fundamental safety functions – primarily train separation – are still based on sophisticated realizations of the century-old track-circuit concept. Federal Railroad Administration and the industry are now exploring "positive train control" approaches based on use of locomotive GPS position determination and on-board track geometry data, with digital communication to a control center. Either this or other concepts using wayside transponder beacons for location appear to offer substantial safety and traffic capacity benefits, although the necessary investment would be quite large.

Inter-city Highway Transportation

Many elements of ITS apply to non-urban environments. For example, "mayday" systems – devices that sense an accident, or can be activated by the vehicle operator, and automatically communicate the situation (typically using satellite or cellular systems) to an emergency response agency. By incorporating a GPS receiver, this system is able to provide precise location information to the responder.

Route navigation systems, typically based on GPS access with a CD-ROM map database, are already quite popular in Japan, although still relatively costly. While perhaps of greatest value in urban areas, particularly when accompanied by effective routing algorithms, they can also be of considerable value to the intercity traveler. In addition to navigation functions, such systems can contain information on restaurants, hotels, services, etc, as well as local traffic problems or – of particular interest on rural roads – warnings of approaching adverse weather conditions. More localized meteorological information can also be of great value to highway agencies in assuring that warnings and countermeasures are deployed quickly, effectively, and efficiently.

The typical modern automobile is already highly digitized, with approximately 20% of its value associated with electronics, and total on-board distributed computing power said to be cumulatively greater than that on NASA's space shuttle. To a large degree, these devices simply represent improved ways to achieve traditional functions, and their presence is relatively invisible to the driver. But innovations now being actively explored offer more dramatic change. Various night vision technologies, originally developed primarily for defense applications, may be a powerful means of coping generally with the high rate of nighttime accidents, and with the decline in night vision inherent to the aging process.

Maritime Transportation

Navigation has always been particularly critical in the marine environment for both safety and finances. GPS systems and steady improvements in oceanic meteorological sensing and forecasting provide the foundation for efficient routing between ports. Continual communication among corporate headquarters, customers and ships similarly enables economic optimization in routes, schedules and cargoes. Highly automated port facilities contribute to minimization of the time spent loading and unloading, further improving ship utilization. High-precision augmentations of GPS in harbors and other critical regions (used in concert with digital charts), radar for collision avoidance, and radio communications provide the basic framework for marine safety, which includes avoidance of environmentally damaging spills.

Transport IT and Changing Demand for Transportation

The changes described above suggest the degree to which information technologies can complement and facilitate transportation functions, yielding more attractive cost and service characteristics that often affect, to some degree, the choices made by users and customers. However, less obvious, but potentially of greater direct impact, is an IT-driven shift in the underlying spectrum of demand for transportation services based on substitution of telecommunications for physical movement. This is often referred to as "tele-substitution."

Mobility and access needs and desires can be met in many ways. The place at which products are produced is similarly subject to a wide range of choices. The need for transportation is often discretionary, and, when it is needed, alternative means can be found. The brief discussion that follows suggests some of the many ways in which IT substitutes for transportation, stimulates or eliminates the need for it, alters user travel or mode choices (including facilitation of public-sector travel demand management strategies), or substantially changes the nature of the transportation services needed. The extent (if any) of substitution of movement by communication is still speculative. Evidence from France, and from US in the recent past suggests that in the last two centuries, both transportation and communication have been complementary growing in parallel – with little support for substitution (see Fig. 5). Since these effects are just beginning to emerge and are not readily quantified, this discussion is necessarily anecdotal and descriptive in nature.

Local Mobility and Access

For many people, mention of information technology impacts on transportation most often conjures up the image of telecommuting – replacing the trip to one's workplace by digital tools and communication links that permit working from home, with

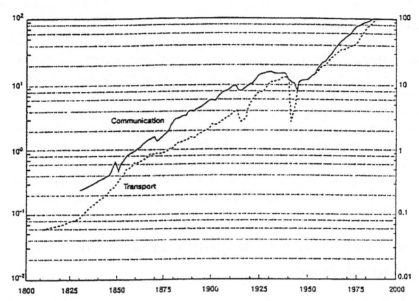

Figure 5a. Growth of passenger transport and communication in France (in passenger-kilometers and total number of messages transmitted), index 1985 = 100. Source: Arnulf Grübler, The Rise and Fall of Infrastructure: Dynamics of Evolution and Technological Change in Transport (Heidelberg, Germany: Physica-Verlag, 1990), figure 4.5.2, p.256.

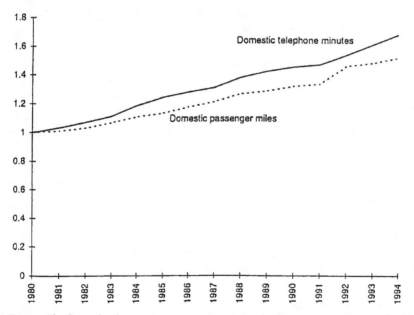

Figure 5b. Growth of passenger transportion and telecommunication in the United States, 1980-1994, index 1980 = 1.00. Source: Federal Communications Commisssion, 1996. Statistics of Communications Common Carriers, 1995/1996 Edition. Washington, DC; U. S. Department of Transportation, Bureau of Tranportation Statistics, 1996. National Transportation Statistics, 1997. Washington, DC.

shopping, entertainment, social interactions and other activities also accomplished largely through digital links. But even if this relatively extreme view is not realized, there is a core reality to the expectation of significant potential changes in personal and corporate activities – with likely parallel alterations in the provision of mobility and access, particularly in the urbanized areas which account for 75% of the population and almost 2/3 of national passenger car mileage.

At present, telecommuters comprise only a few percent of the workforce (estimated at 4 million in 1995, or about 3%), and the most common situation is that individuals still travel to the regular workplace two or three days per week. The growth rate for this style of work appears to be rapid, although difficult to measure, and it remains to be seen at what point it may either accelerate or level off. Transportation effects are limited, since trips to and from work only represent about a little over a quarter of urban travel in any case, and often include other errands that would require local travel in any event. Thus, for the foreseeable future, direct impacts of "pure" telecommuting are expected to be modest.

However, the more general phenomenon of home-based work, with telecommuting as a subset, has much broader implications. Establishment of a fully functional home office now requires only a relatively modest investment for capabilities including fax terminal, copier, Internet (with a World Wide Web site) and e-mail connection, voice mail, and multiple telephone lines. These tools, plus a powerful desktop computers running commercial small-business software can support a substantial business activity in a fully professional manner, with operating costs far below the level associated with rented quarters. The net effect is a marked increase in ease of entry into the marketplace.

A variety of societal trends are encouraging self-employment: voluntary or forced early retirements, contracting out of services formally performed by employees, need to tailor work to responsibilities to care for family members, aversion to large and impersonal work situations, a need or preference for flexibility in work style and hours, the wish to pursue several types of work simultaneously, or the desire to make one's choice of residence independent of workplace. Information technology enables and stimulates realization of goals such as these.

It has been estimated that approximately 8 million employees work from home, either as telecommuters or as field representatives, salespeople, or service technicians with no formal office. There are also about 24 million self-employed home-based workers. All of these categories are being enabled and reinforced by technology, demographic and economic trends; all show steady and rapid growth.

While eliminating trips to an external workplace, this kind of arrangement may generate trips to customer locations and to suppliers. However, the home-based worker is more likely to be able to schedule travel in off-peak hours for business as well as other activities, thereby his or her personal mobility. Transportation implications include the reduction in work trips, as well as some degree of congestion relief in periods of peak travel. The broader and more speculative impact is the potential to increase urban sprawl and migration to the more-rural suburban areas as a consequence of improved mobility.

No matter where employed, many people are attracted to the convenience of shopping from home, whether via catalog and telephone, television, Internet, or other means. This phenomenon has already contributed significantly to the growth of businesses well attuned to it, including those that can provide the rapid and efficient delivery services upon which the entire process depends. This approach is now being introduced by many supermarkets, so that groceries can be ordered without leaving one's home, for next-day delivery. A related development that markedly improves access to many services as well as goods is the relatively low cost to many businesses of providing 24-hour phone service, now being supplemented by posting catalogs and accepting orders via the Internet.

The World Wide Web also offers direct access at any time to many documents, forms, tickets and reservations, and general information resources that otherwise might require significant effort, during business hours, to obtain.

Information technology also can affect access and mobility in a very different way by providing the means for implementation of demand management and congestion mitigation measures by public authorities. Given the political will and necessity, IT can greatly facilitate the management and control of motor vehicle access to urban areas through road pricing (adjustment of electronically-collected tolls) or prohibition of entry for some or all vehicles.

Long-Distance Mobility and Access

The telephone has provided a partial alternative to some types of business travel, especially in the form of conference calls that involve numerous participants, but has seldom been considered a serious alternative for most meetings. The video form – with television cameras and monitors at each location, providing the visual contact that is so important to human discourse – has been available in some fashion for decades, but problems of cost, availability and performance have limited its use. However, technology advances have now begun to generate serious interest in "virtual" meetings. The steady decline in equipment and telecommunications costs, accompanied by significant performance and ease-of-use improvements, is producing a growing willingness consider video conferences as a possible replacement for some single-location meetings.

Although not a replacement for physical presence, video meetings, if effectively implemented, have several important strengths beyond economy. They are easier to schedule and can involve more people, often from more places. Systems are now becoming available that, at moderate cost, can enable even a conventional desktop personal computer to operate as a video terminal. Should this capability reach sufficient acceptability in cost and performance to become relatively standard, the accessibility of business colleagues for digital "face-to-face" meetings would become substantially greater. Personal relationships could become significantly easier to establish and maintain, even while travel budgets shrink. (Realization of this scenario could have significant impact on airlines, dependent on business travel for almost half of their passengers and about two-thirds of their revenue.)

Information Technology to Provide Access Beyond the Scope of Transportation

The preceding discussion addressed ways in which information technologies affect the need for physical mobility to provide access and thereby substitute for transportation or make it unnecessary. However, tele-substitution functions – business, medical, educational, shopping, etc – can also apply to situations in which the distances, travel costs, or other impediments are so great, that physical transportation is not considered at all In short, the desire for access simply would not be met at all, or would be satisfied in a less adequate manner. Telecommunications and associated technology can effectively remove distance or locational remoteness as a constraint on access to education, libraries, and some medical care services and consultation,

For example, the Internet permits people with common interest in virtually any subject to gather (electronically) for discussions, debate and exchange of information – no matter how esoteric the subject or geographically dispersed the participants may be. An absence of like-minded individuals in a person's local community need not be an impediment to actively following a particular interest, whether professional or amateur, intense or casual. As valuable as this type of access is in general, it is of special importance to people who, for whatever reason, may be unable to travel physically and would otherwise be precluded from the activities and services enabled by information technologies.

Information technology is also creating opportunities well beyond the realm of simple telecommuting or home-based self-employment. Some types of work – such as telemarketing, invoice processing, and software development – do not require physical presence at a particular location. Businesses requiring these services now can and do seek them anywhere in the world, based primarily on economic considerations. At the same time, individuals and whole communities gain a significant degree of access – that would not exist without IT – to very distant commercial and employment markets in any activity that does not entail substantial physical movement of resources or products.

Goods Movement

For the most part, freight transportation does not lend itself to tele-substitution, but there are significant exceptions to this generalization. In particular, when the commodity to be moved consists primarily of information, telecommunications can *substitute* for transportation, rather than playing the more common role of *complementing* it. There has long been a substantial degree of synergy and interchangeability between transportation and communications where information is concerned.

In some cases, the substitution represents a meaningful replacement of ton-miles of transportation, as when newspaper content is transmitted across a continent or abroad for local printing and delivery, as contrasted to shipment of tons of newsprint. That process has now been extended via the Internet to delivery to the final consumer, at home or at work, who may or may not choose to print out articles of interest. Library reference books and journals may travel to users electronically rather than physically. The often-bulky user manuals that have in the past accompanied software for personal computers are increasingly replaced by compact disk versions, if not provided entirely digitally over the Internet. A movie can be often be obtained digitally, from a cable service, rather than physically picked up at a video store.

The Transformation of Production by Transport IT

The rapidly growing investment in US in transport IT has meant not only the supply of new transportation services, and the changing of transportation demand as noted above; it has also revolutionized the type and level of competition and the structure of management in transport firms and in production firms who are their customers. Use of IT has meant the improvement of the quality and variety of services to customers and suppliers. IT is promoting much broad restructuring and strategic changes in these industries – facilitating the entry of small firms into new markets and creating new economic activities such as global sourcing arrangements, networking of firms, and faster response innovation systems. IT applications at the level of enterprise and activity, by changing the production processes, products, and lines of business, can trigger job redesign, restructure work flows, or alter the place of work – thereby changing employment patterns.

Since IT has been often used in transport and other firms not only to automate and speed up human processes but also to improve reliability, flexibility, and cost, there is widespread expectation of productivity payoffs in the economy. Yet many macroeconomic studies report no such substantial gains in productivity. This is the 'productivity paradox'. In the words of Robert Solow, "the computer is everywhere except in productivity statistics".

Several factors underlie the fact that macroeconomic productivity statistics do not mirror the improved quality and quantity of services and profitability one observes at the enterprise and industry levels. National statistics on industry productivity do not capture aspects of service quality in terms of how well an airline company or a logistical company serves its customers. Further, the benefits of IT use by transportation firms may accrue to and be reflected in the productivity's of their customer industries (typically manufacturing) than those in transportation industries. Again as noted earlier, the effects of transport IT on productivity can not be disentangled easily from the effects of other factors – e.g. management and worker quality, non-IT technologies etc. Consequently, it is fair to view these macroeconomic studies of IT as taking a narrow view of IT impacts on transport firms and their customers (NRC 1994).

Paul David(1990) offers a historical perspective to the productivity paradox by drawing attention the long time lag between the advent of the dynamo – a broad technology with pervasive effects like the computer – a century ago and its impacts on the economy, reflecting the time needed for the new technology to penetrate the economy widely.

For transportation firms IT use is more *necessary than optional*. Airlines. trucking firms, railroads, logistical companies can not function without IT given the vast transaction- processing systems and control systems they need to operate. Further, the use of new sophisticated IT can offer innovative transport firms important strategic leadership opportunities. Consequently, in what follows the economic impacts of IT are explored at the enterprise, activity, industry, and workforce levels.

Enterprise Level IT Effects

Applications of IT focused initially on consolidating back-office operations and lower the costs of paper-intensive activities such as purchasing and accounting, and proceeded to lowering costs of logistics and customer response. Now IT use is intended *to offer flexibility in rapidly evolving business environments, preserving or raising market share, and improving the quality of services and customer interactions.* Examples of such use of IT are a) the CRS(computer reservation systems) and their strategic use not only for cost reduction and revenue enhancement but also for offering new customer services in airlines (described in previous sections), b) the ability to fuse different transport databases, analyze them in order to develop "transport-governance systems", which underlie the ability of logistical companies to offer seamless supply chain management services to production firms(thus lowering production costs of production) worldwide – thereby blurring the boundaries of transport and production firms[1]. Such benefits from IT accrue to the transport firms as well to their customers and generally to the production and consumer sectors.

In such knowledge-rich transport and logistical firms, which set the competitive standards for their industries, the key firm assets are increasingly knowledge assets. These knowledge assets are in two forms: the knowledgebase about customers and markets embodied in the firm's IT and supporting systems; and the knowledge and

---

[1] For example, a logistical firm such as Skyway equipped with the full range of transport IT manages the (timely, reliable) frequent supply (whose requirements may vary periodically on short notice) of all retail US outlets of a fashion clothing chain such as Gap from production centers scattered all over Asia— balancing the knowledge of the location, content and condition of transportation assets of all modes (every 15 minutes) with the fast varying demand locations of Gap products over the US.

professional skills acquired by the workers and management during the development of flexible response capabilities and innovative logistical and production solutions for customers, and during the use of IT in improving the decisionmaking inside the firm. Entirely new economic activities and subindustries can emerge from these knowledge assets.

Industry Level IT Effects

Among the various industries, transport is one of the more substantial users of IT. As noted earlier, this has been spurred on by transport deregulation, economic restructuring, and the complementaries between IT and transport technologies.

In the air transport industry, use of IT has a major impact on *business and operational aspects – in sales, marketing, load management, logistics planning, safety systems, air traffic control etc.* The consequences on operations are enormous. Some of these have been described above: computer-based flight simulators for training, highly automated air traffic control systems, collaborative airplane design, etc. The key point to note is that in air transport and intermodal logistics – vital to the fast growing knowledge-intensive industries – transport IT is associated with increasing complexity and customization, as industries become more transaction-intensive, and increase the scale and variety of their linkages with other enterprises. In this process, transport IT is altering the existing relationships among end-users, manufacturers, wholesalers, and retailers. Intermediries are being bypassed; direct sales to end users are being made by producers – all of this may generate entirely new lines of business comprising of new and different industries.

The current ferment in and the broad effects of transport IT on operations of and linkages among industries is reminiscent of the era a century ago, when the railroad and the telegraph transformed the wholesaling and retail sectors and new retailers (the department store, the mail-order house, and the chain store) arrived.

Activity Level Effects of IT

As transport IT arrives at a time of intense global competition and economic, the firms look beyond incremental change. They begin to redefine their goals and images as old operations get modified and new activities develop for the firm. In this environment there is a clear organizational need for flexibility, a capacity to respond or change quickly, and the ability to experiment often. Workers need to know how to construct queries, analyze information, interpret information, and draw inferences and act. New perspectives are needed for organizational learning, understanding and assessment (NRC 1994).

IT comprises of a set of tools that affect what workers do and how they do it. In this context, IT can help redesign the work process so as to promote in the worker and management flexibility, quick responsiveness, and experimental attitudes. IT facilitates the separation, the reorganization, and recombination of work activities without regard to space and time. The outcomes of such redesign of the workplace are the increased knowledge levels of workers, the decentralization of decision authority (the "flat" organization) that allows more autonomy and broader tasks at lower organizational levels, and network organizations. The levels and extent of supervision will also change.

Combined changes in job content, tools, organization and infrastructure will require managers to develop new skills and attitudes. Training this different labor force is likely to be redesigned and continuous.

It is reasonable to conclude that the IT-facilitated capability to reconfigure activities in a myriad of arrangements will profoundly affect the nature and location of jobs, the organization of firms, and the structure of US industry.

Bibliography

Branscomb, L.M. and Keller, J. H., 1996 *Converging Infrastructures: Intelligent Transportation and the National Information Infrastructure*, MIT Press.

Chandler, Alfred, 1973 "Decisionmaking and Modern Institutional Change" Journal of Economic History, 33, March, pp.1-15

David, Paul A. 1990 "The Dynamo and the Computer: An Historical Perspective on the Modern Productivity Paradox" American Economic Review, May, pp.355-361

Hopkins, J., 1996 *The Role of Information Technology in the Transportation Enterprise*, Issue Paper prepared for the Symposium on Challenges and Opportunities for Global Transportation in the 21st Century, October.

Jagoda, A. and M. de Villepin, 1993 *Mobile Communications*, Wiley Publication,

Jordan, Daniel R. and Thomas A. Horan, 1997 *Intelligent Transportation Systems and Sustainable Communities: Initial Findings of a National Study,* Paper submitted to the 1997 Transportation Research Board Annual Meeting, January.

Lakshmanan, T.R. and M. Okumura, 1995 "The Nature and Evolution of Knowledge Networks in Japnese Manufacturing", Papers of the Regional Science Association, 74, 1: 63-86.

Machlup F, 1962 *The production and distribution of Knowledge in the United States,* Princeton University Press, Princeton, N J.

(NRC) National Research Council 1994 *Information Technology in the Service Society,* National Academy Press, Washington D.C.

National Research Council, 1996*The Unpredictable Certainty – Information Infrastructure through 2000*, National Academy Press.

Pace, Scott, et.al., 1995, *The Global Positioning System: Assessing National Policies*, Critical Technologies Institute, RAND.

Porat MU, Rubin MR 1977 *The Information Economy, 9 volumes,* Government Printing Office, Washington, D.C.

Scientific American, 1990, *Trends in Transportation: The Shape of Things to Go*, May, pp.92-101.

418

Tapscott, D., 1996 *The Digital Economy,* McGraw-Hill

The Death of Distance, 1995, *The Economist,* Sept. 30.

The Hitchhiker's Guide to Cybernomics 1996 A Survey of the World Economy, *The Economist,* Sept. 28.

U.S. Department of Energy, 1994 *Beyond Telecommuting: A New Paradigm for the Effect of Telecommunications on Travel,* Report DOE/ER-0626, Sept.

U.S. Department of Transportation, 1993 *Transportation Implications of Telecommuting,* April.

List of Contributors

Editorial group

Martin J. Beckmann, Brown University, Economics Department Box B, 64 Waterman Street, Povidence, R.I. 02912, USA

Börje Johansson, Jönköping International Business School, Jönköping University, P. O. Box 1026, S-551 11 Jönköping, Sweden

Folke Snickars, Department of Infrastructure and Planning, Royal Institutet of Technology, S-100 44 Stockholm, Sweden

Roland Thord, Thord Connector, Lützengatan 9, S-115 20 Stockholm, Sweden

Contributors

Jean-Pierre Aubin, Doctoral School for Mathematical Decision Making, University of Paris-Dauphine, Place du Marechal de Lattre de Tassigny, F-75775, Paris Cedex 16, France

Christopher L. Barrett, TSA-DO/SA, Mail Stop M997, Los Alamos National Laboratory, Los Alamos NM 87545, USA

David F. Batten, The Temaplan Group, P. O. Box 3026, Dendy Brighton 3186, Australia

Antoine Billot, Department of Economics and CERAS-ENPC, Université Pantheon-Assas, Paris 2, 92 rue d´Assas, 75006 Paris, France

David E. Boyce, Transport and Regional Science, Department of Civil and Material Engineering, University of Illinois at Chicago, 842 W. Taylor Street (mc 246), Chicago, IL 60607-1023, USA

John L. Casti, Santa Fe Institute, 1399 Hyde Park Road, Santa Fe, NM 87501, USA

Kei Fukuyama, Department of Social Systems Engineering, Tottori University, Tottori 680, Japan

Herman Haken, Institute for Theoretical Physics and Synergetics, University of Stuttgart, Phaffenwaldring 57/4, D-70550 Stuttgart, Germany

Björn Hårsman, INREGIA, Box 12519, S-102 29 Stockholm, Sweden

Walter Isard, Department of Economics, Cornell University, Uris Hall 436, Ithaca, NY 14853, USA

Bruce N. Janson, Department of Civil Engineering, University of Colorado at Denver, Denver, Colorado 80217-3364, USA

Kiyoshi Kobayashi, Department of Civil Engineering, Graduate School of Engineering, Kyoto University, Kyoto 606-01, Japan

Heli Koski, Department of Economics, University of Oulu, P O Box 111, 90571 Oulu, Finland

Robert E. Kuenne, Department of Economics, Fisher Hall, Princeton University, Princeton, NewJersey 08544, USA

T. R. Lakshmanan, Bureau of Transportation Statistics, Department of Transportation, 400 7th Street. S.W. Washington, D.C. 20590, USA

Der-Horng Lee, Institute of Transportation Studies, University of California at Irvine, Irvine, California 92697-3600, USA

Anna Nagurney, Department of Finance and Operations Management, School of Management, University of Massachusetts, Amherst, MA 01003, USA

Peter Nijkamp, Department of Economics, Free University of Amsterdam, De Boelelaan 1105, 1081 HV Amsterdam, The Netherlands

Olle Persson, Department of Sociology, University of Umeå, S-901 87 Umeå, Sweden

Tönu Puu, Department. of Economics, University of Umeå, S-901 87 Umeå, Sweden

John M. Quigley, Graduate School of Public Policy, 2607 Hearst Avenue, University of California, Berkley, CA 94720, USA

Christian Reidys, TSA-DO/SA, Mail Stop M997, Los Alamos National Laboratory, Los Alamos NM 87545, USA

Donald G. Saari, Department of Matematics, Northwestern University, 2033 Sheridan Road, Evanstone, IL 60208-2730, U.S.A

Stavros Siokos, Department of Mechanical and Industrial Engineering, University of Massachusetts, Amherst, Massachusetts 01003

Tony E. Smith, Regional Science Program, University of Pennsylvania, Philadelphia, PA 19104, USA

Wei-Bin Zhang, Institute for Futures Studies, Box 591, S-101 31 Stockholm, Sweden